MORE THAN A FARM ORGANIZATION

FARMERS UNION
IN MINNESOTA

BY
DON MUHM

MORE THAN A FARM ORGANIZATION

FARMERS UNION IN MINNESOTA

Published by
Lone Oak Press
304 11th Avenue Southeast
Rochester, Minnesota 55904-7221

DEDICATION

This book is dedicated to F. B. Daniel.

Throughout his life, and the span of time incorporated in this book, F. B. Daniel has been in close association or personally involved with the activities, events, organizations, people and resulting conditions portrayed in this book.

Without his contribution, resulting from this association, and countless hours of hard work and total commitment to this project, this book would not have been written.

ACKNOWLEDGEMENTS

The Board of Directors of Minnesota Farmers Union Foundation does recognize and thank, sincerely, the many individuals and organizations who generously contributed to the compiling and the publication of this history. Their concern and generosity made it complete.

The names of all who contributed pictures, clippings. information and effort are too numerous to list individually. We truly hope that all who helped and are not identified will know that our thanks is for them as well.

Likewise, generous financial support on the part of both individuals and organizations made the writing, printing and publication of this history possible. Gifts were made in a variety of ways and over a period of time, making appropriate recognition more difficult. We do express our sincere appreciation to all contributors here shown and to any whose names we have failed to list.

INDIVIDUAL CONTRIBUTORS

Marion Grasdalen
Lloyd Peterson (in the name of Harry Peterson)
Harold and Delores Windingstad
Orion and Idella Kyllo
Bob Bergland
Jim and Pam Deal
Loren and Marion Sorg
Don Knutson, Dennis Forsell
   & Members of Green View Board
The Schufte Estate
Dave Frederickson, Dennis Sjodin
   & Members of Minnesota Farmers Union Board
James Frederickson
Dennis and Elaine Forsell
Wava and Levern Larson
Dorothy Froehner
Clarence Hillestad
Judge Clint and Trix Wyant
Cy Carpenter

MEMORIALS - IN THE NAME OF:

Ray Grasdalen
Ernest Hokenson
Dorothy Froehner
Maurice Melbo
Clarence Hillestad
Leslie Schmidt
Milly Frederickson Lang

ORGANIZATIONS

Farmers Union Marketing & Processing Association
Dakota Electric Association
Wright-Hennepin Electric Cooperative
North Star Electric Association
Northern Electric Cooperative
Goodhue Electric Cooperative
PKM Electric Cooperative
First State Bank of Sauk Centre
First State Bank of Castle Rock
Peoples Bank of Kimball
State Bank of Warren
Agribank FCB
Town and Country Bank of Almelund
Perham State Bank
Stearns County Farmers Union
Mille Lacs County Farmers Union
Lac Qui Parle County Farmers Union
Freeborn County Farmers Union
Dodge County Farmers Union
Milaca Farmers Union Local
Paynesville Farmers Union Local #290
Morris Farmers Union Local
Providence Maxwell Farmers Union Local

Recognition and thanks is extended by the Board
to the following individuals and organizations
for their encouragement and support:

Nora Christianson
Archie Bauman
Leo Klinnert
Verner Anderson
Bob Anderson
Junice Sondergaard
Milt Hakel (deceased)
Bob Denman
Dick Johansen
Bob Handschin (deceased)
Ethel Wiseth
Willis Eken
Marion Fogarty
Bessie Klose
Dennis and Joanne Sjodin
Howard Peterson
Joe Larson
Ellsworth Smogaard
Lyle Reimnitz
Elnora Swenson
Donna Klatt Swenson
Mark Johnson
Harry Moreland
Don Erp
Jodi Olson
Lee Egerstrom
Minnesota Historical Society
SW Minnesota Historical Center at Marshall
Jackson County Historical Society
Kandiyohi Historical Society
Canby Public Library
Parks Library, Iowa State University
Bob & Hilda Rickert

And finally, the Board wishes to thank chief
writer Don Muhm and publisher Ray Howe, whose
interest in, knowledge of and appreciation for
family farming and rural community resulted in
much helpful advice and a better book.

MFU Foundation Board meeting 1997. From left to right: Shari Ciccarelli, F.B. Daniel, Dennis Sjodin, Cy Carpenter, Orin Kyllo, Leona Jordahl, Russ Rudd, Elmer Deutsch, Curt Wegner. Not pictured: Brenda Velde, Harold Windingstad, Marion Fogarty

Photo: Bob Nandell

Author Don Muhm

was born on a tenant term near Kanawha, Hancock County in north central Iowa. He is a graduate of Iowa State University and a life-long agricultural journalist who has spent his entire career as a farm editor for daily newspapers, including the Omaha World-Herald and The Des Moines Register.

During his 33 years as farm editor of The Des Moines Register he won numerous national farm writing and reporting awards, and has been honored by many midwest end national farm and commodity organizations. He is the only person to be named national farm editor of the year three times by the Newspaper Farm Editors of America.

The Des Moines Register named him Iowa Farm Leader of the Year in 1993, only the seventh person to be so recognized, and an agricultural journalism scholarship in his name has been established by Iowa State University.

Since his retirement in 1993, several organizations have honored him, including the Iowa Farm Bureau which in June, 1994, gave him its highest honor, its Distinguished Service to Agriculture award. Other recognitions have come from the National Farmers Union which in 1994 presented him with the Milt Hakel Journalism award; the National Corn Growers Association, the American Soybean Association, the National 4-H Council, Iowa Beef Breeds Council, the Iowa Farm Safety Council, the Iowa Association of Electric Cooperatives and the Iowa Association of Soil and Water Conservation District Commissioners have all made special awards to Don.

He and his wife Joann have three sons and six grandchildren .

Contents

DIRECTORY   9

    Chapter 1 – In the Beginning ...   9
    Chapter 2   The Roots of the Farm Movement...   9
    Chapter 3 – Tough Times Grip Rural America...   9
    Chapter 4 – A "Cooperative Giant" is Born...   9
    Chapter 5 – Building a Giant Supply Co-op...   9
    Chapter 6 – Post-War Era Brings Challenges...   9
    Chapter 7 – The Christianson Years...   9
    Chapter 8 – New President and a New Home...   9
    Chapter 9 – A Time of Change & Litigation...   10
    Chapter 10 – Battles for the Rural World...   10
    Chapter 11 – Minnesota Farmers Union Today...   10
    Chapter 12 – It All Begins With The Local...   10
    Chapter 13 – The Green View Affiliation   10
    Chapter 14 – Another Look at Farmers Union...   10
    Chapter 15 – Epilogue: A "Thank You"...   10
    Chapter 16 – Minnesota Farmers Union Timetable...   10

1. ...IN THE BEGINNING...   11

    Preface   11
    No one should take any of these things for granted.   12
    Not the Best of Times...   13
    With This Flow of Immigrants Also Came an 'Idea'   14
    A Contrasting Picture of Farming in Minnesota in 1930   14
    Minnesota Targeted For Chartering in 1929   15
    A Look at an Early Farmers Union Local   17
    A Look at Minnesota in the mid-1920s   18
    Farm Land Prices Fell Sharply After World War I   19
    Organize "The Patrons of Husbandry" – The Grange   20
    Troubled Times Always Spawn New Organizations   21
    Farmers Union History Made in Jackson Co.   22

More Jackson Co. Farmers Union History 23
A Look at the Farmers Union Logo 25
Education – The Base of the Triangle 25
One Leg of The Triangle is Cooperation 27
Legislation – The Second Leg of the Triangle 27
The Farmers Union Tools – the Plow, Rake & Hoe 27
Which Part of the Logo is Most Important? 28
Three Examples of Successful Legislative Action 28

CHAPTER 2 **31**

The Roots of the Farm Movement... Early Leaders 31
John C. Erp, first president Minnesota Farmers Union 31
Erp Elected State Farmers Union President 33
"The Struggle... To Obtain Equality... For All Classes... " 33
Delegates in 1929 Vote For Canby Oil Co-op 34
Calendar of Farmers Union Locals 34
Minnesota Loses Its National Charter 35
A Look at Editor A. W. Ricker 37
C.C. Talbott's Untimely Death 38
Stalwart Members/Workers – Lura & Oskar Reimnitz 38
Youth Programs in the Christianson-Lura Reimnitz Era 41
Special Are Those of Vision... Wilford E. Rumble 42
Early Farm Leaders Seek "Justice" 44
A True Prairie Populist – George Sperry Loftus 46
Other early leaders: 47
In 1941 Minnesota Was Different From Others 49

3. TOUGH TIMES GRIP RURAL AMERICA... **57**

A Decade of Challenge and Turmoil 59
Minnesota Farmers Union Leader Has a Plan 60
A New Organization is Formed 60
Governor, FDR Quiz Bosch About Protests 62
A Near Revolt by Thatcher, Ricker Occurs 63
Premiums Come From "Tin Can Campaign" 63
New Farmers Union Elects Kuivinen 64
Ricker Was Chairman of Minnesota Organizers 65
Yellow Medicine First To Be Rechartered 65
Delegates Pick Kuivinen for President 66
A Profile of Einar Kuivinen 67
Kuivinen's Farmers Union 68
Kuivinen – "Always Willing To Help" 69
Kuivinen's Views on Farmers Union 70
Minnesota Farmers Union Backs FDR 71
B. Franklin Clough Elected MFU President - 1945 71
First Woman Elected To State Board 72
Rural Electrification & Telephone Service (Lack of) 72
Kuivinen Comments on Communism in "The Glimpse" 73
Kuivinen: It's "The People Versus The Trusts" – Again 74
Farmers Union Seeks 100 Percent of Parity 75
A Look at Cooperatives, 1947 Style 75

Kuivinen Loses Presidency to Wiseth   76
Kuivinen's Farmers Union Chapter Ends   77

4. A "COOPERATIVE GIANT" IS BORN...   **79**

Brief Biography of M.W. "Bill" Thatcher: 80
M.W. Thatcher: A "Prairie Radical"   80
A "Team" – Farmers and Cooperatives   84
"Our Greatest Asset... Greatest Strength... the People... "   85
The Case of "Elevator M"   86
The "Battle of the Tin Cans"   88
FDR Prods Wallace Into Action   90
Anti-Cooperative Attack Backfires   92
Thatcher Proposes "Food For Peace" in 1943   94
Thatcher's GTA Made Use of Radio, TV   95
"Mercy Wheat" Sent to War-Ravaged Europe   96
Thatcher Sought An "Agricultural Relations Act"   97
Thatcher Called Farmers "Soldiers of the Soil"   98
Phenomenal Growth for GTA: More Than 600 Members   99
A Real Champion for Rural America – M.W. Thatcher   100
Thatcher's Recollections of The Early Years   101
Thatcher's Views of a Family Agriculture   103
Humphrey Calls Thatcher – A "Political Maverick"   103
Thatcher's Goal – "Welfare of the Farm Family"   104
Thatcher Responds to "Red Smear"   106
A 30-Year Cooperative Summary   107

5. BUILDING A GIANT SUPPLY CO-OP...   **115**

Early Efforts to Organize an Exchange   116
A "Modern" Machine – the "Co-op Tractor"   118
Grocery Marketing – A "Worthwhile" Experiment   119
Syftestad Warns of New Anti-Co-op Foes   120
"Cenex – Where the Customer is the Company"   121
The Syftestad Story, and a "Word-War" with Thatcher   122
Steichen Praises Farmers Union as "Greatest Hope"   124
Central Exchange Growth Continues   124
It Was a Time of Change   126
It Was "A Close Vote" To Change Name   127
Total Cenex sales in recent years: 128
The Farmers Union 'Link' is Attacked   129
A Look Back at the Roots of CENEX   130
Top Exchange/Cenex Managers/Presidents/CEOs   131

6 ...POST-WAR CHALLENGES...   **135**

Roy E. Wiseth, FU president, Minnesota politician   135
Death Comes in a Minnesota Farm Field   136
Lessons of Life Came Early   136
A Distinguished Visitor With An Offer   137
Strong Sentiment for Family Farmers   137
Farmers Union Tradition Continues   138
Needed (Financial) Help Came From Co-ops   139

Wiseth Sees Change, Determination   140
Wiseth Urges Support for "Brannan Plan"   140
Wiseth Urges End of Farm Bureau-Extension Tie   141
Wiseth Speaks Out Against Communism   141
Wiseth Sees Growth and Progress Ahead   142
Something New – Bus-Trips to Washington, D.C.   143

7 THE CHRISTIANSON YEARS   145

A Brief Biography of Edwin Christianson: 145
The Day the Haymaker Became Manager   145
Thatcher, Christianson – Something in Common   147
Economic Lessons He Never Forgot   148
Long, Busy Days at the Gully Co-op   149
"Adventures in Good Citizenship"   151
Only One Regret: "We Didn't Do Enough"   152
Special Reception for New, National Vice President   153
Christianson Helps Initiate "Green Thumb"   153
A Look at Nora Christianson   154
His Own Words: "The President's Corner"   154
Christianson Criticizes "The Communist Smear"   157
Concern About GI Farm Trainee Program   159
Then, No More Words, Messages... or, Leadership   160
The Legacy of the Great Gully Cooperator   161
Summary of Edwin Christianson's MFU years   162

THE GALLERY   165

8... NEW PRESIDENT, NEW HOME   191

Short Biography of Cy Carpenter: 191
An Ardent Advocate of Family Agriculture   192
A Farmers Union Dairy Co-op   193
He Came From Farm Country   193
A Personal Look at Cy Carpenter   195
A Fateful Time of Transition   196
A Call for a "National Food Budget"   198
"Between You and Hunger" – A Food Policy Conference   199
Carpenter: Patton the "Most Significant" Leader   201
The National Presidency 1984-88   201
Dark Days For Farmers   201
White House Actions are "Immoral, Illegal"   203
One of the Finest Hours – The "Hay-Lift"   205
Dedicate New Farmers Union Headquarters   207
Meetings, Meetings and More Meetings   207
A Somewhat Different Farm Scenario   208
A Farm Partnership – NFU & NFO   209
There Were Legislative Successes   211
Some Messages from Cy Carpenter   212
Carpenter Puts Priority On Helping Farmers   212
It's More Than A Farm Organization... Much More!   214
Carpenter's Ally & "Equal Partner" – His Wife, Fran   215

Battling Against Corporate Invasion of Agriculture 215
Something New – "Sip-and-Scheme" Coffee Sessions 217
Carpenter Supported "Big Three" Co-op Merger 218
In Retrospect: The Career of Cy Carpenter 218
First Drop in Farmland Values in 22 Years 219
Carpenter Criticizes Ag Teacher Cuts 220
Carpenter Calls For Reagan Farm Chief To Resign 220
GAPP With GATT? 221
Never One for Retirement, Except... 222

9 ...A TIME OF CHANGE, OF LITIGATION... 223

No Education Fund Policy, No Organization At All? 224
Many Uses For "Co-op Education Fund" Dollars 225
They Sought More "Business-like Use" of Funds 225
Secrecy Prevails as Voting Materials Are Prepared 227
Daniel on Special GTA Actions: "Damned Misleading" 229
GTA Employee Admits Lobbying Vote 231
Concerned GTA Members Organize 232
Newspaper Headline: "Group to fight GTA in court" 233
GTA's First Loss in 31 Years 235
Education Fund History 236
"Loyalty" – The Key to Progress, Growth 237
"Holy Cow" in Polish? 238
Syftestad: "Co-op Education Both Desirable & Necessary" 240
A Television Media Attack 240
"Some Felt (5 Percent)... Was Too Much Money" 241
There Were "Some Darned Good Men" Back Then 243
More Change – New Name For Farmers Union *Herald* 245

10...MANY BATTLERS FIGHT FOR FARMERS... 247

Brief Biography of Bob Bergland: 247
The New Co-op's Success Got Attention 248
A Period of Progressive Growth 250
Bergland Sounds Warning – Surpluses Are Likely 254
Bergland on Politics: "Only Natural" 254

11... FARMERS UNION TODAY... 257

The Willis Eken Era, 1984-91 257
Eken: Farm Bill Offers "Little Help" 260
A Farm Rally Draws 10,000 261
Baumann: "We never learn anything" 261
Groundswell's Demands 261
The Farmers Union-NFO Marketing Pact 262
Eken a Farmers Union "Torchbearer" 264
The Dave Frederickson Era 266
Frederickson Speaks His Mind 267
Eken Resigns; Frederickson Takes Over 268
Frederickson: More Than Prices Involved 268
"Make People the Most Important Part" 269
About "The Farm Lobby" 269

About the Threat to "Family Farmers"...   269
About Farmers Union...   270
Birthdays Celebrated On The Road...   271
"Work With Others"– A Frederickson Goal   271
The Top 10 Farm Cooperatives   272
Farmers Need "Safety Net"   273
One Farm Where There Were Three   274
First Congresswoman Helps Idea Become Reality   276
A Modern View of Farmers Union   279
Another Definition of "Farmers Union"   279
Cy Carpenter's View...   280
Offices and locations of Minnesota Farmers Union: 280

12... IT ALL BEGINS WITH THE LOCAL...   281

A 30-Year Look At Farmers Union   281
The 30-Year Co. Record   284
The "Core" – The Farmers Union Local   287
A Look at Cottonwood Co.   289
The Swensons of Cottonwood Co.   291
A Half-Dozen "Torchbearers" From A Family   293
About Clifford and Elnora Swenson...   293
Three "Torchbearers" Honored in 1943   295
The Farmers Union Education Council, 1942-72   297
Some Special People...   298
Clint Hess – "Destination Washington"   298
Archie Baumann - Dedicated Worker, Cartoonist...   299
A Pair of Farmers Union Workers... the Grasdalens   299
State Vice Presidents...   300
Minnesota Farmers Union Executive Committee... 1997   300

13... GREEN VIEW – A FARMERS UNION SUBSIDIARY...   303

The Birth of Green View   304
Background of Green View   305
Major Developments Related to Green View in 1969   305
The Origin of the name "Green Thumb"   306
Green Thumb state directors: 310
Green View state executive directors: 310

14 ...ANOTHER FACET: INSURANCE   311

The Farmers Union Claims Department   316
The "Mile High" Club   317
Borghorst Retires; Frederickson Takes Over   318
The Insurance Program Today   319
"Behind Our Success... "   319
Prediction – "We'll be around a long time... "   320
Another Farmers Union Cooperative – FUMPA   321
Support From FUMPA For Co-op Education   322
The Role Of Women In Farmers Union   324
The Pattern For Women: Certain Jobs Only   324
Then a New Idea, "City Cousins"   325

She Researched Other Farm Groups	326
An Experiment With a Petition	326
A Look At Secretary Bessie Klose...	327
Women on State Board...	329

15... EPILOGUE ... A "THANK YOU"... 331

Farmers Union: "We Were One Big 'Family'... "	332
The Challenge... Capturing the Past	333
Acknowledgments:	333
Cy Carpenter	333
F. B. Daniel On Cooperatives	334
In F. B.'s Own Words... .	336
Milton D. Hakel, Sr.	337
Farmers Union Staff	338
Credit to Farmers Union *Herald, Minnesota Agriculture*	338
Advice from Tony T. Dechant	339
Green View, Inc.	339

16... MINNESOTA FARMERS UNION TIMETABLE 341

Presidents of Minnesota Farmers Union:	357
Presidents of the National Farmers Union:	357
Presidents of National Farmers Union and years served:	358
History of County Charters	358
Timeline of key state involvement with Farmers Union:	360
Minnesota State Officers by Year Elected:	361

BIBLIOGRAPHY: 363

Books:	363
Newspapers and Magazines:	363

GROWTH OF FARMERS UNION IN MINNESOTA 365

Minnesota Farmers Union Membership by Year	369

INDEX 371

TABLE OF FIGURES 385

AFTERWORD 391

DIRECTORY

MORE THAN A FARM ORGANIZATION...
THE FARMERS UNION IN MINNESOTA

CHAPTER 1 – IN THE BEGINNING ...

Minnesota is young and inviting. Pioneers and settlers from Europe invade the state, seeking a brighter future in "the New World".

CHAPTER 2 – THE ROOTS OF THE FARM MOVEMENT...

A restlessness amid clouds of uncertainty shrouds the countryside, and farmers band together to begin the battle for change, and to challenge the marketing structure. Early leaders gain a Farmers Union charter in Minnesota.

CHAPTER 3 – TOUGH TIMES GRIP RURAL AMERICA...

Harsh economic times set stage for rebellion, including a "Farm Holiday" in which Minnesota farmers and Farmers Union members participate in farm-strikes road blocks, "penny auctions" at farm foreclosure sales, etc. And a new Farmers Union is born in Minnesota. Government responds with historic farm programs.

CHAPTER 4 – A "COOPERATIVE GIANT" IS BORN...

One of the greatest cooperative figures in history, M.W. Thatcher, moves front and center in the battle for farmers in the marketplace where farm co-ops encounter not only discrimination but staunch opposition. But Thatcher is to chart a historic course and change forever how most farmers market millions of bushels of grain through a productive "partnership" with Farmers Union.

CHAPTER 5 – BUILDING A GIANT SUPPLY CO-OP...

Early farm leaders joined together to buy in bulk farm supply items, coal and food in the first Farmers Union "exchanges". This interest led to the creation of the "Farmers Union Central Exchange" and what today is a multi-billion-dollar enterprise know as "Cenex".

CHAPTER 6 – POST-WAR ERA BRINGS CHALLENGES...

Minnesota farmers and others donate millions of bushels of "Mercy Wheat" to help feed war-ravaged Europe. A Goodridge farmer becomes state president; Farmers Union encounters a "Red Smear" campaign.

CHAPTER 7 – THE CHRISTIANSON YEARS...

Farmers Union membership soars to record levels under the leadership of Edwin Christianson, "The Great Gully Cooperator". It is a time of unequaled growth and progress for Farmers Union, and farm cooperatives.

CHAPTER 8 – NEW PRESIDENT AND A NEW HOME...

A stroke disables Ed Christianson and Cy Carpenter moves into the presidency. It is a new time, and a new fight for the "Family Farmer". And for a new home for Minnesota Farmers Union.

CHAPTER 9 – A TIME OF CHANGE & LITIGATION...

Farmers Union members take to the court to battle what they consider "a betrayal" in their long and rewarding relationship with regional cooperatives.

CHAPTER 10 – BATTLES FOR THE RURAL WORLD...

Farmers Union members and leaders gain elective office, or earn government appointments, where their fight for rural Minnesota and America takes on a broader dimension. People like Bob Bergland, Alec Olson, Coya Knutson, and Fred Marshall.

CHAPTER 11 – MINNESOTA FARMERS UNION TODAY...

Strong support for family farmers and a "family-type agriculture" continues under the presidencies of two former Minnesota lawmakers, House Majority Leader Willis Eken and former State Senator David Frederickson. Both were raised in Farmers Union families, and have seen change and challenges develop for rural communities, schools, churches, and farm families.

CHAPTER 12 – IT ALL BEGINS WITH THE LOCAL...

Strength of Farmers Union comes from the grassroots local, the core and heart of the organization. Featured in this chapter are three representative county units – Becker, Cottonwood and Isanti, and several Farmers Union stalwarts and leaders.

CHAPTER 13 – THE GREEN VIEW AFFILIATION

A special, productive relationship that has grown out of the Federal Green Thumb Program – Minnesota's Green View, Inc. It has established excellent work projects for the State of Minnesota, its parks, its forests and its highway rest areas. Green View is a time-tested affiliate of Minnesota Farmers Union.

CHAPTER 14 – ANOTHER LOOK AT FARMERS UNION...

Providing insurance and marketing services for Farmers Union members is an enterprise dating back many years. This chapter identifies some of the early leaders and provides information about the origin of these affiliated companies.

CHAPTER 15 – EPILOGUE: A "THANK YOU"...

This closing segment expressing gratitude to the early leaders and pioneers who through dedication and effort established the foundation for what exists today in rural Minnesota – a respected, enduring organization which continues to work for a brighter future for farm people and family agriculture.

CHAPTER 16 – MINNESOTA FARMERS UNION TIMETABLE...

Here are nuggets of history about the farm organization movement nationally and in Minnesota spanning most of the 20th Century. Featured are people, events and highlights, including some of the challenges encountered, and battles waged for farmers and the farm cooperatives they fostered.

1. ...IN THE BEGINNING...

"It should not be forgotten that settlers coming to America left behind them a feudal pattern of land holding... many were attracted to the "New World" by the dream of becoming land-owners and farmers in their own right... "
Milton D. Hakel, author, 1983, *Who Needs a Farmers Union and Why?*

"Farmers should be encouraged to enhance their bargaining power through cooperative marketing... a large part of the 'farm problem' – which in fact consists of many problems – must be solved by the individual and cooperative efforts of the farmers themselves... "
U.S. Sec. of Agriculture W. M. Jardine, *Yearbook of Agriculture*, 1927.

Figure 1 Horsepower in the 1920's

PREFACE

Too often it seems most people take for granted what's in place, those things their ancestors and predecessors had to work hard to achieve, and in many cases almost literally fight for, against what must have seemed at times hopeless odds. The newer, younger persons sort of forget, or indeed, were never fully cognizant of, the magnitude of the efforts, the kind of battles and all which preceded their world of today, and which they seem to think will never reoccur or confront us or them again. Or, perhaps, they feel there's no use attempting to struggle and oppose something you probably can't overcome anyhow.

Sadly, too many appear to have forgotten, or ignore completely, the lessons of history. Instead, the written record should provide hope that there is something there from which we all can learn out of the pages of the past. Lessons, perhaps, that can be used to enhance today, and hopefully improve tomorrow and the days ahead.

That is the purpose behind our efforts and those of others made to research, prepare and publish farm organization history. What's printed within these pages perhaps represents really only a brief period in the overall scheme of things, a short time really in the history of our world as experienced by those we salute, revere, and acknowledge in this book, focusing on a limited but unique period in the history of agriculture, family farming, and rural community life.

At the same time we trust the reader will absorb some of the flavor, some of the significance, and some of the concerns that caused people to come together, to tackle problems much bigger than what a person could combat alone, all in the quest for a better, brighter tomorrow for rural people. What they did, what they sought, what they accomplished, where they succeeded, where they fell short of their goals, is a part of the story we hope to relate.

The primary goal is to portray a period when Minnesota was much younger, and to try to tell the story of people who embarked and shaped their history, and charted a course for their children, and for their children's children.

Many but unfortunately, not all of those who contributed to this grassroots effort, will be mentioned in these pages. There simply is no way to fully credit those who incubated, developed, and supported with their time, talents and resources this significant farm movement that took place in the 20th Century. Our records and personal recollections are not infallible, and we apologize in advance for failing to mention certain people, developments or events that should rightfully be a part of this history.

Indeed, there were those who were highly visible by their actions, their words or their roles – the "giants" of the times, the gifted speakers, dedicated and persistent leaders, dreamers, planners and, most importantly of all, builders – persons who received much public attention and notoriety for what they did or for what they tried to do. We know at the same time that there were the loyal, less-identified legions who quietly and without fanfare or widespread recognition, convinced in their purpose and mission, provided the foundation and support needed to make the kind of progress and impact their leaders hoped to achieve.

What exists in today's Minnesota world didn't just happen. This influential farm organization, for example, didn't always have the respect or stature it enjoys; the big, successful farm cooperative that today is such a profitable multi-million-dollar enterprise didn't just emerge upon the landscape, open their doors for business and start marketing grain or selling supply items, and paying dividends.

NO ONE SHOULD TAKE ANY OF THESE THINGS FOR GRANTED.

What we have, use, enjoy and are proud of today likely represents the product of countless hours, days and years of planning, dreaming, meetings and conferences, sweat, tears, hundreds and hundreds of minds and thoughts, and thousands and thousands of travel miles to attend all kinds of business sessions with boards of directors and managers, fellow-members, partisans, opponents, and others.

One story passed on in the Farmers Union ranks is how early organizer C.C. Talbott of North Dakota ran out of funds traveling and recruiting members. He went home, sold a bull, pocketed the proceeds, and hit the road again seeking more new members. Another Talbott tale pertains to how two men routinely were used to call on a potential recruit at his farm – a local person whom everybody knew, to introduce the visitor-organizer to the farmer prospect.

And, when the farmer-prospect was hard at work harvesting his crop or doing some chore and didn't want to lose valuable time visiting, one of the two visitors would volunteer to take over the farm work for the host farmer. That system proved pretty effective in gaining new members.

In this book we try to give credit and recognition to many of those who contributed to growth, progress, and change, and who in short made it all come about without seeking personal financial gain. We acknowledge that there may have been shortfalls, that all targets and goals were not attained, and that in some instances the results wrought by change and changing times produced undesired results. At the same time it is hoped that what was begun years ago has provided a foundation on which today's generations, as well of those of tomorrow, can build, grow, and prosper.

NOT THE BEST OF TIMES...

Many of those who first settled the Upper Midwest came from places distant from the new land where they immigrated. These pioneers came to America from all corners of Europe – from Germany they came hoping to establish a "bauernhof" (farm), from Norway looking forward to working new "jord" (soil or earth), from Denmark to become a "landmand" (man of the land), from Sweden seeking new "lande", and from Holland to homestead a "boer" (farm) of promising "akkers" (fields) and to find a "het/vanbelofte" (promised land), and from Finland – the "Suomi" – to be "maamies" (farmers) in the new "maailma" (world).

Hundreds and hundreds left "the old country" to come to the new "landen in Amerika", and to create "pockets" of new farming communities settled by Danes, Norskies, Finns, Germans, Dutch, and all.

The one common denominator involved in their migration westward was very basic – land, thousands and thousands of acres of new land. The resource so much limited in the "old country", and so available in the new "landen" to the west, across the seas, that the American Government would provide "free" land, a homestead to those who would live on it for only five years.

It was new land as well, virgin prairie or woodlands and meadows and valleys, which had never felt the prod of a plow or the nudge of a hoe. Not the tired, worn-out land farmed for countless generations in what these settlers called "the Old Country".

As Farmers Union author/editor Milton D. Hakel wrote in a 1983 booklet *Who Needs a Farmers Union – And Why?*: "It should not be forgotten that settlers coming to America left behind them the whole feudal pattern of land holding which prevailed in Europe. Many were attracted to the 'New World' by the dream of becoming land-owners and farmers in their own right... "

Some came to this new country not just for land, but for other reasons as well – to avoid persecution or the military, or to escape the struggles with over-farmed, worn-out European land, personal troubles, high taxes, whatever. But to the West they came, by boat, by wagon, by horseback, by train, and all, sharing the hope and dream that a new, more rewarding chapter in their lives would begin in The New World of Minnesota, the Dakotas, Montana, Iowa, Wisconsin, Nebraska, and the Northern reaches of the United States.

WITH THIS FLOW OF IMMIGRANTS ALSO CAME AN 'IDEA'

With this flow of immigrants who became the true prairie pioneers on the midlands of America came an idea born a few years earlier in Europe – the concept of working together to achieve a common goal, through the process of organizing and cooperating.

However, as so often seems to be the case, and simply put, it may have not been the best of times for the people of agriculture and rural Minnesota when the tide of immigrants, turned New World pioneers, arrived in their new country. Many saw a need to share the burdens of establishing a new life, settling on the barren plains, prairies and woodlands. They felt that they needed mutual support, to organize, and share in opportunities and in responsibilities, a feeling of "closeness of community" – the sentiment that if any needed help, all felt the obligation to offer assistance in any way they could.

In time, the "new landen" became the right environment for the people to ally themselves even closer, in new organizations, like "alliances", "equity" associations, "leagues", and so on, as a means of fighting social and economic injustices. One of these was Farmers Union, very early in the century in Texas, and then, in the early 1920s when the first known locals sprang up in southwestern Minnesota.

Out of this cooperative environment was to sprout the seeds of a giant grain cooperative which would battle for more dollars for the farmer-producers in a way not before seen even in Europe where the cooperative idea had its roots, and become the world's largest marketer of the produce of the land.

Through their cooperatives these farmers would also be able to do many things, like being able to purchase carloads of sugar, salt, coal, foodstuffs, twine, lubricating oils, tires and such at discount prices and more cheaply, retaining precious dollars essential to creating a new life in the "new country of Amerika".

A CONTRASTING PICTURE OF FARMING IN MINNESOTA IN 1930

Debate existed concerning a farm program proposal that was to transfer control of agricultural commodity marketing from private to cooperative ownership in 1930. The "Marketing Act" was the product of President Hoover's "Farm Board" established to help improve the farm economy. He had wooed the farm vote pledging to farmers that he would create a "Federal Farm Board", a vehicle that would allow farmers to gain control of marketing, and hopefully, pricing their produce.

To achieve the marketing objective, a "Farmers National Grain Corporation" was to be established. It would be farmer-owned and farmer-controlled. And, M.W. Thatcher would leave his Farmers Union Terminal Association to manage the new "national" grain corporation in Washington, D.C. Emil Syftestad became the manager of Farmers Union Terminal Association, which included the Farmers Union Central Exchange, the procurement office for Farmers Union locals, and a growing number of fledgling local cooperatives.

Writing in the Farmers Union *Herald*, editor A. W. Ricker predicted a bitter fight from the grain trade interests to any effort that provided a bigger share of marketing to the budding farm cooperative movement. "Will the owners of the marketing machinery fight to keep the ownership of their property?," Ricker wrote. "The answer is, they will. How much private capital is tied up in such ownership of marketing machinery? The answer is, billions of dollars... "

About the same time Ricker saw all kinds of problems, a completely different scenario was described in a Minneapolis *Tribune* editorial which Ricker reprinted in his newspaper:

"All through central, and southern Minnesota, and to a remarkable degree in North and South Dakota and Montana, from the Mississippi River to the Rockies, a sound farming industry is established which enriches its patrons and spreads fertility and prosperity all over the countryside.

"Here we have no 'hard times', few 'busted' banks or 'frozen credits', in the land of kine and swine, the lamb and ham, the wool and bull, the honey bee and hennery, the land flowing with milk, honey and money... "

It was pointed out that some Farmers Union partisans saw The Minneapolis *Tribune* as "a willing spokesman for Mill City (Twin Cities) grain traders and processors", and not sympathetic toward what was really happening in rural Minnesota just months ahead of the Great Depression era.

Ricker was an early Farmers Union advocate. In 1925 he wrote in his Farm Market Guide newspaper that Farmers Union was "the largest and a promising organization" with livestock marketing capability on 10 major livestock exchanges, was successful in insurance activity, and had established carlot purchasing of a wide variety of goods for its members.

Ricker was head of the Producers Alliance and a party in talks of a merger involving the Alliance, the Equity Cooperative Exchange, and the Farmers Union. Each organization had its strengths. The Equity was experienced in marketing, and had some marketing facilities; the Alliance was skilled in organizing farmers and was headed by a well-known dynamic leader, C.C. Talbott of North Dakota; Farmers Union boasted both recognized legislative and leadership qualities nationally.

At a meeting in Fargo, ND, where all three organizations were represented in January, 1926, Paul Moore of the Equity issued a caution related to the financial troubles of the marketing concern: "The mother of your marketing organization was the little local out at the country crossroads. You let that local die. And when you get into trouble and need help, you looked around and found you had forgotten to keep in existence, the thing which made you."

MINNESOTA TARGETED FOR CHARTERING IN 1929

When North Dakota was granted a charter at the 1928 National Farmers Union convention, the word went out that Minnesota was next. Officials knew that Minnesota would be a difficult state to organize because there was strong opposition in the state, especially the grain traders in Minneapolis who earlier had fought the entry of earlier farm organizations such as the Producers Alliance, the Non-Partisan League, and the Equity Cooperative Exchange.

In addition to the expected opposition to a new organization, there was the unhappy history where efforts to establish marketing pools for wheat, egg/poultry and potatoes had been unsuccessful. Also, representatives of Land O'Lakes, Central Cooperative Association and Twin Cities Milk Producers felt the task of organizing farmers cooperatively was pretty much completed. Finally, there was another reason – the Farm Bureau. In some areas, it was growing in membership and had the advantage of a ready staff of government employees who were both county Extension Service staff members and paid organizers for Farm Bureau.

Dispatched to lead the new organizing effort was D. D. Collins, a member of the National Farmers Union board of directors from Colorado. Collins knew that there was a pocket of Farmers Union supporters in west central Minnesota, and that there had been early locals established in southwestern regions of the state near Jackson. He chose to headquarter in Canby, the community where John C. Erp, an early organizer and a member of the board of directors of the Farmers Union Terminal Association, resided. (Erp later would be elected vice president of the Farmers Union Central Exchange when it was incorporated in 1931.)

The Minnesota chartering convention was held in Willmar Nov. 12-13, 1929. National Farmers Union President C.E. Huff presided. John Erp was elected president of the new Minnesota Farmers Union – a post he would hold until the state unit lost its charter in 1938. Other officers elected: Hemming Nelson, Kandiyohi Co., vice president; Carl Lundberg, Lincoln Co., secretary-treasurer. Elected to the state board of directors were A. J. Blodgett, Wright Co.; Paul E. Aelke, Jackson Co.; Nels Pederson, Chippewa Co.; and Carl Winters, Yellow Medicine Co.

There were major issues and concerns which brought immediate attention from the new organization, including the threat of oleomargarine to the butter market. A special report on the impending oleo-butter "war" came in a letter written by Harold A. Mattson of Local #103, Dawson Co., and printed in the Farmers Union *Herald*. Wrote Mattson about a meeting held Mar. 13, 1930, at the home of Olaf Mattson:

"We discussed the oleo problem. We took a rising vote of those who were in favor of disposing of oleo and not using it themselves, also having the storekeepers dispose of it and not sell it. There was not one person left sitting of the hundred or more present... "

Later, as drought and poor markets gripped farmers, dissention and discontent developed. One source of disagreement pertained to the Farmers Union Livestock Shipping Association, which had enacted policies that many patrons saw as unfair. One of the critics was Minnesota president John C. Erp who angrily surrendered his own stock certificate along with that of the state Farmers Union.

Then, the Minnesota membership became seriously divided concerning not only the regional cooperative marketing problems, but also concerning continued affiliation with the National Farmers Union. For one thing not fully explained in records, Minnesota halted remitting memberships to the national office – something quite disturbing to the national staff.

Further complicating matters was a call issued by Robert Miller of Underwood for a second state convention. The Otter Tail member set up a meeting in the armory in Montevideo, and invited national representatives to attend.

The disagreements seemed to relate to support for regional cooperatives, federal farm programs and policies, and membership in Farmers Union. Adding to the confusion was the fact Miller's second Farmers Union received recognition and support from all three regional cooperatives – the Terminal Association, Central Exchange and the livestock co-op.

In the end, Minnesota lost its charter. However, almost immediately efforts to re-charter began in earnest.

A LOOK AT AN EARLY FARMERS UNION LOCAL

An insight into the early years of Farmers Union activity in Minnesota is possible through the examination of an old record book of Rustler Local No. 9 of Wright Co. This volume lists the actions, concerns, and programs of one of the first Farmers Union groups organized in the state.

The book begins with a report that the applications of "15 male and 13 females" were accepted as charter members of a new local at a meeting in Cokato on Aug. 23, 1920, organized by O. M. Jandra. He officially represented the Farmers' Educational and Cooperative Union of America, according to minutes penned by A. A. Anderson, newly-elected secretary-treasurer that evening.

F. E. Tomlinson was elected president, with Algot Bergman named vice president. Other officers chosen at that first meeting were Otto Johnson, doorkeeper, and A. O. Norton, conductor. The group of 25 also chose a name for their new Farmers Union unit – the Dewey Local. However, a year later the members voted for a new name from a list of a dozen candidates, and chose Rustler Local. It won narrowly over two other candidates – Justice and Mud Lake.

The minutes kept by Albert A. Anderson referred often to men as "Brother" so-and-so, while women members were identified as "Sister". The minutes of the Dec. 16, 1920 meeting mentioned selecting delegates to both a "county meeting" and to attend the "Maple Grove Union" meeting on Dec. 17. So this is the first record found by researchers of this history of Farmers Union in Minnesota concerning early activity at the county level involving other Farmers Union locals.

In January, 1921, a committee was formed to organize a "farmers' stock shipping association", and F. E. Tomlinson was named "purchasing agent for Dewey Local No. 9". And "orders for corn and feed" were taken at the bi-monthly meetings in Jan., 1921, with "hayseed and flour" orders noted the following month. In March, auto tires and lubricating oils were ordered for members.

The new local also showed interest in local governmental matters, as evidenced by a committee organized to meet with the local "town board" concerning "improvement of our public highways" at a Mar., 1921, meeting. And there was mention made of procuring "a warehouse" at the first April meeting of Dewey Local.

When the local phone company raised its rates, the members circulated a petition complaining about the hike and telling the company to disconnect the telephones of the undersigned parties "within five days after receiving this notice". Then later the minutes show that any Farmers Union member not disconnecting from the local Cokato Telephone Co. "will be expelled from the Union".

Then, toward fall, the local arranged for shipments of coal, salt, oats and potatoes for the members. In Feb., 1922, minutes suggested that members "Order goods from 'the Union Store'," without identifying what store had goods available, or where this "Union Store" was located. Later, orders were taken for sugar, hayseed and twine, and mention was made in the minutes of "the county Union store".

In Mar., 1922, Jacob Thompson was appointed director on the board of the Cokato Farmers' Union Exchange, and a "Ladies' Committee was organized, with Mrs. Lewis Johnson as chairman. In May a motion was approved "to investigate why members do not attend (Farmers Union) meetings".

In August the local took orders for peaches. At the Oct. 20 meeting the members voted to organize a new telephone company, and organized "a husking bee to be held at the home of Sister Lucy Farrell".

On May 18, 1923, the Rustler Local No. 9 voted to withdraw from the Cokato Farmers' Union Exchange, and demand the "release" from the Exchange of Rustler Local secretary A. A. Anderson. No reason for this withdrawal was given in the minutes. Then, later on the same day the members voted to start a new business that would be called "Cokato Farmers' Union Produce Company".

The last entry in the record book kept by Albert A. Anderson noted that "an investigating committee of three be appointed to look into the Cokato Exchange business", but didn't say what was behind this action.

A LOOK AT MINNESOTA IN THE MID-1920S

There were 875,749 people living on about 192,000 farms in Minnesota in 1925, when the state boasted a total population of 2.6 million. Just the previous year there had been a 5 percent increase in the state's farm population.

On the dark side, farmland prices had slumped sharply following the boom years of the world's first major war which broke out in the late teens of the 20th Century. For the new generation of farmers, including many who were first-generation Americans, to see land lose value was something new and novel. For example, average farmland values in Minnesota had more than doubled in one decade, from 1910 to 1920, according to the U.S. Department of Agriculture.

But then farmland values that averaged $45.60 an acre in 1910 in this state, and peaked at a record average of $109 an acre in 1920 slumped to just under $80 an acre five years later. The same thing happened in neighboring North Dakota. Average land values declined by 25 percent, slipping from $41 in 1920 to $29.75 in 1925. In Iowa, farmland dropped even more – by one-third during that same 5-year period, soaring upward from $96 an acre in 1910 to $227 a decade later.

At the same time, the farm markets were as unpredictable as the weather. Northern hard red spring wheat bids on the Minneapolis market averaged $3.09 a bushel in May, 1919. Before the first World War in 1913, the price on that same market averaged only 94 cents a bushel. And by 1927 the average paid was $1.47 – or less than half of the wartime price.

In Oct., 1925, the Farm Market Guide reported both the market price and the estimated cost of production for major crops grown in the northwest states: Corn – 68 cents per bushel ($1.41 cost of production); wheat – $1.17 ($2.42); oats – 38 cents (78 cents); barley – 60 cents ($1.24).

There were about 4.5 million acres of both corn and oats raised in Minnesota in the late 1920s. Corn farmers harvested an average of 34 to 35 bushels per acre, and then sold for as low as 64 cents a bushel in 1927 or 20 cents less than the average price of 84 cents received during the 1914-20 period. Oats usually yielded less per acre, and sold for less, too, than corn. Oats averaged 31 cents in 1925 in Minnesota when a "bumper crop" of 42 bushels per acre was harvested.

Barley was seeded on 1.5 million acres of land in 1927 in Minnesota. The crop that year averaged 30 bushels per acre and sold for 75 cents a bushel – a pretty good price compared to the 44-cent barley market only four years earlier.

Minnesotans milked 1.5 million cows in the late 1920s, second to Wisconsin where nearly 2 million head produced milk. Nationally there were 22 million dairy cows then, with an average value of $77.43 per animal. The price of milk, delivered to the front door of the Minneapolis consumer ranged from 10 cents to 12 cents a quart. And, Minnesota was the nation's No. 1 butter producer (268.5 million pounds in 1926) and second to Wisconsin in the output of cheddar cheese. The state also ranked third in egg production, trailing Illinois and Iowa.

Pork production was big business in Minnesota, too, in the late 1920s, with 3.7 million head on the state's farms. There were 3.1 million hogs marketed at the South St. Paul stock yards, then the nation's third largest livestock market (trailing Chicago and East St. Louis). Of the state's 2.7 million head of all cattle, 1.6 million head were marketed at the South St. Paul yards.

There were 628,000 horses and 14,000 mules on Minnesota farms back then, with the average mule worth about $12 more per head than what a horse sold for. The USDA reported a Minnesota mule had a value of nearly $80 a head, compared to the average value of $67 for a horse, $77.43 for a milk cow, and $10.22 for a sheep.

FARM LAND PRICES FELL SHARPLY AFTER WORLD WAR I

Commodity prices had fluctuated in line with production and other factors, although in general market prices collapsed to unprofitable levels after the World War I boom period. All of this led to little confidence in the prospects for the farm economy. In his report to the U.S. President in 1927, Agriculture Secretary J. M. Jardine, an Idaho native and Kansas educator, noted "the farm problem". It was actually, he added, "many problems". "It must be solved by the individual and cooperative efforts of the farmers themselves," Jardine told President Coolidge in his late 1927 report. "Many farm products are no longer marketed in the locality where they are produced. They are sold in distant markets in competition with products from other parts of the country and even from other nations."

Secretary Jardine also pointed out the emergence in the early 1920s of "a number of large-scale cooperative associations formed almost overnight", adding, "the members had for the most part no experience in cooperation... problems of marketing were strange to their directors... and their managers, even though experienced businessmen, were not familiar with the peculiar difficulties and responsibilities involved in the management of a cooperative organization... ."

Many of the members of the new cooperatives "believed that farm products could be sold on the basis of monopoly control of supply rather than by service to customers... "

Also, help from the Halls of Congress was more uncertain, in the wake of Presidential vetoes that had killed hoped-for farm support legislation that had been finely tailored by sympathetic and understanding Midwest lawmakers familiar with the problems confronting farmers, and eager to offer some legislative solutions which might ease the economic burdens.

In 1929 Iowa-born Herbert Hoover was in The White House. Theodore Christianson of Dawson was the governor of Minnesota, and Charles E. Adams of Duluth was lieutenant governor. Ready to campaign successfully for governor was Floyd B. Olson of Minneapolis, with K. K. Solberg of Clarkfield as lieutenant governor. The former governor of Missouri, Arthur M. Hyde, was the U.S. Secretary of Agriculture, and a newly-appointed Farm Board was supposed to come up with programs and proposals to rectify the increasingly gloomy state of affairs for agriculture.

For the farmer, the moment of truth comes in the market, and what his or her produce commands in bids and prices from the buyers who perennially seem better organized than are producers and farmers. There is no better barometer of rural prosperity than that found in the marketplace, where value is placed on the fruits of one's labor.

Just as bad was the fact that farm land – the basic economic yardstick measuring the farm economy – which had declined sharply in the post-war years, and continued its costly skid eroding rural real estate equity well into the Great Depression era of the early 1930s. For example, land in Minnesota in the mid-1920s sold for a fourth less than it had five years earlier.

The bullishness that precipitated such prosperity for farmers had its roots in the years ahead of the first Great War – a time from 1910 to 1914 when things generally were in good financial balance between the farm and industrial economic sectors. However, after the war ended, the financial bloom began to fade – slowly at first, but disturbingly steady year after year.

Truly, it was an era crying out loudly for grassroots leadership even though there were farm organizations already representing farmers and looking out and speaking out for the people of rural America. Both Farmers Union and Farm Bureau had emerged on the farm scene earlier in the century, and both were fledglings in the farm lobbying arena where The Grange had about a half-century of experience promoting better things for the people of agriculture.

ORGANIZE "THE PATRONS OF HUSBANDRY" – THE GRANGE

One of the chief founders of The Grange as it is commonly called was Oliver Hudson Kelley who located on a farm near Elk River in Minnesota, and later turned it into a showcase landmark.

Kelley was a writer, and had worked briefly as a reporter for the Chicago *Tribune* before moving to Minnesota where he farmed and also was agricultural editor of the Sauk Rapids *Frontiersman*. His farm became a showcase. While a post office clerk in Washington, D.C., he and six others founded the "Order of the Patrons of Husbandry, the National Grange".

Later he moved his family to the Grange headquarters located first in Washington, and later in Louisville, Ky. His farm became a tourist attraction when it was taken over by the Minnesota Historical Society in 1981 and operated as an "interpretive center" depicting the events and character of the country from which Kelley became a national farm leader.

Historians cannot pinpoint exactly when or what factors caused yet another new farm organization to enter the agricultural arena. Certainly any student of agricultural history can easily identify the periods when certain groups or organizations came on the farm scene, along with some of the factors which existed at the time that might have aided its emergence.

For example, the National Farmers Organization (NFO) originated in the agricultural picture in the mid-1950s on the heels of severe drought and extremely poor prices for livestock – particularly the price of hogs which plummeted unprofitably from Korean War highs in the mid-$20s to near 10 cents a pound by the mid-1950s.

Then, 20 years later the American Agriculture Movement developed after surplus production and other factors caused economic woes for many farmers in the late 1970s.. A "strike" tactic, similar to the "market holding actions" of the National Farmers Organization in the 1960s and the "Farm Holiday" tactic of the 1930s, was one chief suggestion of the AAM leaders. All farmers had to do was not produce, and prices would increase to the often sought cost of production plus a "reasonable" profit level sought by so many so often in agricultural history.

So "STRIKE!" was one of the AAM's early advocacy proposals. And indeed, a "Farm Strike" tabloid newspaper made its debut shortly later in 1979 edited by Elmer Carlson, a Farmers Union member from Audubon, IA.

Then, another new group emerged during the "farm crisis" years of the 1980s, when farmland values dropped sharply year after year for a half-dozen years, when farm foreclosures and bankruptcies abounded forcing thousands of farmers off the land, and the doors closed at scores of rural, Midwest banks, and even the farm credit banks encountered red-ink bottom lines. It was, many said, the worst financial problems ever for the farmer and his lending institutions since the Great Depression era.

So emerged in the Farm Crisis Era of the 1980s yet another farm organization – the "Farm Unity Coalition" movement, indicating something new – a more united front where several organizations would merge into an action group to do what none perhaps could do singly and alone.

The "coalition" approach was not new or novel to farm groups. The Farmers Union during the Nixon years had sparked a short-lived effort to form a consolidated farm front – bringing together the major general farm organizations (with one notable exception –the Farm Bureau) and national commodity organizations. A coalition meeting was held in St. Louis, Mo., on April 6, 1970. It was an impressive gathering. But sadly, however, this group was never to assemble like that again. The closest resemblance to that "farm coalition" meeting was a "Farm Crisis Rally" organized by the Farmers Union and the NFO in Ames in Feb., 1985, on the campus of Iowa State University. And 60 years earlier, in 1925, efforts were made to unify farm organizations at a meeting of 24 farm organizations (including a number of state Farm Bureaus) in Des Moines, IA. Out of that meeting emerged a coalition group which became known as the Corn Belt Committee (Officially, its name was The Grain Belt Federation of Farm Organizations).

As far as some observers are concerned, it was agricultural history repeating itself again in the 1950s, the 1970s and 1980s. That's because, it seems, whenever tough times prevail, you can expect another farm movement to occur – even though obviously there already seem to be plenty of farm organizations around representing, and looking out for farmers and rural people. The thrust of such times seems to be a movement born out of frustration – the fact that tough times for farm people developed despite the presence of farm organizations.

Some of those who criticize existing organizations for not doing enough to help farm people often seem to be those who did little or nothing to augment the efforts of those groups earlier, and then attack them for not doing more to prevent the current farm crisis or challenge at hand.

It's like during the rough economic times folks desperately look for, and hope for new leadership, a new face or personality which might possess new solutions and answers to the oldest farm problem of all – good, prosperous times for farmers. And many times the dream is that the best solution will come out of sharing – sharing troubles, and ideas of how best to solve them.

FARMERS UNION HISTORY MADE IN JACKSON CO.

Some of this solid cooperative philosophy was voiced in Jackson Co. in southwest Minnesota by Charley F. Wendel of Wakefield, one of 20 charter members of the first local organized in the autumn of 1918 – the Rost Center Local.

Wendel had this comment on the beginnings of the Farmers Union: "The beautiful thing we have seen in the Farmers' Educational and Cooperative Union of America is the way the men and women have seen the tasks that somebody ought to do and have accepted them and bent their shoulders to the tasks.

"Now is the time for honest and loyal leaders and members to pull themselves together and show the present and rising generations the stuff they are made of. Our children will live after us and suffer the penalties of our mistakes and harvest the rewards of our honesty and courage."

The movement spread like wildfire quickly throughout southwestern Minnesota, and especially in Jackson Co., where records show there were 27 Local Unions existing by 1922, with 2,000 members.

In Feb., 1922, five locals – Silver King, Des Moines Valley, Star, Hummel, and Banner, held a joint meeting to order rail carloads of flour, salt, fruit, and machinery for members. However, there was no mention of this business meeting in the Jackson *Republic* which on Feb. 17, 1922, reported in a newspaper headline, "Farmers' Co-operative Union Holds Successful Meeting".

The Jackson newspaper article said 75 Farmers Union members from the Valley, Silver King, Star and Hummer locals attended the meeting. For some unexplainable reason, this news account failed to list the Banner local in its report.

The article continued: "Several very good addresses were given by different members... at noon a bounteous lunch was served by the ladies to which all did justice... after lunch, various business and farm problems were discussed, after which all departed for their homes, having spent a most enjoyable day." But there was no mention of the historic decision to organize a cooperative store which would be in business in Jackson Co. for nearly 50 years.

Arrangements were made to rent a building in the county seat town of Jackson in which to store the cooperatively-purchased merchandise. Then in 1923 a meeting was held in Jackson to establish a Farmers Union Exchange there.

Within six weeks $6,000 was subscribed and the Farmers Union Exchange was incorporated in Oct., 1923, with Fred Hample as president and Charley Wendel as secretary. The Exchange opened for business in Mar., 1924 with Louis S. Bezdicek as manager. A second store was established at Lakefield in west Jackson Co. that same year.

What Farmers Union members in Jackson Co. did was not novel. Dewey Local in the Cokato community in Wright Co., which was chartered on Aug. 23, 1920, with 28 members, in November that same year ordered a carload of sugar. It also organized a Farmers Union Shipping Association, and initiated a fund-raising effort for a Wright Co. Farmers Union store. Even though such early organizing efforts occurred in several areas of Minnesota before and after 1920, there wasn't enough membership to warrant a state charter until 1929 after the Northwest Organizing Committee began making great strides in building membership in North Dakota, Minnesota and Wisconsin.

MORE JACKSON CO. FARMERS UNION HISTORY

"We handle everything needed by the farmer and his family."
Slogan of The Farmers' Exchange of Jackson, MN., 1923.

Jackson Co. Farmers Union president Robert Rickert prepared a brief history of the organization in 1995, which varies somewhat from other records regarding the origin of the movement in southwest Minnesota.

In Rickert's brief, six-page hand-written account, he reports how the Jackson Co. Farmers Union was first organized in Kimball Township in 1922 by a group of farmers who were interested in banding together. Mrs. Paul Goede has the charter for Hummel local in Kimball Township dated Jan. 24, 1922. Fred Hample was president, and Andrew O. Johnson was vice president. The secretary-treasurer was Tom Albrecht.

It is thought the county was first organized in 1923 with approximately 1,200 members. H. K. Hinds was the first Jackson Co. chairman. Charles Wendel served as secretary-treasurer. These members were very "cooperative-minded", Rickert writes and quickly organized Farmers' Union Exchange stores in Jackson and Lakefield.

A special meeting was held Sept. 15, 1923, in Jackson to elect a temporary board of directors. This board then would launch a drive to get members to subscribe stock for the two, new cooperative stores. Elected to the board were John Schluter (who was temporary chairman), Charles Wendel (secretary-treasurer), William Benda, F. O. Lee, Martin Anderson, Martin Miller, and Fred Hample.

Two weeks later a second meeting was held, and officials announced that 192 shares at $25 each had been subscribed, or a total of $4,800 raised. On Oct. 22, 1923, at a third meeting, it was reported that 48 more shares had been subscribed and now the drive had netted $6,000. Two more meetings followed quickly, including the first regular stockholders meeting on Oct. 31, 1923, where action was taken to incorporate the Farmers' Union Exchange of Jackson. Fred Hample was elected chairman, and Charles Wendel was elected secretary – a post he would hold for 25 years.

The first paid manager was Louis Bezedicek, employed in 1924. The next manager was Dan Durbala. After Durbala retired, Harry V. Johnson became manager. He was succeeded by John Mellum. The Jackson "co-op store" was first located in what now is the Jackson City Hall. A new store was built in 1941.

A copy of the constitution and bylaws of the Farmers' Union Exchange of Jackson is available at the Jackson Co. Historical Society in Lakefield. It states the Exchange had capital stock of $25,000, divided into 1,000 shares. The slogan for the Exchange printed on the brochure states, "We handle everything needed by the farmer and his family."

The stated objective of the Exchange was printed in the brochure. It stated in part, "The objects of this association shall be: 1. The conversion and disposal of all agricultural products. 2. Buying and selling coal, lumber, flour, farm supplies and other merchandise including livestock for the mutual benefit of its shareholders and other Union members and such other benefits as the shareholders may direct."

Membership eligibility was confined to membership in "the F. E. And C. U. of America" (Farmers' Educational and Cooperative Union of America), according to the brochure, with no person able to own more than 40 shares, and only one vote per member.

The brochure spelled out details how the Exchange would operate, the number of directors (seven), terms of office, and pay ($3 per meeting and 10 cents per mile going to meetings).

Not as much detail was available for the Lakefield "co-op store". For example, while it isn't known for sure, Rickert believes the Lakefield "co-op store" was put into operation the next year after the Jackson store opened, or in 1924. The original Lakefield store was located in what was known as the Bertels Building, and then moved to what later became a hatchery. The "new" Lakefield store was built in 1947 along Highway 86 and today houses the Jackson Co. Historical Society and Museum. Store managers over the years included John Koster, John Wendt, Walt Geiser and John Mellum.

Early patrons recalled that peaches were a popular co-op store seller, with members paying only 99 cents a crate while other local stores charged $1.35 for the same type of peaches marketed in the same size crate.

Another big-seller was twine. But sometimes patrons had to pay up-front before the twine would be shipped from the prison where this product was manufactured. Involved in the twine business were Paul Aelke, William Bender, and Joe Slebodnik.

By 1926, farm machinery was added to the store's inventory. And the first tractors were being sold in 1930. Old-time patrons recall also that David Olson kept an inventory of petroleum supplies at his Jackson Co. farm for members until the Farmers Union Central Exchange in St. Paul began marketing such products through local cooperatives. Both of the "co-op stores" are now defunct. Popular was the annual Farmers Union picnic held near Windom at Fish Lake, according to records kept by the Historical Society.

Farmers Union was reorganized in Jackson Co. in June, 1950, thanks to efforts by Russel Schwandt, who was the state organization director then. Leo Freking of Heron Lake was elected county president and served as chairman until 1955. Alex Milbrath was president 1956-58, while Ray Salzwedel, Sr., served 1959-63. Clarence Granztke was elected chairman in 1964. In 1966 Robert Rickert became president, and held that office until 1981 when Shirley Rubis was elected. In 1987 Marvin Luckt was elected. In 1990 George Paulson began serving as county chairman.

In Mar., 1950, a new Farmers Union local named Hunter-Heron Lake local, was organized in the Lakefield creamery hall. Its name comes from the two Jackson Co. townships the new local represented. In time, it became known as the "Double H Farmers Union". The first president was Harm Ackerman. Ole Jeppeson was elected secretary-treasurer.

In Aug., 1950, the first annual Jackson Co. Farmers Union picnic was held at Kilen State Park. There were 625 in attendance, with music provided by the Bohemian Accordian Band. Activities included hog-calling and nail-pounding contests and a ball game. Speakers included the state president, Roy E. Wiseth of Goodridge.

A LOOK AT THE FARMERS UNION LOGO

From its very beginning in 1902 the goal of the Farmers Union (officially it is the Farmers Educational and Cooperative Union of America) has been simply this: a better life for all Americans, and especially for those families who live, work and play in America's rural communities.

The three main thrusts of the organization are reflected in the Farmers Union Triangle logo – the symbol showing the familiar trademark of the united plow, hoe and rake encompassed by a double equilateral triangle area presenting three words as legs of this triangle: "Cooperation, Legislation, and Education".

Figure 2 Farmers Union Logo

EDUCATION – THE BASE OF THE TRIANGLE

The base of the Farmers Union Triangle is "Education". It is the vehicle used to encourage member-participation in the organization's activities, in its cooperatives, and in its local, state, and national legislative programs. Education helps prepare the membership to take part more effectively in the important matters of cooperative organization and in legislation. For many years the educational programs of Farmers Union received the financial support of the large, regional farm cooperatives that marketed their grain and which sold them farm input and supply items. This relationship existed because legendary leaders like M.W. Thatcher of the Farmers Union Grain Terminal Association and others felt that educational programs represented a good investment in building and maintaining a cooperative organization. The bylaws of this cooperative provided that 5 percent of the gross savings would be provided to state Farmers Union organizations on the basis of cooperative support generated within the respective state.

To Thatcher and his noted peers, editor A. W. Ricker, and farm leader C.C. Talbott of North Dakota, the three "legs" – education, cooperation, and legislation – all were equally important and vital.

"When Equity (the old Equity Cooperative Exchange and predecessor of Farmers Union Terminal Association) fell, we recognized that information and education had to be an important part in the building of any cooperative organization," Thatcher said in a 1948 speech.

Indeed, Thatcher's emphasis on "education" as a vital part of building a farm movement or a farm cooperative is a sentiment he long had fostered. It dates back to when he was a member of the "Northwest Committee", one of three pioneer leaders who set out to organize farmers in the late 1920s. His two cohorts on this Northwest Committee were a pair of Farmers Union stalwarts, Farmers Union *Herald* editor A. W. Ricker and C.C. Talbott, then president of the North Dakota Farmers Union.

This Committee was authorized by the National Farmers Union board of directors at a St. Paul meeting in Mar., 1927, and given the charge to organize new states in the north central region of the United States, including Minnesota, and to qualify those states for chartering in the NFU.

In a Committee report printed in the Feb. 2, 1930, Farmers Union *Herald*, emphasis was clearly put on creating "a farmer-organization", controlled by farmers, and financed by farmers. The authors (Thatcher, Ricker and Talbott) said that it was vital for leadership to be "trained", and that "there must be education... directed and controlled by the farmers themselves, otherwise it is certain to be corrupted... "

Further, the Northwest Committee said in its 1930 report that "any attempt at organization of farmers is met with hostility, for the simple reason that you cannot organize farmers to buy and sell cooperatively without stepping on the toes of the commercial interests... but the moment you attempt to organize farmers to do something that will save them money, you are treading on toes whose owners will fight, and fight with any weapons at hand – falsehood, slander, and even abuse... "

The report also suggested the importance of an organization having its own newspaper, asserting that "the papers which these farmers read are victims of a system which makes the advertiser the 'boss' of the newspaper... the editor dare not disobey his master... "

Then the report added, "There can be no solution of the farm problem until farmers are organized so that they may read their own papers, be directed by their own leaders, and have machinery through which they may do their own business in their own way... marketing machinery, without understanding how to use it, or the educated will to do so, may do the farmers no good... there must be education... directed and controlled by the farmers... "

ONE LEG OF THE TRIANGLE IS COOPERATION

Farmers Union people have laid the foundation for, and have built, more cooperatives than any other group or organization in the nation. These pioneering leaders recognized that no single factor in the daily activity of farm people could be more important to the farm family, or to its future.

Cooperatives comprise the most effective economic vehicle available to farmers, while representing an organized effort promising enormous social implications as well. While the number of new farmer cooperatives being started has declined in recent years, the development of existing cooperatives and support for necessary changes in cooperatives continues to be a cornerstone of Farmers Union activity. Like farming itself, cooperatives have been confronted with periodic challenges and change.

LEGISLATION – THE SECOND LEG OF THE TRIANGLE

A keen, active interest in legislation has been demonstrated over the years by Farmers Union, locally, at the state capital, in Washington, D.C., and elsewhere. Progressive legislation is essential for farm families and for farm cooperatives to survive and grow, early leaders emphasized.

The thrust of Farmers Union legislative activity is three-fold in purpose: 1) To educate all of its members; 2) To communicate concerns, proposals and ideas to non-members and the general public; and, 3) To acquaint public leaders in government and elsewhere with the concerns and recommendations originating within Farmers Union and its membership.

Closely linked with legislative activity are research and information activity which can foster education and knowledge concerning the legislative positions and actions sought by Farmers Union. This is accomplished in several ways, via media news releases, bulletins, letters, backgrounder-type brochures provided to all interested parties, news conferences, meetings, and personal contact made by the organization's leaders and membership.

Such activity has one mission – to alert and acquaint people with the matters at hand, the concerns of farmers, the scope and dimensions of existing problems and challenges and the possible potential impact and implications of these events and consequences on Minnesota farm families, the public, and the society in general, as well as any proposed remedies or legislative solutions that might be considered.

M.W. Thatcher often said, and with conviction: "Farm prices are made in Washington". That's because he firmly believed sound, progressive legislation and government programs could be an important economic ally for family farmers.

THE FARMERS UNION TOOLS – THE PLOW, RAKE & HOE

There is another Farmers Union symbol just as well-recognized as "the triangle" – the Farmers Union's logo showing three items found on every pioneering farm: the plow, the rake and the hoe.

The symbols represent a closeness to the land, the source of food and sustenance. The plow is the dominant "tool" in the Farmers Union insignia. It is the time-honored symbol of agriculture, and how before the planting and promise of the seed must first come the seedbed preparation. And the plow featured in the logo is a single share, horse-drawn type implement – which reflects a family-sized farm.

The hoe symbolizes the nurture and care given to the crop, the removal of weeds and enemies of what is being grown. It is a recognizable tool which represents that an investment of sweat and muscle is a part of producing food.

The third object is the rake. Like the hoe, it is a mainstay farm item which is mindful of hard work, but which also reflects the end result of plowing and planting and caring for a crop – the harvest. The rake symbolizes that time of year when one gathers up the reward from what was planted.

WHICH PART OF THE LOGO IS MOST IMPORTANT?

The three sides of the Farmers Union triangle were discussed by the late, legendary farm-cooperative leader, Edwin Christianson, in a speech given July 20, 1967 at a meeting of Farmers Union Central Exchange officials, patrons and staff:

"Farmers Union was founded very wisely on a triangular base of education, cooperation, and legislation, and certainly it must be recognized that these tools have proven useful and successful when we have used them with a purpose and a determination.

"I would not attempt to say which one of the three sides has been most important; but it is quite clear to me that except for contributions of farm cooperatives, not much would have been possible."

Then the respected farm leader who was state president for 22 years added, "A partnership of farmers in and through their cooperatives is going to be every bit as important in the years ahead as in the past... "

THREE EXAMPLES OF SUCCESSFUL LEGISLATIVE ACTION

There were several examples of how Farmers Union's legislative program can benefit farmers and others during the 1970s after Cy Carpenter of Stearns Co. became the organization's seventh Minnesota president.

One opportunity came in 1972 when the Farmers Union became aware of a serious shortfall in providing educational benefits to returning military servicemen and women eligible for such assistance and training under the nation's G.I. Bill. Carpenter and his Farmers Union colleagues discovered that such G.I. Bill farm training programs were not available to most of the Vietnam veterans in the state. A vocational specialist was hired to research and document the situation.

Legislation was drafted, and a concerted effort made to have Minnesota provide sites, instructors, and facilities, to comply with the Veterans Administration Veterans Program. Then, too, there was lobbying and this situation was brought into sharp focus. This led to funding of the G.I. Farm Management Program, and the number of farm management classes available in Minnesota increased from a mere two to 92!

Then, five years later, Farmers Union found out that the U.S. Veterans Administration had begun using a new policy which denied farm management training and related educational benefits to young farmers if their net income exceeded a yet undefined minimum. After two years of meetings, letters back and forth, testimony before Congressional subcommittees and four trips by Farmers Union representatives to Washington, D.C., the Veterans Administration backed down. And then the Veterans Administration invited recommendations from Farmers Union concerning farm training programs and its controversial eligibility requirements.

A third example of legislative leadership was more dramatic. It came in 1976, a year that will be remember for the siege of hot, dry weather which gripped the state and shriveled up crops and created one of the most severe shortages of hay in the state's history.

Farmers Union saw the threat emerging head-on, and marshaled an effort to not only locate desperately needed forage supplies but also to secure reduced rail rates plus federal assistance to import the hay into the drought-hit regions of Minnesota where the future of entire dairy herds were in jeopardy.

Not only did it organize the highly-publicized, successful "hay-lift" that moved precious tons and tons of hay to where it was needed, Farmers Union also arranged a revolving fund put together by Minnesota Farmers Union and the three Farmers Union regional cooperatives (each loaned $25,000 to be used for purchase and transportation of hay). Farmers paid for the hay on delivery, plus freight. The Farmers Union set up a revolving fund, collecting money, and paying expenses as well as arranging for the location, purchase and delivery of the hay.

In the end, this Farmers Union program effectively capped the price of hay, saving farmers thousands of dollars. All of these drought emergency loans were repaid by grateful dairy farmers, who witnessed more stable and equitable hay prices than what they normally might have expected in a severe drought year and who were thankfully better able to maintain their herds instead of being pushed into a forced, emergency liquidation and a drought-depressed down market for their animals. However, because of litigation relating to poor quality hay shipments, the hay-lift turned out to be a money-losing proposition for the chief sponsor and instigator, Farmers Union.

However, twenty-plus years later you'll find many Minnesotans recalling proudly "the year of the drought" and how the Farmers Union was able to engineer a successful, modern-day "good neighbor" program and thwarted what could have been disaster for dairy farmers and others.

The spirit of the people was akin to years earlier when in rural Minnesota the Farmers Union movement caught fire. A time when people showed interest in establishing a new farm organization.

Early immigrant Minnesota farmers had struggled gallantly to establish, and to maintain, farm and home amid heavy debt brought on largely by the World War I boom that caused land prices to soar, and then decline sharply and steadily in the post-war period into the depression years of the 1930s. At the same time farmers experienced depressed commodity prices, high taxes, rampant foreclosure action and had almost no public voice speaking for and fighting for farmers.

Because they had virtually no protection in public law, farmers felt the sting of exploitation frequently, at the hands of ruthless railroad officials, grain merchants, livestock traders, bankers and others. Such pressures compounded the battle to survive, and hang on, in the hope the plight of farm families would improve.

Succeeding generations have trouble comprehending how those plucky, first-generation Minnesota farmers and their families were able to survive and persevere. How, without modern communication systems, were they able to join together as good neighbors, in a new "union" and withstand all of the many injustices they faced as pioneers, and newcomers in the "new world". With the flow of immigrants also had come a concept – cooperation.

Perhaps in some circles, and to some people, these things weren't novel or new. But they found new supporters and new participants in the period of harsh economic times following World War I.

CHAPTER 2

The Roots of the Farm Movement... Early Leaders

"You cannot balance the budget, you cannot make the farm relief legislation work, you cannot save the banks, the railroads, the insurance companies, and other commercial and industrial enterprises with a dollar that buys: four bushels of wheat from a Kansas farmer, ten bushels of corn from a Nebraska farmer, or, twenty pounds of cotton from a Texas farmer... "
National Farmers Union President John Simpson,
Letter to President Franklin D. Roosevelt, Apr. 2, 1933.

"The task of the Farmers Union is more than to get a few more cents for his products, or save a few cents on coffee; it is the struggle to obtain equality for all classes as guaranteed in the Constitution of the United States... "
D. D. Collins, convention chairman
First Convention of Minnesota Farmers Union, Willmar, MN, Nov. 1929

John C. Erp, first president Minnesota Farmers Union

"He was gone a lot... he saved a lot of farms for people... "
Lola and Donald Erp, speaking about their father, John C. Erp

Figure 3 John C. Erp

Born: May 17, 1885, Gibson City, IL.
Died: April, 1960 Burial, Canby, MN
Education: Attended rural school, Gibson, IL.
Employment: Life-long farmer, IL (1909-10), Canby, MN, 1911-46.
Community service: Organizing Farmers' Educational & Cooperative Union in Minnesota; state president of Minnesota Farmers Union when first chartered in 1929 until the organization lost its charter in 1938.
Married: Anna Stroh, Bloomington, IL, Aug. 30, 1908. Eight children, Rozella, Alfred, Viola, Irene, Edwin, Lloyd, Lola, and Donald. Anna Stroh Erp was born June 5, 1887, in Anchor, IL, and died Feb. 27, 1979, in Canby.

Perhaps it was history repeating itself. Just as at the turn of the 20th Century when agriculture was besieged by despair, and the troubled times spawned a new farm organization in Texas.

The formation of the Farmers' Educational and Cooperative Union of America was, in the words of its first president, Charles S. Barrett, "A protest of the man and woman in the field against the uneven distribution of opportunities and privileges" for farm people.

Farmers Union was to foster, sponsor, conceive and develop concepts of parity for farmers, emphasize cost of production as an important measure of return on investment, legislation as a means to bring economic justice to the marketplace as well as to promote farmer-owned cooperatives that pioneered in grain marketing. It was to lead in calling for natural resource conservation, seek social security for farmers, urge the development of national and international commodity trading agreements, the formation of the International Federation of Agricultural Producers, and other progressive developments.

Such things grew out of hard, difficult times. Just as nearly a generation later, the same financial scenario existed when the Farmers Union movement made inroads into the Upper Midwest, the Northwest, and Minnesota.

When a biographer asked two of the siblings of John and Ann Erp what they recalled about their father and his years as the first president of Minnesota Farmers Union, both at first hesitated. "We were very young," was the answer given by both Lola Erp Ouverson and her brother, Donald.

But then one provided this observation from those days many years ago growing up on a western Minnesota farm during the depression of the 1930s: "He (her father) was gone a lot." But then the other added something more significant about John Erp, and those years: "He saved a lot of farms for people ... "

Exactly how John C. Erp was able to "save a lot of farms for people", his offspring could not recall, being just young farm kids at the time their father was making agricultural history. Nor could researchers 60-some years later come up with records that provided an insight into the times and the activities of the late Mr. Erp and his new farm organization he had helped organize.

His story begins as a Freeland Township farmer in Lac Qui Parle Co., where he and his wife located near Canby – 18 years before he became Minnesota's first Farmers Union president.

The Erps as a young, married couple had moved to Minnesota in 1911, three years after their marriage and a brief farming career in their native state of Illinois. While it is not known exactly when John Erp joined Farmers Union, The Canby *Press* in its Dec. 13, 1929, edition identified him as secretary of the Lac Qui Parle Co. Farmers Union organization. Herman Laabs of Freeland Township was elected president of the Freeland local, while Gunder Anderson of Hantho became its secretary-treasurer.

Erp "called the roll of each local and received a verbal report of the paid up membership in each and also the amount of grain, coal, feed, flour and twine business completed during the year", reported the Canby newspaper which at the top of the front page called itself the "Official Farmers Union Paper", and on each side of its name, The Canby *Press*, proclaimed the publication to be "A Progressive Newspaper".

After delegates voted to hire a manager for the Madison Livestock Shipping Association, it was announced that a "jubilee meeting will be held in Canby in honor of the new state (Farmers Union) president, John Erp".

ERP ELECTED STATE FARMERS UNION PRESIDENT

A month earlier in 1929, a double honor had come to 44-year-old John Erp at the first meeting of the new – and first – state board of directors of Minnesota Farmers Union held in Willmar:

First, J. C. Erp was elected state president of the new farm organization he had helped organize in the late 1920s. Then, Erp was elected delegate to the National Farmers Union's 27th annual convention scheduled later in Nov., 1929, in Omaha, NE.

There were 339 registered convention delegates meeting Nov. 12-13, 1929, in Willmar. The crowd was so much larger than expected that sessions had to be moved from the Masonic Temple to the City Auditorium, according to a front page report by A. W. Ricker in the Farmers Union *Herald* (a bi-monthly newspaper founded two years earlier).

In addition to electing Erp, delegates also elected Hemming S. Nelson of Kandiyohi Co. state vice president. Nels Pederson of Chippewa Co. was elected treasurer, and Carl Lundberg of Lincoln was chosen secretary. Other state directors elected were Paul Oelke of Jackson Co., and Carl Winters and Oscar Brekke, both delegates from Yellow Medicine Co.

Minnesota membership was reported as 6,940 members, well in excess of the 5,000 required by the National Farmers Union to qualify for a state charter. Most of the membership then could be found in 14 counties: Yellow Medicine (555 members), Kandiyohi (526), Lac Qui Parle (520), Jackson (482), Big Stone (446), Lincoln (389), Wright (320), Chippewa (284), Nobles (222), Swift (221), Otter Tail (210), Renville (168), Lyon (119), and Marshall (108). Those 14 counties had 4,570 members, or about two-thirds of all who then belonged to the Farmers Union in Minnesota in 1942.

In a historical article published in 1988, F. B. Daniel reports that the Farmers Union organization in Minnesota apparently began in 1920 in Wright Co. when G. H. Blodgett "brought the Union with him from Iowa". The movement also developed about the same time in Jackson Co. in the early 1920s.

"THE STRUGGLE... TO OBTAIN EQUALITY... FOR ALL CLASSES... "

The Canby *News* in its Dec. 20, 1929, editions reported on the "Jubilee Meeting" in the Canby City Hall celebrating John C. Erp's election as Minnesota Farmers Union president, and the election of two other men from the area to state offices.

Presiding at the celebration honoring the new state Farmers Union leaders was Farmers Union organizer D. D. Collins. The newspaper reported on Collins' speech to the crowd:

"The first 20 minutes (of the meeting) were used by the chairman (Collins) in stirring recitation of the battle to secure equality for the laborer and the farmer. One of his outstanding statements was, 'The task of the Farmers Union is more than to get a few more cents for his products, or save a few cents on coffee; it is the struggle to obtain equality for all classes as guaranteed in the Constitution of the United States... ' "

The newspaper report also indicated a lighter side to that first state convention, noting that newly-elected Minnesota Farmers Union President Erp had refused to wear overalls at that meeting. And when officials presented Erp with a new pair, suggesting that he "put them on as the insignia of his loyalty to his overall friends", Erp again refused.

After this second attempt to get President Erp to wear overalls at the meeting failed, the matter was put to a delegate voice vote. It was an unanimous "aye" vote favoring Erp to don the overalls. Still, he balked. But when a standing vote was taken, the new president had to comply – much to the amusement of the audience, according to the Canby *News* account. Then followed, the newspaper added without elaboration of any kind, "Mr. Erp's short address that was full of meaning and seriousness".

The newspaper reported that Nels Pederson, the state Farmers Union board member, had a message that day "that reached many hearts who could say 'Amen' to his convictions ably expressed in his short talk". Carl Winter, state secretary and another member of the board of directors, and like Erp, from Canby, was said to have "made a good speech, reciting some of the difficulties in organizing farmers... his talk was well-received... ", the newspaper article continued.

The colorful newspaper report also included this paragraph: "A word of appreciation was spoken to the families, wives, and children, who are willing to care for the work of the farms, while husband and father fill the front line ranks in the war for justice... "

DELEGATES IN 1929 VOTE FOR CANBY OIL CO-OP

The major news of the convention was buried in the news columns, toward the end of the article. It was the fact delegates voted unanimously to establish a bulk oil station in Canby, with this explanation, "the farmers believe it would be better to keep the ten to fifteen thousand (dollars) profit that goes east, and spend it on shoes, groceries, etc., right at home".

The Canby *News* article ended with this editorial comment: "The meeting adjourned and a happy bunch of farmers went home, to recall the several hours of laughter and jubilee. May their laughter continue until the day of smiles and justice become the fruits of the (Farmers) Union endeavor ."

CALENDAR OF FARMERS UNION LOCALS

A special Farmers Union page was published as part of The Canby *Press*. It listed a calendar of monthly meetings for local organizations by county, and also the names of local presidents and secretaries.

Here is the listing of locals from Dec., 1929:

Yellow Medicine Co. – Forteir, Burton, Wergeland, Sandness, Norman, Normania, Oskosh, Hammer, Florida, Stony Run and Swede Prairie.

Lincoln Co. – Alta Vista, Ash Lake, Hansonville, Hendricks, Marble, Royal.

Chippewa Co. – Big Bend, Havelock and Crate, Tunsberg, Woods.

Lac Qui Parle Co. – Baxter, Freeland, Garfield, Hamlin, Mehurin, Providence,

Ten Mile Lake. Co. president – Herman Laabs; county Secretary – J. C. Erp.

In a historical article, F. B. Daniel writes, "You are not necessarily wrong if your memory says Minnesota Farmers Union dates back to the year 1942. That is the year of our current charter.

"The state lost its 1929 charter in the 1930s when the pressures and trials of the Great Depression, the drought, and the constant agitation of the grain trade and the Minneapolis Chamber of Commerce divided its members into seemingly irreconcilable opposing camps."

Daniel and others credit a group identified as "The Northwest Committee" as playing a major role in the Farmers Union movement into the Upper Midwest and into Minnesota. This "Committee" was created by the National Farmers Union board of directors at a meeting in St. Paul on Mar. 12, 1927.

Figure 4 L to R: M.W. Thatcher, A.W. Ricker, C.C. Talbott

Three stalwart farm leaders were appointed to this organizing committee:

M.W. "Bill" Thatcher, general manager of Farmers Union Terminal Association (which he fashioned out of the financially-troubled Equity Cooperative Association of St. Paul).

C.C. Talbott, president of North Dakota Farmers Union.

A. W. Ricker, editor of the then new Farmers Union *Herald* (which was the successor news publication to Ricker's *Farm Market Guide* and Equity Cooperative *Herald*, publications sponsored by the Farmers Union Terminal Association).

This committee's success was immediate– North Dakota was chartered in 1927, with Montana and Washington-Idaho organizations chartered the following year. At the national convention in 1928, the Northwest Committee promised to bring another state – Minnesota – into the Farmers Union fold within a year. And that they did.

There were 2,796 signed up members in Minnesota that year, records show. National Farmers Union board member D. D. Collins set up an organizational campaign headquarters in Canby in Lac Qui Parle Co. and announced a goal: 5,000 members by year's end. It was to be an objective that was exceeded by nearly 2,000 (6,940). So Minnesota, described as a hard state to organize, and Oregon both gained charters in 1929.

Minnesota was a difficult state in which to organize Farmers Union for several reasons: First, it was a grain and livestock state, both enterprises highly dependent then on non-farm marketing agencies to purchase their crops, milk, and meat animals. This meant dealing with well-financed interests such as the Minneapolis Chamber of Commerce noted for its resistance to the rebellious policies advocated by groups like the Producers Alliance, Farmers Union and others where they intended and did seek revolutionary changes in marketing practices.

Herald Editor A. W. Ricker acknowledged this vested resistance to farmers organizing for change when he wrote in his newspaper: "Considering the fact that practically every farmer who joined the Farmers Union in Minnesota has done so after having been given every reason why he should not do so by our opposition, is evidence of the fact that the member joined because he was convinced of the soundness of the Farmers Union program... "

Also, officials reported that it cost $13,000 for National Farmers Union to organize and charter Minnesota – more than any other previous state chartered. By comparison, it took only $4,500 each to activate charters for Wisconsin and Montana. Those were amounts expended by the Northwest Committee and did not include thousands more invested by Farmers Union Terminal Association, or by grain, livestock and insurance departments in recruiting and organizing Farmers Union memberships and locals.

From the start of the Farmers Union movement, emphasis was put on "education" as an important activity. In the Farmers Union *Herald* in Feb., 1930, this point was made clear:

"There must be education, and that education must be directed and controlled by the farmers themselves. Otherwise it is certain to be corrupted," charged *The Herald*. In the last analysis, farmers must come to realize that they will never emerge from bondage until they become a class unto themselves, fight their own battles, and follow their own leaders. By their own leader, we mean farmers who become leaders through organization training.

"To be of any value, a farm organization must be just that – a farm organization. It must be financed by farmers, controlled by farmers, and led by farmers. And farm leadership is not possible unless such leadership is first trained through organization... "

Further, *The Herald* editor appealed for a "farmer" newspaper – something the publication itself unquestionably was: "There can be no solution to the farm problem until farmers are organized so that they may read their own papers, be directed by their own leaders, and have machinery through which they may do their own business in their own way."

Farmers Union dues, the editorial continued, cannot cover the costs of educational activity needed to promote organization and cooperation. "These funds to organize and educate must come from the only other place where funds may be obtained – namely out of the earnings of sales organizations, insurance, commodity purchasing associations, local cooperatives, etc... "

A LOOK AT EDITOR A. W. RICKER

A brief biography of A. W. Ricker:

Born: Dec. 15, 1869, Johnson Co., Iowa.

Died: Feb. 11, 1955, San Gabriel, CA.

Education: Graduate, Iowa City High School

Employment: Iowa newspapers; *Pearson's Magazine*, New York City; Minnesota *Star*, *Farm Market Guide*, Farmers Union *Herald*.

Community Service: Consultant and advisor, Farmers Union; speaker at numerous farm events.

Honors: National Farmers Union distinguished service award (presented posthumously at national convention, 1955).

Figure 5 A.W. Ricker

The author of the Farmers Union *Herald* editorial appealing for a "farmer's paper" and for cooperatives to provide educational funds likely was Iowa-born and educated A. (for Allan) W. (for Warren) Ricker (who was called "Rick"). He was born on a farm near Iowa City in Johnson Co. on Dec. 15, 1869.

After completing high school in Iowa City, he began working for newspapers – a trade he pursued the rest of his life – for special interest liberal publications and trade journals. It was while working for the Minneapolis *Star*, the Non-Partisan League paper, that he met M.W. Thatcher who was auditing the publication's books at the time. Then later, Ricker became secretary of the Producers' Alliance and edited its paper, the *Farm Market Guide*. It was through the Alliance that Ricker met Charles C. Talbott of North Dakota, and later introduced Talbott to Thatcher – a meeting which was to lead to a strong, productive relationship.

And when the three groups – the Farmers Union, the Producers' Alliance and the old Equity organization – joined together in 1926, it proved to be a fruitful union. This threesome would develop a productive, warm relationship that became the "team" which organized Farmers Union in the Upper Midwest and Northwest, and spawned the birth of what would become super-cooperatives.

Ricker is said to have described the relationship this way: "The Alliance (Talbott) furnished the organizers, the Farmers Union (Ricker) the good name and a good history, but it was the old Equity (Thatcher) philosophy and spirit which provided the stickers in our organizations."

Not only did A. W. Ricker faithfully cover Farmers Union news events for 20 years, he also would swear in new Minnesota state officers at conventions, be a speaker, and an advisor and consultant to the farm organization. Sometimes his reports read more like editorials than news dispatches, and were not without obvious bias and subjectivity. Editor Ricker, indeed, wrote in *The Herald* with an obvious bias that clearly informed the reader that the writer "is one of us", and that his audience represented "the underdog" in the nation's economy, and that the only salvation would be cooperation with enough political clout to influence legislation.

C.C. TALBOTT'S UNTIMELY DEATH

M.W. Thatcher, the general manager of the Farmers Union Grain Terminal Association, who had been a speaker at the 1942 state reorganizing convention, again was a featured speaker the following year at the 2nd annual Minnesota Farmers Union convention program held in Montevideo in Nov., 1943. Other speakers that year included A. W. Ricker, editor and publisher of the Farmers Union *Herald*, and E. H. Everson, president of the South Dakota Farmers Union.

Figure 6 C.C. Talbott

The third member of the dynamic Farmers Union organizational trio was C.C. Talbott of North Dakota. His promising career was cut short by death April 8, 1937, of injuries sustained in an automobile accident on Mar. 26 on U.S. Highway 10 near Steele, ND. At the time of his death, Talbott was president of North Dakota Farmers Union and president of Farmers Union Central Exchange. He was headed to Bismarck for a radio broadcast when the accident occurred. Talbott's son Glenn, only 36, would immediately become the North Dakota president and a distinguished leader, just as someday would a Talbott grandson, Stanley Moore.

STALWART MEMBERS/WORKERS – LURA & OSKAR REIMNITZ

Brief Biography of Lura Mae Chapman Reimnitz
Born: Oct. 12, 1907, Kent, IA.
Died: Oct. 30, 1974, Willmar, MN.
Education: Graduate, Gregory High School, Gregory, S.D.
Employment: Secretary, 1943 Minnesota Farmers Union state convention; special assistant to insurance director, 1944-45; education field worker, 1945; state

secretary-treasurer, 1945-50; state education director, Minnesota Farmers Union, 1950-72.

Marriage: Oskar Reimnitz, Mar. 9, 1929, Huron, SD. Two sons, Lyle of St. Paul, MN, and Dale of Cambridge, MN.

Honors: Recognized for 25 years of service to Minnesota Farmers Union.

Brief biography of Oskar E. Reimnitz
Born: Aug. 19, 1901, Wells Co., ND
Died: June 21, 1985, Willmar, MN

Employment: Farmed for a number of years; worked for rural electric cooperative in Willmar area, 1945; organizer, Minnesota Farmers Union; employee, Farmers Union Oil Co. of Willmar for many years.

Community Service: State convention rules committee, 1945; state FU executive Committee, 1946.

Honors: Minnesota Farmers Union "pioneer", 1975 convention.

Figure 7 Laura & Oskar Reimnitz

Easily one of the most dedicated, long-time Farmers Union workers was Lura Reimnitz of Kennedy in Kittson Co. She and her husband Oskar were early members, staff members and leaders.

During her long career, Lura Reimnitz was to play an important role in the lives of thousands of young Minnesotans through educational programs, the Farmers Union camping activities, and the operation of state conventions and other meetings. Her leadership featured versatility and an ability to handle a wide variety of Farmers Union chores.

Lura's years of service to Minnesota Farmers Union began shortly after the organization was re-chartered. She served as secretary at the second state convention held in Alexandria in 1943, and began working on a part-time basis for the newly re-chartered organization later that year. Her duties included contacting high school science and history classes, setting up youth groups, day camps and helping with state camps.

In 1944 Lura was appointed "insurance specialist" and assistant to insurance director Edgar Nordstrom. Then a year later she began working part-time as education field-worker, and became a member of the state education council. Later in 1945 she became state secretary-treasurer and assumed day-to-day management of the state office then located in New York Mills. She also was a member of the state convention credentials committee.

Her best-known role, though, began in 1950 when she became education director for the Farmers Union. She was to hold that position for 22 years – a period of unprecedented and phenomenal growth for the organization. Thousands of Farmers Union juniors enrolled in special classes and programs during Lura Reimnitz' tenure as state education leader, with those boasting special dedication and study activity climaxing their youth careers recognized with the prestigious Farmers Union "Torchbearer" award in special ceremony held at each state convention.

At the same time Lura Reimnitz played a major role in the development of Farmers Union "Ladies Fly-In" trips to Washington, D.C., including serving as a group leader and tour guide for the delegation.

Her husband Oskar was a volunteer worker for Farmers Union in the early years after the organization was re-chartered. Then in 1945, he became a state organizer with C.C. Griffith. He served on the state convention resolutions committee in 1945, and then became a member of the Farmers Union state executive committee in 1946. In 1947 he left the Farmers Union staff to become an employee of Kandiyohi Rural Electric Association, and then to work for Willmar Farmers Union Oil Company. He worked there for many years.

Oskar and Lura Reimnitz became members of Farmers Union in the early 1940s. He was a native of Wells Co., North Dakota, while Lura was born in Union Co., Iowa. In a 1976 interview, Oskar reflected on his long career working with, and for, farm cooperatives. "People have to help themselves," Reimnitz said. "You can really get involved trying to explain the situation with cooperatives. The best way to understand a cooperative is if you get a dividend check. Cooperatives today are different. They have grown and expanded; services have improved. Back in the early 1920s no laws were available to help them like now."

Over the years, the Willmar couple served as guides and tour leaders for many Minnesota delegations to national conventions, on "Fly-ins", and other trips, as well as to meetings of the International Federation of Agricultural Producers, an organization supported solidly by Farmers Union since its inception.

Two major activities which Lura Reimnitz supervised were the Farmers Union summer camping program, and the "Torchbearer" ceremony held as a part of the annual state convention. Since its inception, more than 1,400 young Minnesotans have earned this prestigious award, including former state Farmers Union president Willis Eken and National Farmers Union President Leland Swenson.

In a special report to the state board of directors dated Aug. 28, 1972, Lura told how her camps held from June 1 to Aug. 5, were attended by 378 junior youth (ages 6 to 14), 114 senior youth (8th Grade through high school), and 26 ladies (she recommended that the "Ladies Camp" be discontinued). She noted the popularity of camps, and growing competition for campers adding, "at one time Farmers Union provided the only camps for farm youth... today schools, churches, lodges, schools, sports, band, music camps are common... "

At that time the National Farmers Union required youths to attend two state senior camp sessions to qualify for "National All-States Camp" near Denver in the Rocky Mountains. Minnesota had 63 eligible young people who qualified for the All-states Camp; however, only 21 attended. Again, Lura Reimnitz said she felt young people had other interests which were keeping more youth from attending what she described as "once-in-a-lifetime" camping experiences.

At the end of her career, Lura Reimnitz left an important legacy – a 30-year history of Minnesota Farmers Union which she prepared especially for Edwin Christianson, the state president she worked with for 20-some years. This document proved to be an important resource in the preparation of this history.

YOUTH PROGRAMS IN THE CHRISTIANSON-LURA REIMNITZ ERA

"He Loves His Country Best, Who Strives to Make It Best"
Junior Motto, 17th Annual Convention, St. Paul, 1958.

"We live in an age when opposing world powers are capable of destroying human lives by the tens of millions literally overnight," featured speaker Flossie Harris, educational director for the National Farmers Union told her Minnesota audience. "The world's great powers compete for a military ascendancy which none dare employ because it appears clear that none would win or, indeed, survive in a total war with present weaponry."

Harris' message that November night was that "we need to help tomorrow's adults learn, understand, and cherish the basic values of human freedom". Her listeners were those attending the "Junior Night Program" at the 17th annual convention of the Minnesota Farmers Union.

The convention theme was, "Pioneers For Progress In A Changing World". It was the age of Sputnik (the Soviets had successfully put a dog into orbit the previous year), the space race, and the nuclear-"Cold War" threat. The speaker reminded her listeners of such rapid, far-reaching changes in her address.

The sessions were held in the Theater Section and Stem Hall (used by the convention for the first time because "it was necessary to get a larger meeting room", Lura Reimnitz told the audience) in the old St. Paul Auditorium, with the "Junior Night Program" held on Sunday evening Nov. 23.

The special night reflected the skilled hands of State President Edwin Christianson and the State Educational Director Lura Reimnitz. It was what a veteran Farmers Union member and observer described as "typical" in the days of the Christianson-Reimnitz leadership team, and a time when interest in the organization spurred record membership.

Most of the stars of that evening production were 33 young people designated "Torchbearers" – an honor bestowed as a "national award" for completing five years of "required Farmers Union project and study work". Five of the "Torchbearers" came from Beltrami Co. – the largest delegation from any of the 13 counties that were represented by the "Torchbearer Class" of 1958. They were Karen De Vries, Alvin McClellan, Erma McClellan, Connie Northrup, and Voyle Wingren.

These counties each produced four "Torchbearers": Roseau – Anna Flick, Marie Sjostrand, Walter Thompson, and Wanda Urtel; Swift – Patty Kvam, Walter Munsterman, Betty Sanders, and Glenda Strommer; and, Yellow Medicine – Mona Dandurand, Lois Danielson, David Greseth, and Larry Stensrud.

Also receiving special recognition that night were junior members completing their sixth, seventh, and eighth years of project work and study, and three 1958

"Torchbearers" who were in England: Sheldon Haaland (Yellow Medicine Co.), Donald Swenson (Cottonwood Co.), and Glen Anderson (Lac Qui Parle Co.). The six-year honorees were James Hanson of Fertile, Veldo Jones of Hawick, and Curtis Swenson and Eugene Engler, both of Westbrook. The lone seven-year honoree was Rodney Efterfield of Big Woods Local in West Marshall Co. Eight-year recognitions were given to Elvern Gunderson and Hiram Sirjord, both of Bejou, and Betty Jean Ronholdt of Murdock.

The schedule of this special evening program began with Lura Reimnitz inviting the audience to join in the singing of the Farmers Union song, "Hail Our Union". After making a few remarks, she introduced State President Christianson.

Christianson pointed out it marked the first time the Junior Members had been a part of the regular convention schedule, with their "Junior Night" held before the state meeting. Also, he added, the convention was being held later in November to better accommodate the corn harvest schedule for farmer-members. And, it was the first time in his reign that National Farmers Union President James G. Patton was not on hand to present the national honors to the group of new-named "Torchbearers".

Lura Reimnitz directed the Torchbearer Ceremony, which features a procedure where each honoree comes to the front of the meeting hall and lights a candle. Then the group of inductees sings in unison, "Follow The Gleam".

At the 1958 program, Velda Jones of Kandiyohi Co., a 1957 award-winner, recited the Torchbearer Pledge, while another 1957 honoree, Avis Fricker of Polk Co. presented "Hold High the Torch". Betty Jean Ronholdt, a 1955 Torchbearer, sang, "I Would Be True" while all of the new class sang, "Hail Our Union". Then, after Ronholdt performed, "May The Good Lord Bless and Keep You", a recessional parade of newly-designated Torchbearers from the stage followed with all leaving the platform carrying their lighted torches.

A coffee-and-cake reception was held afterward, with Ed Christianson announcing a special resolutions committee meeting at 11:30 p.m. yet that evening, with an 8 a.m. breakfast session the next morning to begin the convention's second day.

SPECIAL ARE THOSE OF VISION... WILFORD E. RUMBLE

"His handiwork will live as far into the future as any man can see... For farmers and cooperatives a true and effective friend... "
Alfred D. Stedman, Nov., 1971, writing on Wilford E. Rumble in FU *Herald.*

Figure 8 Wilford Rumble

Those familiar with the remarkable career of Wilford E. Rumble referred to him as "the architect of cooperative law".

This Minnesotan "fashioned and built the legal foundations at the state and federal levels which today support the work of cooperatives far and wide", wrote St. Paul newspaper farm editor Alfred D. Stedman in an obituary Nov. 8, 1971, in the Farmers Union *Herald*. Rumble died in a St. Paul hospital on Oct. 26 at the age of 80.

"The good effects of his (legal) handiwork will live as far into the future as any man can see," wrote Stedman about the attorney of the Twin Cities law firm Doherty, Rumble and Butler.

"Nearly six decades ago (early in the 20th Century) Wilford Rumble was beginning his scholarly studies of cooperatives and their economic and legal problems."

That was back when, Stedman reported, "farmers could, and did, go to jail for organizing to sell or process their own products or buy tools or supplies for themselves through their own cooperatives". There were no state or federal laws protecting farmers and cooperatives. It was Rumble who saw the need to advise cooperatives "then springing up like mushrooms" in Minnesota and elsewhere, according to Stedman.

"He (Rumble) was in the thick of it – counseling co-ops, planning and executing court battles, defending in lawsuits, staving off imprisonment until new laws could be passed, getting co-op leaders out of jail, explaining to legislators why in justice and reason there must be pioneering changes in the law to safeguard the works of co-ops...

"In words that legislators quickly understood, he told over and over again how farmers were being squeezed between the millstones of low farm prices and high farm costs, and he spelled out the letter and reasoning of new the laws that were needed... he and others fashioned the foundations for cooperatives through the Minnesota Cooperative law of 1916 and the federal law (the Capper-Volstead Act of 1922)... "

When the great challenge came from the National Tax Equality Association in the 1940s, Rumble and others, including M.W. Thatcher (see Chapter 4), provided strategic leadership through the prolonged and bitter battle. The Tax Equality group sought to stall, and perhaps destroy the mushrooming cooperative movement, by amending income tax laws so that cooperatives would be liable for income tax on refunds paid to their members.

Rumble's method, Stedman wrote, "was not oratorical or written bombast, but clear and simple telling of the facts and earnest and persuasive explanation of their meaning". He portrayed farmers as "all capitalists operating their own farm factories" and not as radicals or revolutionaries. They formed cooperatives, he said, "as a kind of partnership to get higher prices for their crops and lower prices for their supplies". At the end of the year, he would explain, the cooperatives paid members in cash or stock what was due them. These refunds were not profit or income to the co-op, but were income owed members as owners of the cooperative, Rumble would explain.

He spoke out against the cooperative critics, saying, "There is no merit in their position, and no intricate or complicated technical argument is necessary to meet it... whatever the cooperative receives by way of income belongs to its patrons". He repeated this message often, and before the Congressional committees when they considered major cooperative legislation.

However, it was not until Rumble and cooperatives won their fight to safeguard the tax status of cooperatives in 1962 that there would be an end to the battle with the anti-co-op tax group which began its fight more than 20 years earlier. So, it wasn't until then that the existence and future of cooperatives moved beyond question and doubt.

It was during the 1940s that Rumble and his associates won other challenges to permit farmers and their cooperatives to transport their own commodities across state lines and retain exemptions from costly transportation fees, permits and safety regulations.

During his career Rumble represented several farm groups – the Minnesota Farmers Union, the Farmers Union Grain Terminal Association, the Farmers Union Central Exchange, the Land O'Lakes Creamery Association, the Central Livestock Association, the former Twin Cities Milk Producers Association, Midland Cooperatives and even the Minnesota Farm Bureau Federation at an earlier time.

It was his depth of knowledge and understanding of agriculture, his warm feeling and concern for farm people and other traits which enhanced his effectiveness as a legal advocate and spokesman for farmers and their cooperatives, according to Alfred Stedman.

"The lucid logic of his language, and the direct forcefulness of his speech made him a great favorite as a speaker in agricultural conventions," Stedman wrote. "The wide respect in which his integrity was held... rubbed off to a degree on the cooperative movement and helped give cooperatives a public respectability and stature that might have been hard for them to otherwise attain in the face of the harsh and widely-publicized assaults by some business competitors. In his death, farmers and cooperatives have lost in Wilford Rumble a true and effective friend."

EARLY FARM LEADERS SEEK "JUSTICE"

In Farmers Union history, credit for interest in legislative activity by farmers is given to Charles Barrett, the Georgian who became national president in 1908.

Barrett, who was to hold that office for 20 years, was among the first to recognize and demonstrate the need for farmers to create and utilize legislative capability. A. C. Davis served as national secretary 18 years, or nearly as long as Barrett was NFU president.

The National Farmers Union is not truly a "national" organization, but instead a federation of state Farmers Unions. This relationship has over the years provided an arena marked by more than a little internal dissention at times as state units sometimes tend to be independent, and even rebel against policies and positions held by the parent organization. In 1928 at the national convention, internal differences led to the election of C.E. Huff of Kansas as national president. This seemed to create what some described as "a new birth" to Farmers Union, underscoring its commitment to the three-phased program (education, cooperation, and legislation) that had helped build the organization for so many years.

However, Huff took office at a time of controversy, particularly the struggling Federal Farm Board and the new, untested Farmers National Grain Corporation. As a result, John Simpson of Oklahoma was elected national president at the 1930 convention held in St. Paul.

It was in the late 1920s that the Farmers Union *Herald* became the successor to the *Farm Market Guide* and the Equity Cooperator *Herald*. The editor was A. W. Ricker, the Iowa native and one of the key figures in the "Northwest Committee" which engineered the "amalgamation" that played such a pioneering role in the emergence of the Farmers Union in the Upper Midwest states. *The Herald* began publishing in Mar., 1927, with a 50-cent subscription price that was supplemented with subsidies provided by the three Farmers Union regional cooperatives (Farmers Union Terminal Association, Farmers Union Central Exchange and Farmers Union Livestock Commission Company). There was no definite publication schedule, with the newspaper's size and publication dictated by the money available from the three cooperatives.

Ricker was seen as a savvy editor, and an advocate for farmers and cooperatives. His newspaper and his leadership played a major role in the development of one of the most successful farm organization efforts ever. He shared a genuine and dedicated interest in working to better the position of farmers by setting the stage for them to organize and help themselves through the three-phased Farmers Union program of education, cooperation, and legislation.

The Farmers Union stressed three areas in particular to attempt to improve the lot of family farmers, rural community life, and the nation:

First, it sought equal opportunity and treatment for farmers and not favors or special privileges. Just as the Producers Alliance had championed "cost of production plus a reasonable profit", this was the goal of the new Farmers Union. Some 30 years later the National Farmers Organization would preach the same agricultural gospel. In the 1990s the terminology is for farmers to experience "a level playing field" in competing for markets and fair returns for their production.

Secondly, when they individually could not achieve justice in the marketplace, they came up with an alternative – the cooperative. And then they fought for equal treatment for their cooperatives with competitors in the marketing of farm production.

Finally, being among the first to recognize the need for seeking helpful and corrective legislation, and having gained the ability to create legislative capability, the Farmers Union all too often wound up bloodied in the very process in which it excelled. That's because what legislation it sought originally often became amended and changed to the point where the final product barely resembled the original proposal. And then, in the administration, it generally emerged further distorted.

A True Prairie Populist – George Sperry Loftus

"George S. Loftus was like the pioneer who tore up the virgin soil with a breaking plow... under the green of buffalo grass was many a hidden boulder... he broke that prairie planting the seeds of cooperation and unity... "
Farmers Union *Herald*, May, 1943.

To the grain growers of the Northwest early this century, their economic lives were being controlled by powerful forces beyond their control, the grain traders of Minneapolis and its hated Chamber of Commerce which in reality was the regional grain exchange, or primary market for what they harvested.

It was a much-disliked system that would be challenged by one of the architects of modern-day cooperatives, George Sperry Loftus, a former farm boy, railroad company executive, elevator operator, dynamic orator and arch-foe of the Minneapolis grain trade. He became the first to battle the Minneapolis Grain Exchange on behalf of farmers, and when denied a seat on that Exchange, with the help of Otto Bremmer of the American National Bank of St. Paul and St. Paul businessmen, he started a St. Paul Grain Exchange.

Loftus' background provided extensive knowledge of the grain merchants and how they operated, and a deep-seated bitterness against their unethical, cheating practices buying grain from the producers. His convictions led him to become a strong supporter of an anti-monopoly progressive, Wisconsin Senator Robert M. LaFollette.

When Loftus was appointed manager of Equity Cooperative Exchange in 1912, the Minneapolis grain merchants recognized it as almost a declaration of war, according to the book, The Greenwood Encyclopedia of American Institutions, Farmers' Organization. In 1914 the Equity organized the St. Paul Grain Exchange to compete head-on with the Minneapolis grain institution, and lobbied for North Dakota's Legislature to authorize building a state-owned terminal elevator. When the North Dakota state elevator matter lost, Equity decided to build its own facility in St. Paul.

The Equity elevator was completed in 1917, a year after Loftus died, in time to profit from World War I demands. Equity earlier had expanded its activities into livestock marketing, and wool sales and cooperative buying of agricultural necessities and supplies. None of these operations, though, was as financially successful as the grain terminal.

Loftus, with his great skill of oratory, had done a superb job goading farmers into building an important marketing tool. After his death, and the end of the wartime boom period, Equity tried unsuccessfully to organize growers into a "wheat pool". In Jan., 1923, it was deeply in debt and headed toward receivership. It was then that the astute manager M.W. Thatcher took over, and borrowed on the experience of George Sperry Loftus to take the failed cooperative and transform it into the nation's largest and most successful grain cooperative in history.

Then, a later Equity's successor organization, Thatcher's Farmers Union Grain Terminal Association, bought St. Anthony & Dakota Elevator Co. on May 1, 1943, in a $2.7 million deal which then represented the biggest acquisition in the cooperative's history. Involved were 135 grain elevators and 38 lumber yards in six states (Minnesota, the Dakotas, Montana, Nebraska and Iowa).

Observed the Farmers Union *Herald* in its June, 1943, edition: "George Loftus was the pioneer who blazed the way for the cooperative movement... M.W. Thatcher has been the builder, his imagination and vision have made a reality of a pioneer's dream... Loftus cleared the way in the wilderness... (which) Thatcher has followed to build well and strong... "

Thirty years earlier, in 1913, the newspaper said, "George S. Loftus was storming the prairies to arouse the producer to the evils of the then existing grain marketing system and to convince him of the need of a farmer-owned grain marketing organization if he was to hope for a fair return for his labor...

"Loftus was like the pioneer who tore up the virgin soil with a breaking plow... under the green of the buffalo grass was many a hidden boulder which, catching the plowshare, threw the long moldboard out of the furrow and jolted harshly the man whose hands guided the plow... but he broke that prairie to plant the seeds of cooperation and unity and that which he has cultivated has not returned to its former wild state... "

OTHER EARLY LEADERS:

Charles Barrett - President of National Farmers Union (1908-28) and key figure in the "Amalgamation"of 1927.

C.E. Huff – A Kansan, and NFU president, succeeding Charles Barrett of Georgia in Feb. 1928 and serving until June, 1929.

Frank Livingston – Former teacher, early and longtime Thatcher aide and public relations specialist..

J. Edward Anderson – an early activist from Buffalo and a member of FU select Washington legislative committee in 1935.

T. J. Kelly of Renville – a representative for Minnesota on Corn Belt Committee (Originally Grain Belt Federation of Farm Organizations).

D. D. Collins of Colorado – NFU board member and first crew chief in Minnesota membership drive. Established headquarters in Canby, home town of John Erp in 1928.

John C. Erp – first Minnesota FU president, elected in 1929, was vice president of FU Exchange, and a member of FU Terminal Association board of directors.

F. J. McLaughlin – He was born in Dubuque, IA, but moved to the Dakotas in 1926, and is recognized as one of a group of men responsible through the fabled "Northwest Committee" for bringing the Farmers Union movement into North Dakota, Minnesota, and Wisconsin. He became an employee of Farmers Union Terminal Association in 1929, and two years later joined the staff of the Farmers Union Central Exchange. He was the stepfather of Thomas H. Steichen who was general manager of the Central Exchange from 1957 to 1963.

C. J. Mitchell – another former organizer for the Producers Alliance in South Dakota, and first credit manager for Farmers Union Central Exchange.

Emil A. Syftestad – succeeded M.W. Thatcher as manager of FUTA in 1931; first general manager of FU Central Exchange, 1931-1957.

Charles Egley – long-time manager of Farmers Union Livestock Commission Company, and an organizer of Farmers Union locals. The livestock company played a significant role in providing early financial support for Farmers Union when most meat animals were sold at the "central", or terminal, markets.

Milo Reno – he wore several "hats" – president of the Iowa FU, an organizer of the FU Livestock Commission Co., president of Farm Holiday Association, an "Amalgamation" figure, and an early FU policy-maker.

Paul Moore – Pres., Equity Exchange; "amalgamation" player, FU organizer.

John Bosch – from Atwater, he was a Minnesota Farmers Union executive board member, president of the Farm Holiday Association and a mediator of internal strife of the Minnesota Farmers Union during the depression era of the 1930s.

(NOTE: The following people represent "the second wave of leadership" in Minnesota Farmers Union, rising to statewide prominence and leadership following the revocation of the state charter by NFU and the subsequent rechartering.)

Einar Kuivinen of New York Mills – the first and third Minnesota FU president after rechartering.

Bob Handschin – For 33 years he worked as director of economic research for Farmers Union Grain Terminal Association, retiring in May, 1976. Prior to joining GTA he was a legislative representative for the National Farmers Union in Washington, D.C. for five years. For several years Handschin was editorial page editor for *Co-op Country News* and its predecessor, the Farmers Union *Herald*. A 1932 Phi Beta Kappa math graduate at the University of Illinois, he worked at a wide variety of jobs before becoming a career Farmers Union cooperative specialist. He died in April, 1997, at the age of 86.

Jewell Haaland – A faithful cooperator, and former chairman of the board and a member of the board of directors of Farmers Union Grain Terminal Association for 21 years. He was a patron of Cottonwood Oil Co., the world's first cooperative of its kind founded in 1921. He died in Feb., 1997.

Richard Johansen – key staff worker at Farmers Union Grain Terminal Association for many years and press secretary for its general manager, M.W. Thatcher. He was a skilled writer who authored many of Thatcher's texts.

Joe Nolan – His career was spent heading up the petroleum division for the Farmers Union Central Exchange, from 1931 when he became employed, to his retirement in 1962. Nolan saw the division grow from a group of only six employees to a work force of 300 in sales, clerical and technical personnel, and participated in the acquisition of the Central Exchange's own refinery in Laurel, MT, and a one-third interest in a cooperative refinery in McPherson, KS.

George C. Lambert – He was appointed a receiver for the Equity Cooperative Exchange in 1923, and became one of those who helped organize the Farmers Union Terminal Association which took over the Equity operation. Lambert became secretary-treasurer and general counsel for the new cooperative. He was one of the early leaders seeking Congressional authorization for deeper channels to accommodate barge transportation on the Upper Mississippi River.

Howard Peterson of Benson, devoted member of Farmers Union and champion of cooperatives. He served in many capacities: GTA fieldman, coordinator of member/visitor activities at GTA (succeeding Frank Livingston), county president and member of FU state board.

James E. Manahan – He was an early Minnesota Congressman and is credited with helping save the Equity Cooperative Exchange from its arch rival, the Minneapolis Chamber of Commerce. As a receiver for Equity, Manahan led an investigation which led to an injunction that provided the Chamber of Commerce not interfere in any way with the grain marketing activity of Equity. This helped preserve what then represented the only cooperative grain terminal market in the Upper Midwest.

Luella Jacobson – long-time junior leader, early staff member, camp staffer.

Ralph Ingerson – another key Central Exchange staff member, active in development of oil cooperatives, and the "father" of the revolutionary famous Co-op Tractor developed in 1930s and assembled in St. Paul at one time.

C. W. Stickney – This Clear Lake farmer was the long-time state chairman of the farmer-elected Minnesota state committee overseeing the federal farm programs for nearly 15 years, and was associated with these programs from 1933 to 1953 when he resigned as chairman of the Minnesota Production and Marketing Administration. He had been appointed state chairman of the Agricultural Adjustment Administration in 1938.

D. L. O'Connor – the New Rockford, ND, leader and first president of the Farmers Union Grain Terminal Association board of directors.

Jon Wefald – Minnesota agricultural commissioner who assisted the Farmers Union hay-lift campaign during 1976 drought, and later president of Kansas State University.

Maurice Melbo, Wilt Gustafson and Lyle Heaton – long-time state staff members and Farmers Union bus drivers.

In 1941 Minnesota Was Different From Others

"We can be good Farmers Union members without offensive fault-finding with other organized farm groups... "

Farmers Union Herald, June, 1941.

Minnesota was the only Northwest state where Farmers Union really had any farm organization competition, the Farmers Union *Herald* editorialized in June, 1941. At that time there were "a small Farm Bureau organization in southern Wisconsin... and a small Grange in Wisconsin by the lake shore".

In North Dakota there was reported to be a lone Grange unit, and no Farm Bureau in evidence. The Farm Bureau had organized in a half-dozen Montana counties, while The Grange was active in three counties.

In Minnesota in 1941, however, things were different. While The Grange had some pretty deep historical roots in Minnesota, it lacked presence in those areas where Farmers Union was organized and active. However, thanks to its close links with the Extension Service, Farm Bureau was active in many areas of the state, the newspaper reported. And, it added, many people mistakenly believed that without Farm Bureau there would be no county Extension agent. Evidence that this was not true, the editorial noted, is the fact that all but two counties in North Dakota had Extension agents even though Farm Bureau at that time didn't exist in that state.

The point of the editorial was to suggest that efforts be made to avoid "local conflict" because of farm organization competition. "No doubt as we (Farmers Union) expand and develop in Minnesota, there will be some local conflicts with Farm Bureau officers and perhaps members," *The Herald* observed. "We have instructed our field workers and organizers to make every effort possible to avoid such conflicts."

There are, the newspaper continued, "some goals which all three organizations are agreed on", but also there is disagreement... the Farmers Union organization begins with a local; the Farm Bureau is set up on a county-wide basis and has no locals or local autonomy... the Farmers Educational and Cooperative Union is most of all, as its name implies, an organization interested in developing the cooperative movement... many Farm Bureau members are patrons of our cooperative oil associations and patrons of our Cooperative Grain Elevator Associations... quite a few Grange and Farm Bureau members are patrons of Farmers Union Livestock Commission Co. at St. Paul and West Fargo... "

The editorial concluded by urging "Farmers Union members to set a good example to the members of other farm groups by avoiding controversy as far as possible... we can be good Farmers Union members without offensive fault-finding with other organized farm groups."

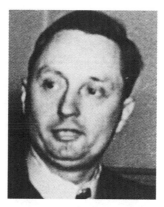

Figure 9 Chas Egley, long-time Farmers Union Livestock
Commission Co. mgr, organizer of many FU locals

Figure 10 Chas Barrett, Pres. NFU (1908-28);
A key figure in "Amalgamation" of 1927

Figure 11 C. E. Huff. Kansan, NFU Pres. 1928-29

Figure 12   A. W. Ricker. Head of Producers Alliance,
FU NW Committee Sec. , editor, Farm Market Guide
& Farmers Union Herald, and an "Amalgamation figure.

Figure 13 C.C. Talbott. Organizer for Producers Alliance, later for
Farmers Union; Pres. ND Farmers Union, member, Exec. Comm.
FU NW Committee, 1st Pres., Board Chairman, FUCE

Figure 14 M.W. Thatcher. Equity Exchange auditor, chairman,
FU Northwest Committee, Manager, FUTA, key staff member
Farmers Nat'l Grain Corp., Pres. & Founder Nat'l Fed. of
Grain Cooperatives, key FU legislation rep., & Gen. Mrg., FUGTA

Figure 15 F. J. McLaughlin. Born in Dubuque, IA, moved to Dakotas in 1926, recognized as one of group responsible through the NW Committee for bringing Farmers Union movement into ND, MN, and WI . Employee of FU Terminal Association in 1929.

Figure 16 C. J. Mitchell. Organizer for Producers Alliance in SD, 1st Credit Mrg., Farmers Union Central Exchange

Figure 17 Emil A. Syftestad. Succeeded M.W. Thatcher as mgr. of FUTA in 1931, 1st Gen. Mgr., FU Central Exchange, 1931-1957

Figure 18 Einar Kuivinen of New York Mills
- 1st & 3rd MFU Pres. after rechartering.

Figure 19 Bob Handschin, dir. Econ. Research
FUTA 1943-1976, NFU legislative rep., 1938-43,
editorialist Co-op Country News & FU Herald. Died 1997

Figure 20 Richard Johansen. Staff at FUGTA,
Press Sec. to M.W. Thatcher. Skilled
writer, authored many Thatcher speeches.

Figure 21 Ralph Ingerson, FUCE staff, developer of the Co-op Tractor in 1930's which was made in St. Paul. Active developer of oil co-ops.

Figure 22 Jon Wefald, MN Ag Commissioner who assisted FU hay-lift campaign during 1976 drought, later Pres., Kansas State Univ.

3. TOUGH TIMES GRIP RURAL AMERICA...

"If you incur a debt when, say, corn is 80 cents a bushel, and then you try to pay for it when it is 2 cents a bushel, it doesn't make any difference how hard you work, you just can't do it... this was the situation (in 1930's)... "
John H. Bosch, Kandiyohi Co., Member, Farmers Union, Farm Holiday Ass'n.

"The bitterness became so violent (National Farmers Union President John A.) Simpson socked (National Farmers Union Editor A. W.) Ricker in the jaw at the 1934 (NFU) convention. . . . . "
Farmers' Organizations, The Greenwood Encyclopedia of American Institutions.

"The Republican 4-H Club: Hoover, Hyde, Hell and Hard Times".
"In Hoover we trusted; now we are busted."
"Give us honest prices so we can pay honest debts."
Signs in a 1932 Des Moines parade led by Batavia Farmers Union Band (Batavia, IA, was the home of Milo Reno, Iowa Farmers Union president and president of the Farm Holiday Association).

To understand the source of farmer rebelliousness in young America, one needs only to understand the tremendous power and influence on early rural society emanating from the railroads and the rail barons who owned and ran them.

Many of the new settlers, and the immigrants from "The Old Country" had been lured to the provinces of the Upper Midwest by the promising propaganda emanating from and distributed by the railroad companies. Colorful, promising accounts of what awaited the newcomers in "the New World" were reproduced, widely advertised and distributed by the railroaders who themselves had been given subsidies and benefits, including land parcels, by the government.

Some of these pioneers even arrived on special "immigrant trains", and often settled on land sold to them by the railroads that hauled them to the west. Also, these newcomers often worshiped in churches built on railroad-donated sites located near the evenly-spaced rail stations which sprung up in the westward movement and settlement of mid-America in the post-Civil War era of the Nineteenth Century. Further, local bankers and businessmen had close financial ties with the large railroad firms which in turn had friendly financial connections with the influential business cliques in Minneapolis and St. Paul, and beyond.

In turn, grain commission companies, railroad companies, and others owned most of the facilities which represented the primary markets for the crops grown by the immigrants, pioneers, and settlers, attracted to the new country by the same railroad firms which now would buy their produce and haul it to distant markets.

Consequently, independent grain firms often found the rail companies stubborn in providing their needs in developing alternate markets for what the new generation of farmers were producing.

However, in addition to the economic alliances that thwarted their efforts to make money, the farmers also found another obstacle – the farm-fresh wheat they sold often was graded lower than what the same grain was really worth, and which later sold for higher prices on up the line and in the commercial markets of the Twin Cities or Duluth, and beyond.

Even the tailings (the material sifted out of the original wheat to upgrade its quality) were worth nothing to the farmer. That is, until this same "valueless" stuff was sold back to them as cattle feed by the same buyers who initially claimed it was worthless and had no commercial value.

With such questionable grain trade practices rather commonplace happenings in mind, it is easy to see why the farm country of young America was ripe for a rather radical and rebellious leadership and a dose of "prairie populism". But there were those equally ready to fight and put out any prairie fire-type protest of the status quo which threatened their profits.

A sample of the kind of opposition Minnesota farmers likely would encounter trying to organize into effective protest units came early this century because of the activities of the Non-partisan League and its leader, A. C. Townley, a native of western Minnesota who rose to fame in the state of North Dakota where a disgruntled politician once told farmers to quit bellyaching about things and "go home and slop the hogs". This quotation spread quickly and became sort of a rallying cry for farmers to no longer take things the way they were, and to organize and together fight for change.

After the League won control of that state's Republican Party in the 1916 primaries, and Townley set his sights on Minnesota, powerful business interests in Minneapolis and St. Paul took action, according to a historical account published in Farmers' Organizations, The Greenwood Encyclopedia of American Institutions, 1986. The Twin Cities businessmen, mainly a group of merchants and lumbermen who feared the movement promised a dire, negative impact on their financial interests, even organized a counter-group – "the Minnesota Nonpartisan League, Inc."

The main goal of this group was to oppose in any way it could anything Townley's Nonpartisan League happened to favor. The Minnesota establishment reportedly had much cash with which to fight the League, and issued a couple of pamphlets in its 18-month existence. One publication had a patriotic theme, and was titled "America First".

However, Townley's climate for organizing a new farm group received a much-welcomed boost when the North Dakota Agricultural Experiment Station issued a bulletin. Its research showed quite clearly how deceptively the grain merchants were deliberately down-grading wheat bought from growers, and then re-grading the same grain to gain a higher price later when the same wheat was sold. Townley and his League organizers circulated this North Dakota bulletin at their meetings, which added a spark of discontent to the farmers and which added interest in the Townley crusade.

When the United States entered World War I, the mood in rural Minnesota changed. Some of the immigrants – not many years distant from the "home country" they left to come to America – openly opposed the war effort. And because some had joined with Townley and his followers, the Townley movement soon became branded and tarnished as "un-American" or "pro-German". This identity cast a disturbing light on the movement, and tarnished it to the point it never recovered.

When a new organization, the American Farm Bureau, came on the scene, Townley shifted gears. He resigned as president of the Nonpartisan League in 1922, and began to organize yet another new group, the National Producers Alliance, instead.

Coincidentally, during this same period, the first locals of the Farmers Union were organized in Minnesota, in places like Jackson and Wright counties. This organization had been founded in 1902 by Isaac Newton Gresham, a native of Georgia who had been an organizer earlier for the National Farmers' Alliance and Industrial Union in the 1880s in the South. It was chartered as the Farmers' Educational and Cooperative Union of America, with the first local organized in Rains Co., Texas.

At first, the dues were 5 cents a month ($1 for initiation fees), with applicants required to be "white, over 26, of good character, a believer in God, a farmer or farm worker, a country teacher, a preacher, a doctor or mechanic, and editors could join with prior approval".

A DECADE OF CHALLENGE AND TURMOIL

Minnesota's Star Rises, Shines, (and Falls)

At the 1928 National Farmers Union convention in Denver it was announced that Minnesota had been targeted for state chartering as a new affiliate in 1929. The organization work had begun in 1928 when the national board appointed D. D. Collins of Colorado as chief organizer, and after the creation of a group called "The Northwest Organizing Committee". It was felt that Minnesota would have sufficient membership in a dozen counties to become worthy of a state charter late in 1929.

The efforts of many – including D. D. Collins, the organizing crew chief and others – were successful. And the first state president elected was John C. Erp of Canby who subsequently was seated on the national board of directors. (Not all state presidents were automatically seated on the national board in those early years as routinely has been the case in recent years.)

Minnesota followed states like the Dakotas, Montana and Washington – major wheat-producing states. The early leaders felt that there had been an advantage in organizing around a major commodity like wheat where there were recognizable common concerns and common interests, and common problems.

However, before long, a series of events occur which will have an impact on the fledgling Farmers Union in Minnesota. The FU Livestock Commission Co. emerges from the assets of Equity Livestock Commission Co. in 1927. The Farmers Union Terminal Association rises from what's left of the restructured, and financially-troubled Equity Terminal Association. The Farm Holiday Association develops out of the economic gloom gripping rural America. And many Minnesotans become active and involved in this effort to "strike" until farm markets bode a price equal to "cost of production plus a reasonable profit".

MINNESOTA FARMERS UNION LEADER HAS A PLAN

One of the leaders of the "Farm Holiday" movement was John H. Bosch, a member and leader of both Farmers Union and vice president of the Farm Holiday Association. He told of those traumatic times in an interview with David L. Nass of the Southwest Minnesota Historical Center at Marshall which was printed in part in Minnesota History (Mar., 1975).

Bosch was the eleventh of 14 children in his family raised near Prinsburg, MN, born on Dec. 4, 1898. His father was an active Populist, and Bosch joined Farmers Union as a young man (Lake Elizabeth Local) in about 1930. He became county chairman and vice president of the Farmers Union Livestock Commission Co. in South St. Paul. He also was a brother-in-law of B. Franklin Clough, one of the leaders of Minnesota Farmers Union when it was re-chartered in 1942, and elected state vice president in 1944 and state president a year later.

Concerned about the depressed economics for farmers, Bosch said he started thinking and came up with an idea that he presented at his local (Lake Elizabeth) meeting: People could do without cars and a lot of things, but they can't do without food. So, the young farmer figured, why not try to do four things: 1) Have a "holiday" on farm foreclosures, just as the banks took a "holiday". 2) Seek "cost of production" prices for farm products. 3) Abolish the Federal Reserve System which he and some others blamed for fluctuations in prices and credit. And, 4) In the event of another war, tax all profits from war production 100 percent. And, if farmers failed to gain these four things, they would "strike" and withhold their produce from the market.

The Farmers Union local delegates approved Bosch's plan, and urged him to take it to the Kandiyhoi Co. convention where it was also endorsed. The county leaders voted to have Bosch take his plan to the Minnesota Farmers Union convention. After the Minnesota delegates also approved this plan, it then was advanced to the National Farmers Union convention in Oklahoma City. However, at the national level, John Bosch's proposals were coolly received. "They were afraid to become involved," Bosch related. "The Farmers Union was quite heavily involved in cooperatives, and they were afraid that, legally, these cooperatives would become involved in this (the strike proposal)... "

A NEW ORGANIZATION IS FORMED

A historical book, *Cornbelt Rebellion*, tells what happened at the National Farmers Union convention of 1931. Delegates rejected the idea of calling for a farm strike because the organization was engaged in business operations it feared might suffer if it publicly supported such a tactic. Instead, Bosch recalled, a resolution was passed calling a meeting (held May 3, 1932 in Des Moines) to create a new organization not officially connected with Farmers Union but receiving the support of Farmers Union members – and leaders, including Milo Reno, Iowa Farmers Union president.

Reno, a lay minister, was a dynamic and gifted speaker from the tiny rural Iowa town of Batavia. A parade in his honor held in Des Moines in 1932 was led by the Batavia Farmers Union band, and featured signs of the time, including one proclaiming "The Republican 4-H Club – Hoover, Hyde, Hard Times and Hell".

The message, of course, pertained to President Herbert Hoover, his Agriculture Secretary Arthur Hyde of Missouri, and economic conditions that existed for many farm and other folks. Reno would become famous for his leadership in the Farm Holiday movement, including his oft-repeated advice to embattled farmers – "Buy nothing; sell nothing", or "strike" until farmers were paid prices equal to their cost of production plus a reasonable profit.

"This name – the national Farm Holiday Association – was selected because we did have the 'bank holiday', and we wanted a 'farm holiday'," Bosch explained. "And the program I set up became the (Farm Holiday) program... "

Bosch, who lacked the identity of Milo Reno, was elected vice president of the Farm Holiday Association. Then, he recalled, a series of things happened sort of spontaneously and likely well beyond the control of the Farm Holiday leaders.

"Knowing that this meeting was held, and that we were probably going to call a farm strike, there were a half-dozen areas that called strikes of their own – just blocked the highway, and not related to anything," Bosch recalled. "Somebody, maybe he wanted the publicity... or maybe his reasons were entirely honorable and good... there was one around Omaha, one at Sioux City – Marshall had one where they had quite a rumpus (On Nov. 10, 1933, a crowd of farm strikers estimated at 1,100 shut down the Swift & Co. Plant in Marshall. A sheriff's car and local fire trucks were damaged)... " (He did not mention a crowd of 2,000 farmers who gathered near Canby, MN, to pay tribute to Nordahl Peterson, a 26-year-old Farm Holiday picket who had been killed by a shotgun blast fired into the darkness of a picket camp apparently at random by an unknown assailant near Peterson's home town of Canby. That western Minnesota town also was the home community for John C. Erp, first president of the state Farmers Union.)

About the sporadic upsurges of demonstrations, protests and such, including the practice where rope nooses were dangled in front of foreclosure participants and even a District Court Judge (at LeMars, IA, in April, 1933), Bosch offered an explanation:

"Here something develops which would be difficult to foresee. A farmer, by the nature of his occupation, is an individual. His farm is different than somebody else's farm – the soil, the amount of rain he gets or doesn't get, the size of his family... does he feed cattle or raise grain? And the date he has to pay (on his mortgage) is different than everybody else's date. To try to fit these all together is an extremely difficult thing... you begin to set up terrific conflicts between different groups. So what should have been your support becomes your opposition... "

Bosch said the Farm Holiday Association tried a strike, then called it off "because it was spasmodic, intermittent". Then, another date was set where coordination was hoped for, but never really amounted to much. "At least 50 percent of the paid-up membership (of the Farm Holiday group) came from Minnesota," Bosch stated. "So all the rest of the United States had less than half... so we (Minnesotans) were supporting what everybody else was or wasn't trying to do... "

Minnesota Gov. Floyd B. Olson "was 100 percent in favor of what we were trying to do", Bosch said. "So was Bill (William L.) Langer of North Dakota. But none of the rest of the governors had the guts... "

Bosch told how farmers would spread out, covering township after township like "minutemen" in a system to organize crowds for "penny auctions" at farm foreclosure sales, or to protest or intimidate activities of law enforcement officers and bankers serving foreclosure notices on farmers.

The activity seemed to center in Kandiyohi Co., Chippewa, Yellow Medicine, Lac Qui Parle, and adjacent counties, Bosch indicated. Then, there was the day in Montevideo when Bosch remembered "I bet I had about 10,000 farmers there" to try and stop a federal marshal in a foreclosure sale. The marshal was threatened, and called names. But his response was something to the effect that he agreed with the farmers about what was going on, but that he had a job he had to do. This action challenging federal action brought a call from Governor Olson, and an invitation "to see him". When he arrived at the Governor's office, Olson told Bosch he had received a call from President Roosevelt about the farm protests in Minnesota.

Olson said the President thought Bosch was doing a good job representing farmers, but that he said that "You can't buck the federal government", adding, "You're going to spend the rest of your life in the pen", Bosch related. And, when Bosch indicated he planned to continue "Bucking the federal government", and that even if he was put in prison, he wouldn't stay there because he could get 10,000 or 100,000 if needed, to free him. At this point, Bosch said, the Governor asked, "Well, what do you want me to do?" And Bosch suggested, "Call the President."

With both Bosch and Olson on the governor's phone, President Roosevelt reportedly told Bosch he respected what he was trying to do, but repeated what the Minnesota Governor had already told the farm leader, "You can't buck the United States Government." When Bosch told FDR, "Mr. President, we did.", Roosevelt said something like he didn't want to see him spent the rest of your life in the penitentiary, Bosch recalled. When the farmer indicated he planned to continue to do "whatever we have to do to stop them", Roosevelt asked the same question Olson had asked earlier, "What do you want me to do?"

Bosch then said he told Roosevelt of his ideas how to help financially-troubled farmers by different actions. Later, Bosch said he also talked with Roosevelt's Secretary of Agriculture, Henry A. Wallace. Federal programs were rushed into place, including the Agricultural Adjustment Act passed a few months later (May, 1933), a Farm Credit Act (June, 1933), federal bankruptcy laws were amended (June, 1934), conciliation committees were set up at the state level, and on Feb. 23, 1933, Minnesota Governor Olson proclaimed a temporary moratorium upon farm mortgage foreclosures and the state legislature later extended the redemption period on past-due mortgages for two years.

Such activities ended the thrust of the "Farm Holiday" movement, and Bosch told historians that "it is not an exaggeration to say that through what we did... the mortgage moratorium in Minnesota and North Dakota and the debt conciliation commission act... probably saved hundreds of thousands or millions of homes and farms from foreclosure... "

A NEAR REVOLT BY THATCHER, RICKER OCCURS

When National Farmers Union balked at supporting the New Deal program of President Roosevelt, M.W. Thatcher, A. W. Ricker and former National Farmers Union President Charles E. Huff of Kansas became genuinely concerned. They knew farmers liked FDR's New Deal – it was the economic salvation for hundreds of thousands of farmers. Also, they saw the protest so prevalent in rural areas vanish quietly as government cash began to pour into the countryside.

They realized further had the Farmers Union allied itself with FDR, endorsed its programs, the organization with its long, liberal history might easily have become the premier agricultural ally of the New Deal and its popular founder, FDR. However, thanks to the intransigence of NFU President John A. Simpson of Oklahoma, who won a bitter election with Huff in 1930, and his close relationship with the controversial Milo Reno of Iowa, the agrarian "Friend of FDR" role was assumed by the competitor, the American Farm Bureau instead.

This situation festered. Nebraska seceded for a time, and Ricker, Thatcher and Huff talked seriously of abandoning Farmers Union and starting a new "union of cooperatives". They and other Farmers Union members were steadfast "New Deal" supporters. The divisive bitterness became so intense that NFU President Simpson socked Editor Ricker in the jaw at the NFU's 1934 convention.

A month later, Simpson died of a heart attack on the steps of the U.S. Capitol, and was succeeded by Ed Everson of South Dakota – another critic of the New Deal. But the Thatcher-Ricker-Huff contingent bided their time, while doing what they could to develop a better relationship with Roosevelt, and the New Deal, and building their respective business enterprises. Reno died in 1936, and Thatcher and his team gained much-needed momentum. It would be the last time this August twosome, Thatcher and Ricker, who had done so much to organize Farmers Union, would be known to quarrel openly with Farmers Union and its policy.

PREMIUMS COME FROM "TIN CAN CAMPAIGN"

Also in 1927, the Farmers Union Terminal Association launched what becomes a legendary episode. It became know as the "tin can campaign" because wheat growers were invited to send, by mail, samples of their wheat crop for testing for its protein content.

The theory behind this effort was akin to the North Dakota Experiment Station research a dozen or so years earlier – the idea that growers routinely were being cheated, and that their wheat was really a superior, quality product worth much more on down the food chain compared to what buyers had been paying for the farm-fresh grain.

The result of the "tin can campaign" was that farmers for the first time were duly rewarded for the true value of their wheat, pocketing premiums which previously grain commission companies had retained for themselves. This program may have been the biggest single factor ever in motivating farmers to join, and be extremely loyal to, the Farmers Union and the cooperatives which helped them achieve a more honest price for their crops.

However, the times were extremely uncertain and troubled, and unrest began to emerge within the Farmers Union membership across Minnesota, almost in proportion to the plummeting of the farm economy, ahead of the time the nation entered the Great Depression in the early 1930s. Some members were unhappy that Farmers Union had not climbed on the political bandwagon of the new president, Franklin D. Roosevelt, as quickly as had some competing farm organizations and groups. Also, they felt not enough support or attention was being given by Farmers Union to the historic, "New Deal" farm programs.

The unrest and discontent was an underlying factor in the suspension of Minnesota's charter by the NFU. The revocation of the state charter was followed by litigation and by a court appeal of the decision against the Farmers Union unit. But, in 1938, the appellate court upheld the revocation. Minnesota was out of the Farmers Union fold (a fate that would happen to other states later, including Iowa 20 years later).

When the original Minnesota Farmers Union, headed by Erp, refused to patronize/associate with Farmers Union regionals, Robert Miller of Underwood formed a second organization that claimed to be a state Farmers Union. J. Edward Anderson of Wright Co. was its secretary. This organization attracted a goodly following, as evidenced by a picnic in Fergus Falls in 1934 which drew 10,000 people – reflecting the height of farmer sentiment.

However, when Erp (described by one member, Ellsworth Smogaard of Madison, as "one who liked to have things done his way") and his followers began to openly question national policies of the parent organization, the National Farmers Union was forced to take action. And the official first Farmers Union of Minnesota lost its national charter.

Almost immediately, though, a new Farmers Union movement began to emerge – and an important step toward regaining the national charter. At a meeting at Moorhead in Nov., 1940, a band of farmers with the help of national and North Dakota Farmers Union leaders, set their sights on rechartering.

NEW FARMERS UNION ELECTS KUIVINEN

Near the end of her professional career, long-time Minnesota Farmers Union education director Lura Reimnitz compiled a self-styled history of the organization which she described as "just the bare facts" telling some of the things that had happened.

This record she constructed is the primary source of information about the second state Farmers Union which came on the Minnesota scene in a movement that began almost when the official national charter of the first organization was revoked.

"In the fall of 1929 a convention was held at Willmar, MN, at which time the Farmers' Educational and Cooperative Union of America was chartered. It continued until 1937... when the National Farmers Union found it necessary to revoke the charter because of inconsistencies and uncooperative methods of those in leadership positions... ", Reimnitz wrote.

During the following years, "interested individuals, including A. W. Ricker, Dan Collins, Esther Schneider (she served as secretary of the NW Organizing Committee), Ione Kleven (she was educational director) and others worked to establish a nucleus of a new organization", she added. In Nov., 1942, this group called a convention at Fergus Falls and applied for a new Farmers Union charter.

Several counties had gained charters in 1941, according to records. One of these counties was Swift. The first president of this newly-chartered county was Theo Frederickson of Murdock – the grandfather of David Frederickson, the current state Farmers Union president. Theo had three sons who farmed in Swift Co., and who all were active Farmers Union members. The sons were John, Donnell and Rudolph. The latter is the father of David Frederickson who became Minnesota president in 1991. Donnell Frederickson was Swift Co. president from 1956 to 1967, and a member of the Minnesota Wheat Commission for several years.

RICKER WAS CHAIRMAN OF MINNESOTA ORGANIZERS

A. W. Ricker, editor of Farmers Union *Herald*, was in charge of the committee organizing Farmers Union in Minnesota, according to a report on organizational work published in the Nov., 1941, FU *Herald*. Mrs. Esther Schneider was secretary of the committee.

Ricker reported that "Minnesota's organization work in 1940 was confined almost entirely" to communities where there was a Farmers Union Oil cooperative. Then in 1941 emphasis shifted to working township by township with the goal of forming a county organization. But muddy roads interfered, after heavy rains. Most farmers at that time lived on "dirt" roads which became impassable when it rained.

While the Farmers Union *Herald* absorbed expenses of Ricker and his stenographers who handled correspondence in the newspaper's office, additional financial support came from the Farmers Union Terminal Association, the Farmers Union Central Exchange and the Farmers Union Livestock Commission Company. For example, GTA paid for the services of Frank H. Livingston, who worked full-time working with both old and new locals, including training sessions for the officers and membership, and others.

At a monthly meeting of the Homestead Township Local near New York Mills in Oct., 1941, Livingston appealed for support at the Farmers Union local level because "through his local he is aiding the building of the (Farmers Union) organization... which will be the only real hope when the post-war depression again blasts the country... "

Livingston also criticized "American industrialists" who he said defend imports of "competitive agricultural products from Latin America where foodstuffs are produced by peon labor at about 6 cents a day" by blaming high food prices caused by the war.

YELLOW MEDICINE FIRST TO BE RECHARTERED

Long-time Otter Tail Co. member Verner A. Anderson of New York Mills remembered, in writing a local history, how the efforts to gain a second state charter were "centered mostly in Lac Qui Parle, Otter Tail, Dakota, Kandiyohi, Wadena and other interested areas in central and northwestern counties in Minnesota... "

Farmers Union records show Yellow Medicine was the first county to be chartered by the National Farmers Union after the state charter had been revoked. Its charter was dated Jan. 14, 1939. Chippewa was also chartered that same year, but the date was not recorded.

In 1942 charters were granted to five Minnesota counties – Kandiyohi (Jan. 17), Swift (July 15), Beltrami (Nov. 4), and Big Stone and Lac Qui Parle, dates unknown. Six charters were issued in 1943 – Roseau and Clay in January, Mahnomen and Polk in March, and Kittson in November. There was no month listed for Becker Co.'s charter.

"As they re-organized," Verner Anderson recalled, "Many locals asked for, and kept, the same local numbers they had held under the old (original) 'Erp' organization (Farmers' Educational and Cooperative Union of America, Minnesota Branch) which lost its charter in 1937... "

Early local charters, Anderson recalled, were issued in 1941 to Heinola (he was first secretary and Einar Kuivinen was first president), Newton, and Homestead locals in Otter Tail Co. Also, to Menagha in Wadena Co.; to Home Office in Dakota Co. (which was comprised of staff and employees of the three regional cooperatives); to Riverside, North Branch, and Springbrook Jupiter locals in Kittson Co.; to Harmony in Lac Qui Parle; and, to Triple in Roseau Co.

State membership on Nov. 10, 1942, when the new Minnesota Farmers Union was chartered, was 1,907, Anderson noted. The charter was signed by James G. Patton, National Farmers Union president, and by J. M. Graves, national secretary.

Delegates Pick Kuivinen for President

Delegates had only one choice for president of the "new" Farmers Union organization. They elected the only man nominated, Einar Kuivinen of New York Mills, president. However, it was a different matter for the vice presidency, won by E. L. Smith of Montevideo who defeated a bid by Edgar L. Nordstrom of Becker Co. for that office.

Five were elected to the state board of directors that day, including two (B. Franklin Clough of Lake Lillian and Edwin Christianson of Gully) who like Kuivinen would within a decade become president of Minnesota Farmers Union. Other directors elected were Curtis Olson of Roseau (Roseau Co.), William Cassavant of Red Lake Falls (Red Lake Co.), and Lloyd A. Lyngen of Montevideo (Chippewa Co.).

Other officers appointed were Irene Paulson of New York Mills, state secretary, Junice Dalen of Madison, education director; and Edgar Nordstrom of Detroit Lakes, director of organization. President Kuivinen and vice president Smith were chosen to head the labor relations committee. Other assignments included: Cooperatives – Edwin Christianson, Gully; legislation – Curtis Olson, Roseau; junior leader – William Cassavant, Red Lake Falls; insurance – Edgar Nordstrom, Detroit Lakes; and, door keeper – Cletus Tacheny of Blue Earth Co., T. A. Christianson of Polk Co. was elected conductor in a contest with David Vincent of Beltrami Co.

Two women were given important assignments. Freida Eisert of Polk Co. was selected national delegate, with Irene Paulson of New York Mills named alternate.

The headquarters was located in the City Hall of New York Mills. Later it would be moved to Willmar, and then in the early 1950s to St. Paul where it would remain at various locations to this writing.

M.W. Thatcher, general manager of Farmers Union Grain Terminal Association, was a featured speaker at that initial convention of the re-chartered Minnesota organization. According to the Farmers Union *Herald*, Thatcher "spoke for an hour" over a statewide hookup of rural radio stations from a session held in the high school auditorium in Fergus Falls. The convention headquarters was in the River Inn Hotel where a banquet hosted by the GTA was attended by 125. Other convention speakers included Charles D. Egley, manager of Farmers Union Livestock Commission Company ("he gave an hour's address"), E. A. Syftestad and J. L. Nolan, both representing the Farmers Union Central Exchange ("they had two hours on the program... ").

A PROFILE OF EINAR KUIVINEN

"The cooperative movement is at the crossroads. Which way it will go – forward or backward – will depend upon how much we can do... It must be recognized by all groups that the attack on cooperatives is a recognition of the power that lies in the hands of 'the Little People' when they work together for their common good... "

Einar Kuivinen, President, Minnesota Farmers Union, May, 1944.

"Einar was a man who could never say no. He was always willing to help the farmers... in the winter of 1936-37, he was instrumental in organizing REA, going house-to-house signing up members... "

Roy Isaacson, a relative.

Figure 23 Einar Kuivinen

Born: Aug. 28, 1908, Virginia, MN
Died: Aug. 21, 1990, Riverbank, CA
Education: Eighth Grade, Otter Tail Co. School, District 134N.
Employment: Farmer, 1937-1950; trenching contractor, New York Mills, 1950-52;California Highway Department, Fort Bragg, CA.

Community Service: Helped organize Rural Electrification Administration cooperative; First secretary, Heinola Local; First state president of re-organized Minnesota Farmers Union, 1942-44, and, 1946-49.

Married: Florence Mattson of Heinola; one daughter, Shirley Cormier, of Twain Harte, California.

The Finish immigrants had a word – "suomi" – which reflected much of the values held by the immigrants who came to America and northern Minnesota late in the 1800's. "Suomi" translates into a philosophy – "Plant and grow your own food, so you will not take it from the mouth of the needy".

Involved in this was a strong sentiment to do all you could to help those in need of assistance. "There was a closeness of community, if any needed help, everyone felt the obligation to openly offer assistance," explained Verner A. Anderson of New York Mills. "I remember we all lived by an 'open door' policy – no need to knock on the neighbor's door, just walk in! And there was usually coffee and lunch always ready. That was the way we lived."

And the Fins had "sisu" – strong, impeccable character and fortitude, recalled Anderson about his generation which used that "closeness of community" philosophy in organizing farmers into the Farmers Union and into the first generation of cooperatives.

The one man who exemplified these Finnish attributes was Einar Kuivinen, who at a very early age exhibited a rare kind of leadership skill. "Einar was an accomplished public speaker at a very young age," recalled Roy Isaacson, a relative. "He took part in locally produced plays and often he had the lead role. And he was a member of almost every cooperative board of directors, usually serving as chairman. If he wasn't on the board of directors, he was very instrumental in helping to organize the co-ops."

As a result of such a presence – a tall, good-looking young man, and a gifted speaker –Einar Kuivinen seemed like a natural choice for state president, observed an old colleague and farming neighbor who was active in Farmers Union in those early days, Verner Anderson of New York Mills.

Anderson, who helped with the farm chores for Kuivinen when he was on the road handling his presidential chores, described Kuivinen this way: "He was quite tall, had rather handsome features, always wore a Stetson hat, was self-educated – don't think he went beyond country school (eight grades), and had a great speaking ability. We lived only three miles apart, and I used to help him with certain jobs when he was (Minnesota Farmers Union) president."

KUIVINEN'S FARMERS UNION

The organization which selected Einar Kuivinen state president, was technically one of two "Farmers Union" units in Minnesota at the time, according to a report by A. W. Ricker, editor of the Farmers Union *Herald*. "Incorporation rights are held by the old, ousted organization of which John Erp of Canby is president... two lawsuits in Minnesota have not cleared the air as to who has the right to use the name 'Farmers Union' in Minnesota... one District Court decided that title rights to the name 'Farmers Union' belongs to the National Farmers Union... another court ruled that Minnesota courts lack jurisdiction over the

National Farmers Union which was incorporated in Texas... now, another lawsuit may be necessary... What name is used is of small consequence... "

Ricker added that it had been rumored the "old" organization headed by John Erp had allied itself with The Farmers Guild started by E. E. Kennedy, the former secretary of the National Farmers Union and in 1942 reported to be attempting to organize dairy farmers into a union being established by John L. Lewis of the United Mine Workers.

The "new" Farmers Union was then an organization that had 2,775 dues-paying members and 5,782 voting members (men, women and juniors). Ricker said that "two-thirds of the membership lived in the northern and northwestern regions of Minnesota". However, the delegates, who came from 25 counties and 65 Farmers Union locals, elected a president and three board members who lived in northwest Minnesota, and a vice president and two board members who lived in the southwestern region of the state.

Editor Ricker concluded his report of the historic 1942 convention by paying this tribute: "Much credit is due the Farmers Union women of Minnesota for the work they have done the past three years... in this time they have set up one of the best junior organizations in the nation... Mrs. Ione Kleven of Appleton served as state director the past year... she is succeeded by her sister, Junice Dalen of Madison who lives in the same county... "

For the organization, Ricker said, "Everybody realizes that organization work in 1943 will be under the handicap of war restrictions on gasoline... this will make organization work difficult but not impossible... building an organization now will depend less on money, and more on the merit of the organization itself and the zeal of its membership... "

KUIVINEN – "ALWAYS WILLING TO HELP"

When elected state president of the re-chartered Farmers Union, Kuivinen was 35 years old, and widely-known and recognized for being extremely cooperative-minded.

"Einar was a man who could never say no," said Isaacson (his mother was a first-cousin of Einar Kuivinen). "He was always willing to help the farmers. In the winter of 1936-37, he was instrumental in organizing the REA (Rural Electrical Administration), going house-to-house signing up members."

Einar was one of nine children born to Jacob "Jake" Kuivinen (who had emigrated with his parents from Finland), and his wife, Mary Anderson (a native of New York Mills). Einar had made excellent use of his eight years in Otter Tail Co. School District 134N. In addition, Kuivinen had attended a cooperative management course taught at Superior, WI, under the auspices of Consumers Cooperative Wholesale, a regional cooperative. Several of his brothers had taken that management training course and became cooperative managers or employees.

"Einar grew up in a close-knit community of dairy and small grain farmers who were cognizant of each neighbor's needs and therefore promoters of cooperatives for mutual benefits through the use and support of cooperatives," commented Anderson. He also had experience organizing labor organizations in the mining regions of Minnesota where his father, Jacob Kuivinen, had worked and where he met his wife and Einar's mother, Mary "Kolpanen" (Finish for "Anderson").

The couple, married in 1894, and operated a boarding house called "Poikatalo" near Virginia, where "Jake" Kuivinen worked as a miner and later foreman in the mines. They moved to Leaf Lake Township in Otter Tail Co. where they farmed. Einar was only six years old when his father died in 1913. His mother, with the help of her sons, continued to farm.

When the Farmers Union movement came along, Einar Kuivinen was a dairy and small grain farmer in the New York Mills area. He was known to be quite innovative in the use of modern, labor-saving farm equipment. In 1941 he had become very active in Farmers Union, was chairman of the then new Heinola local, and was a leader in what became Cooperative Services of New York Mills.

"Einar was instrumental in the development of a local locker plant and credit union in 1942 when I worked with him organizing Farmers Union," Anderson recalled. "Then he became (state) president, and was except for one year until 1950. But he always maintained the family farm, and sold it later when he became a private businessman (in New York Mills) in industrial equipment like (trenching) backhoes and that."

KUIVINEN'S VIEWS ON FARMERS UNION

In a special report, published in the May, 1943, edition of Farmers Union *Herald*, Einar Kuivinen urged a letter-writing/telegram campaign to persuade Congress to fully fund the Farm Security Administration. He said a letter had been sent to every Farmers Union county president suggesting immediate action.

"By doing these things and uniting the efforts of Minnesota Farmers Union with those of all other home-loving farmers... and bending every effort to get our representatives in our democracy in Washington to perform their duty in behalf of the smaller farmers of Americas, our Union then is really accomplishing the basic Farmers Union principles – the security of the farm family on the land... "

Kuivinen mentioned "an especially good report" received from Mrs. O. E. (Lura) Reimnitz, secretary of the Springbrook-Jupiter Local. She called a special meeting which resulted in 86 cards and personal letters sent to Minnesota Congressmen urging them to work to fund the agriculture agency. "Action like this is really commendable and is one of the most effective ways of building Minnesota Farmers Union," Kuivinen said.

Lura Reimnitz's report included a check for $15 for the National Farmers Union budget fund to finance the campaign. She explained that instead of paying for entertainment at their local meeting, the money donated instead was collected to help the efforts to favorably influence Congress.

The 1943 convention held in Alexandria marked the second speaking appearance by a cooperative leader who would become a featured speaker every state convention – M.W. Thatcher, general manager of Farmers Union Grain Terminal Association (GTA).

E. L. Smith resigned as vice president because he had been elected to the GTA board of directors, and B. Franklin Clough, was chosen to replace Smith. And in June, 1943, the first statewide junior and leaders camp was held at Meri-Mac Court in Detroit Lakes.

Significantly, one of the issues discussed at the second convention was working to separate the agricultural Extension Service from the Farm Bureau, something which would occur about 10 years later. Another policy topic was working to establish a ceiling on farmland values which were on the rise because of World War II.

MINNESOTA FARMERS UNION BACKS FDR

In one of its first political endorsements, Minnesota Farmers Union came out in full support of Franklin D. Roosevelt in his fourth presidential bid in 1944. The backing became official when board member Edwin Christianson of Gully was appointed to represent the state Farmers Union on the "Agricultural Committee for Roosevelt" in the fall of 1944.

The third state convention was held in Detroit Lakes in Nov., 1944. The organization was assessed $10,000 to help finance National Farmers Union insurance activity.

Membership on June 12, 1945 totaled 3,316. And in September, Lura Reimnitz joined the state staff as an insurance assistant, following the heart attack death of Edgar Nordstrom.

Then, at the fourth convention held in Willmar, Nov, 1-3, 1945, Kuivinen declined to seek re-election, telling the delegates meeting at the Lakeland Hotel, "I'm a farmer at heart and this position (state president) is an all-consuming one. I cannot serve two masters at one time. I wish to retain my farm, but when the State Union was set up three years ago, there was no way but that I be state president... since then, I have tried to do the best job I knew how, with the limited experience I had in the cooperative field and working with a farm organization... so after thinking this matter over carefully, I have decided it is best that I resign... "

Two were nominated for state president, including Edwin Christianson of Gully who immediately declined because of his present full-time job which was performed under contract and would not permit him to accept the nomination.

So delegates then unanimously elected state Farmers Union vice president B. Franklin Clough of Lake Lillian, with Edwin Christianson voted vice president. (In the convention minutes, however, the number of votes was recorded. Clough received 5,036; Christianson got 5,055.) The state office was moved from New York Mills to Willmar, dues were increased from $3.50 to $5 per year, and Hubert Humphrey, then mayor of Minneapolis, was a featured convention speaker that year – the first of many times Humphrey would address a Minnesota Farmers Union session.

B. FRANKLIN CLOUGH ELECTED MFU PRESIDENT - 1945

Brief Biography of Franklin Clough:
Born: April 1, 1902, Blue Earth Co. Death: Aug. 15, 1972, Willmar.
Education: Unknown.
Employment: Farmer.
Community service: Member, MFU; state vice president (1944-45);

State president (1945-46). Member, local school board. Board member, K-M Funeral Association; Thorpe Elevator and Shipping Association, Farmers Union Oil Co., Lake Lillian; Alfalfa Drying Plant, Lake Lillian.

Marriage: Minnie Bosch, Atwater, June 17, 1924. One daughter (Donna Klatt Swenson), one son (B. Franklin Clough, Jr.).

The 113 delegates at the 1945 Minnesota convention when B. Franklin Clough was elected president represented a membership of 5,690 and 84 Farmers Union locals from 27 counties.

Clough had been chairman of Kandiyohi Co. since 1942 when the county organization, which originally was started in the early 1930s, was re-established under the new state charter. He held that office until Oct., 1945, or a month before he was elected state Farmers Union president.

Clough was active in the Farm Holiday movement and was "in the front lines in a mass (farmer) march on the state capitol which influenced government action," according to his daughter, Donna Klatt Swenson of Lake Lillian. The march led to a temporary moratorium on farm foreclosures, a two-year extension period of loans and mortgages, and other activities to help farm people weather the depression.

When the government encouraged hemp production during World War II, Clough helped organize and raise funds to build a hemp plant near Lake Lillian. And when local funeral costs proved "excessive", Clough played a pivotal role in organizing K-M Funeral Association, a cooperative which still operates today, Swenson reported.

"My father was a man who didn't just complain: he did his best to find solutions," said Swenson. "When he saw a wrong, he worked to change it."

One of the first changes under the Clough administration was to move the state office from New York Mills to Willmar, where space was leased in four rooms in the Foundation Building in the downtown area.

FIRST WOMAN ELECTED TO STATE BOARD

Then, at the 1946 convention, also held in Willmar, three significant developments took place: First, delegates elected the first female to the Minnesota Farmers Union board of directors. She was Freda Eisert of Euclid. Secondly, they restored Einar Kuivinen to the presidency he had vacated voluntarily only a year earlier and elected Edwin Christianson of Gully vice president. And, officials reported "the checkoff system", used by the co-ops to deduct Farmers Union dues being tried for the first time, had been successful.

RURAL ELECTRIFICATION & TELEPHONE SERVICE (LACK OF)

A picture of what the countryside of Minnesota was like in the years shortly after World War II comes in a review of critically-worded Farmers Union resolutions adopted by delegates at the 1947 state convention held in Willmar. In plain language, the resolutions criticized "a shameful record" in providing electrical power to farms, and called rural telephone communication systems "a disgrace". Here's a condensed version of resolutions adopted in Section II, headed by these words:

"Modernize Our Farming, Farm Homes and Farm Communities":

"Less than half of our Minnesota farms yet have electricity, a shameful record for the so-called public utilities, and a great and unnecessary burden upon half a million people without electricity... We urge all possible speed in advancing loans to REA (Rural Electrification Administration) ... we urge the new Congress to remove restrictions on REA generating plants and transmission lines, so that rural people may be finally freed of the predatory grasp of so-called public utilities...

"Less than half of our farms have telephones and most of those are a disgrace to an America that can produce electronic equipment so abundantly... our telephones belong to the horse and buggy era... much of the poorest housing in Minnesota is on our farms... we need long-term loans at low interest for public housing and for modernization of rural homes...

"We believe Social Security programs should be extended to cover the whole population, including farmers and their families, and farm labor, and that Old-Age Assistance payments should be drastically increased or supplemented to provide a decent standard of living for our old folks... "

KUIVINEN COMMENTS ON COMMUNISM IN "THE GLIMPSE"

"If the members spend 100 percent of their time worrying about Communism, then we'll never get 100 percent of parity... it seems strange that during my lifetime every decent and progressive idea to help the common people has been called 'Communistic'... "

Einar Kuivinen, President, Minnesota Farmers Union, 1949.

One of the innovations enacted during the presidency of Einar Kuivinen was the introduction of a newsletter type of periodical publication called *The Glimpse*. It was a cheaply-produced, mimeographed publication printed on inexpensive paper, in contrast to the sharp, opinionated words reproduced there for Farmers Union members.

In the Nov., 1949, edition, Kuivinen authored a commentary in which he reflected on his seven years as president of Minnesota Farmers Union, in light of his losing re-election bid at the annual convention in St. Cloud to Roy E. Wiseth of Goodridge.

First, Kuivinen pointed to the fact that the organization has "grown from a few hundred members to over 9,000". Then, he complimented the delegates "for adopting a hard-hitting program for peace and prosperity" and went on to "urge that Farmers Union members continue to carry on an active fight against bankrupting our country through support for the tottering monarchs of Europe and China with gifts and arms, which are not helping the common people in these countries..."

Kuivinen continued, "It is hypocritical to believe that farmers can achieve 100 percent of parity under the Brannan Plan as long as the militarists and armament-makers are milking the U.S. Treasury of billions for the armament gravy train. Farmers, along with workers and small businessmen, are going to have to raise their voices still louder in opposition to the highway robbery that is being committed in the name of national defense... The Farmers Union is one of the few organizations that has spoken out in opposition to the dangerous proposals of the late Wall Street Secretary of Defense to convert our country into an armed military camp complete with concentration camps and against the fantastic arms program for Europe, a program that has already proven a complete failure in China...

"It is important for Farmers Union members to pass resolutions in favor of junking disastrous bi-partisan foreign policy on which we are embarked, and to pass resolutions in support of programs that will really stop the oncoming depression, such as the Brannan Plan, public works program, revival of excess profits taxation on big businesses' profits and similar measures... "

Kuivinen then shifted to the "Communist issue", observing, "The positive Farmers Union program for peace and prosperity will never be achieved if the membership begins to fight amongst itself over the 'red issue' as happened in the CIO and other organizations. If the members spend 100 percent of their time worrying about 'Communism', then we'll never get 100 percent of parity."

He continued by charging "it is time to recognize that the 'red-smears' originate with Big Business and the reactionary newspapers and that occasionally some of our 'yellow liberals' resort to the 'red-smear' as a cover-up for their own lack of courage to stand by their principles... it is time to see that a program of anti-Communism will lead our Farmers Union movement into the cesspool of reaction in support of depression and war policies... it seems strange that during my lifetime every decent and progressive idea to help the common people has been called "Communistic"... is it not time for Farmers Union to get into the fight for the poor and under-privileged and claim credit for our efforts instead of letting reactionaries give the Communists all of the credit?"

Kuivinen ended his commentary with these thoughts: "Friends, the need for a strong, militant Farmers Union that is not afraid to be smeared and called bad names will be even greater in the future... with farm income dropping faster than 10 percent a year, and with the armament makers more determined than ever to keep their snouts in the public treasury and with the Marshall Plan and foreign trade collapsing, every Farmers Union member, local, county, state and national organizations will have to join in a terrific campaign both in our communities and in Washington to win support for our program to stop depression and war. We must SPEAK UP before it is too late!"

KUIVINEN: IT'S "THE PEOPLE VERSUS THE TRUSTS" – AGAIN

In a tabloid newspaper, The Minnesota Farmers Union *Reporter*, published by the organization regarding the 1949 session of the Minnesota legislature, the headline read: "The People Versus The Trusts".

The article stated in part, "The Farmers Union proudly takes its stand with the people in the fight to curb the concentration of wealth and power in the hands of the few. This fight is an old one... ever since the Civil War, the demand for 'trust-busting' has been an appealing one to the common people of America. In protest against the hard times caused by the domination of the trusts, the people of this country have built great farm, labor, cooperative and political movements...

"Here in Minnesota, The Grange was founded in the 1880s under Ignatius Donnelly in protest against the robbery of the farmers by the grain and railroad trusts. Then the powerful Populist Movement swept the countryside in the 1890s demanding a broad program of reforms.

"In 1902, ten farmers meeting in a Texas barn founded the Farmers Union movement. Here in the Northwest, the Non-Partisan League flourished after World War I out of which different political movements grew. As part of this tradition of fighting trusts, Farmers Union has grown into one of the major farm organizations of America... "

FARMERS UNION SEEKS 100 PERCENT OF PARITY

Einar Kuivinen in 1949 praised the Agricultural Adjustment Act of 1933, claiming it represented "the first time in our history, legislation attempted to protect the farmers from the ups and downs of the business cycle". The use of the parity formula as the base for price supports, Kuivinen said, "was one of the great milestones in American agriculture".

Now, he added, the Farmers Union wanted the 81st Congress to enact the Brannan Plan, where compensatory payments would be made to farmers if market prices fell below certain levels. However, Kuivinen wanted these payments to be limited to the first $10,000 "to halt the growth of corporate farming".

Then he continued, "The Farmers Union recognizes that farmers are not alone. There are city workers, small businessmen, government employees, dentists, doctors, and teachers, all of whom along with the farmer, keep the economic wheels of progress turning... we know these other people need good jobs so they can buy plenty of milk, meat and bread at fair prices to the farmer... that's why we are friendly to the efforts of organized labor to get decent pay for city workers... that's why we supported legislation to raise minimum wages to $1 an hour for low-income working people... "

Kuivinen went on to say that Farmers Union "plans to move forward in the fight for prosperity and security for farm families, recognizing that we cannot isolate ourselves from the economics of the 'Cold War'... the billions spent for war preparations may bolster the profits of big corporations that get war contracts, but those billions spent for war prevent real programs of electrification, housing, pensions, roads and farm price supports for the people... "

A LOOK AT COOPERATIVES, 1947 STYLE

Farmers Union delegates meeting in convention in Willmar on Oct. 30, 1947, expressed concern in a resolution about the threat to cooperatives of an "unjust tax" proposal before Congress. "Never before have Minnesota farmers' cooperatives faced such unscrupulous and heavily-financed opposition from middlemen as today... Congress will decide whether our cooperatives are to be punished tax-wise, hedged in by restrictions, and denied equal government credit with other businesses... ", read a resolution in part.

The preamble noted "The hundreds of Minnesota cooperatives sponsored by the Farmers Union have had a banner year... the three great regional cooperatives we sponsor – the Farmers Union Grain Terminal Association, the Farmers Union Central Exchange, and the Farmers Union Livestock Commission Co. – have forged ahead, winning more battles in the market place for all Northwest farmers... we are proud of the record made by these Farmers Union cooperatives, a record not excelled by any farm group in America... we call on each Union member to be a 100 percent cooperator, and by his example and influence get his neighbor to do the same... "

KUIVINEN LOSES PRESIDENCY TO WISETH

Years after the fact, questions remain about Einar Kuivinen's loss of the Farmers Union presidency at the organization's eighth state convention in 1949. "In general, our area felt he was unfairly treated," commented a pair of dedicated, long-time members (Leo Klinnert and Verner Anderson) from New York Mills. It was no secret that there were "financial problems" as evidenced by the fact National Farmers Union President James Patton send Tony Dechant to Minnesota as "executive secretary" of Minnesota Farmers Union. Staff cuts were made, and money borrowed from the Farmers Union regional cooperatives, which were developments disturbing to many members and others, including the influential and powerful M.W. Thatcher of Farmers Union Grain Terminal Association.

"Some folks in our area had the feeling that M.W. Thatcher was behind this action (to oust Kuivinen)," the two men added about that era. "Maybe, some thought, it was because of Thatcher's desire to control the educational funds from GTA." Anderson feels that "Thatcher wanted to dictate how those educational funds were to be used".

But beyond this suspicion, and recognizing the stature of M.W. Thatcher, there is little evidence to support this claim. And although it is without foundation, none who knew him could question the tremendous influence of Thatcher within farm circles and the Farmers Union, and beyond – all the way to The White House. And, his strong feeling for a vibrant Farmers Union organization, especially in the home state of the growing cooperative he ran.

Klinnert and Anderson feel confident that Kuivinen lost the presidency (by 269 votes) because Thatcher put his support behind the challenger, Roy E. Wiseth of Goodridge. The official first ballot tally showed Kuivinen received 5,915 while his opponent, Wiseth, received 6,184. A third candidate, Russel Schwandt of Sanborn, a popular MFU staff member, organizer and state secretary, got 3,171 votes.

However, on the second ballot, Schwandt, who held a substantial bloc of southern Minnesota votes, threw his support to Wiseth, assuring his election and the ouster of Kuivinen.

A headline in the West Central Minnesota *Tribune* described Wiseth's victory this way: "Roy Wiseth Defeats Einar Kuivinen As Liberals Out-Vote Leftists". While there was discussion of financial problems at the 1949 convention held in St. Cloud, the ouster of Kuivinen was attributed more to his controversial support of Henry A. Wallace and his Progressive Party in the 1948 presidential race.

Kuivinen was known to staunchly oppose the Marshall Plan as well, which put him at odds with members at the 1948 state Farmers Union convention. Where National Farmers Union President James G. Patton spoke of "unity" in his convention address in Willmar that year, Kuivinen found himself opposed for re-election by a last-minute, dark horse opponent, Reuben Felt of Fahlun Township in Kandiyohi Co. Felt, a member of the state board of directors and a member highly critical of the Kuivinen administration, lost by a big margin, picking up only 1,763 votes compared to 5,107 for the incumbent.

Then, a year later, the Willmar newspaper reported that "unlike a year ago there was no debate (between the pro-Wallace supporters and the liberal contingent) on the convention floor, but there was plenty of discussion of those liberal-leftist differences in the corridors of the convention hall and among delegates in hotel rooms".

The newspaper added that there had been a deal made before the presidential election by the Schwandt segment which was largely from southern Minnesota and the Wiseth supporters who were mostly from northern Minnesota. The two groups ideologically were comparable in opposition to Kuivinen, and agreed that whoever received the most votes on the first ballot would be supported on the second ballot by the loser and his group. That explains how Wiseth, thanks to the support of Schwandt delegates, was able to defeat Kuivinen.

KUIVINEN'S FARMERS UNION CHAPTER ENDS

Kuivinen was seen by Klinnert and Anderson as "having great interest in the well-being of family farms, and always ready to do anything to help family farmers". They recalled the soggy, muddy spring of 1950 when extremely poor road conditions interfered with livestock marketing. When the local Farmers Union Shipping Association needed someone to haul animals from farms to market, the man who responded immediately "to haul the animals through muck and mud was Einar Kuivinen", the men recalled.

In the early 1950s, Einar Kuivinen sold his farm in Leaf Lake Township where he and his wife (the former Florence Mattson of Heinola) lived. They moved to New York Mills where he operated a trenching business for a time. Then the Kuivinens, who had one child (Shirley) moved to Fort Bragg in northern California where he worked for the state Department of Transportation. His old colleague from New York Mills, Verner Anderson, corresponded with Kuivinen over the years, recalling "that he (Kuivinen) always was interested in the success of the people's movements, and interested in the progress on the farm – right to the end."

Death came to Einar Kuivinen in a hospital in Riverbank, CA, on Aug. 21, 1990 – just a week before his 83rd birthday. His survivors included his wife Florence, a brother (Oliver) who lived in Florida, two sisters (they lived in Oregon and Michigan), and two grandsons. An obituary in The New York Mills *Herald* dated Aug. 30, 1990, reported no funeral services had been held "at Mr. Kuivinen's request", and that the body had been cremated with memorials to go to the Alzheimer's Association.

77

4. A "COOPERATIVE GIANT" IS BORN...

"M.W. Thatcher was different from any other man I have known... He didn't visit Presidents; they visited him... A living legend – a man whose voice has always been heard throughout the land wherever there was economic injustice or social indignity... "
Hubert H. Humphrey, Dec. 28, 1976, at Thatcher Memorial Service

"As long as men till the nation's soil, farmers will remember with gratitude your wise and eloquent efforts in their behalf... there is much yet to be done so that equity can be assured for American agriculture, but the groundwork you have helped create in a half-century of cooperative leadership has brought us close to that day."
President Lyndon B. Johnson, in telegram to M.W. Thatcher on his 1968 retirement.

"The co-ops in this Federation (of Grain Cooperatives) have had a tremendous effect... increasing the efficiency in grain marketing... Assuring an honest and competitive market... yet not able to negotiate price... "
Orville L. Freeman, U.S. Secretary of Agriculture, 1967.

"Farm people of America have a vital role to play in fulfilling America's leadership destiny... but farm income can and must be improved if we are to maintain our nation's economic strength... "
President-elect John F. Kennedy in telegram to M.W. Thatcher at 22nd annual meeting of the Farmers Union Grain Terminal Association, Dec. 13, 1960.

"In this economic war, our greatest associate and ally is the National Farmers Union and its state units... "
M.W. Thatcher, 1949.

"It (Farmers Union) is the only organization that started out to get a full share of income for farmers... so they can have the same advantages that the people have in cities... "
M.W. Thatcher, 1951.

"We must never forget that our greatest asset, our greatest strength, lies with the People who conceived this great (Farmers Union Grain Terminal) Association, who built it, who use it, and who believe in it.
"We must never forget those people... bound together, just as GTA is, by a great farm organization, the Farmers Union... their common objectives are the same."
M.W. Thatcher, 1953.

BRIEF BIOGRAPHY OF M.W. "BILL" THATCHER:

Born: May 5, 1883, Valparaiso, Ind. Died: St. Paul, MN, Dec., 1976

Education: Public Schools, Valparaiso, Ind.; Purdue University, two years; Atheneum Business School, Chicago, Ill.

Employment: Accountant, Chicago, 1906-1909; Marwick, Mitchell, Peat & Co., Minneapolis, MN, 1909; Osceola Mill & Elevator Co., 1909-14; Frame Dougherty & Co., 1914; self-employed, Equitable Audit Co., 1914-1922; business consultant, Equity Cooperative Exchange, 1923-31; Washington, D.C., representative, Farmers National Grain Corp., and Northwest Farmers Union Legislative Committee, 1932-37; general manager, Farmers Union Grain Terminal Association, 1938-68.

Reputation: He knew personally every U.S. President from Franklin D. Roosevelt to Richard M. Nixon; presidential candidates made it a point to visit him at his office when campaigning in the Twin Cities.

Honors: Recognized as "First Farm Act Pioneer" for playing a major role in the drafting and implementation of federal farm programs by U.S. Secretary of Agriculture Orville L. Freeman on the 35th anniversary of passage of Agricultural Adjustment Act in 1968.

Figure 24 M.W. Thatcher, about 1926

M.W. THATCHER: A "PRAIRIE RADICAL"

Perhaps the late U.S. Vice President and U.S. Senator Hubert Humphrey of Minnesota described Myron William Thatcher best when he observed in his eulogy at Mr. Thatcher's Memorial Service that "M.W. Thatcher was different from any other man I have known"

One of the facets which made Thatcher different, Humphrey pointed out, was the fact that Thatcher didn't visit Presidents; instead, they visited him in his office at the Farmers Union Grain Terminal Association in St. Paul. "Who the hell does Thatcher think he is?" asked a staff member accompanying presidential candidate John F. Kennedy when informed that Thatcher would not visit Senator Kennedy in his plush presidential suite at a downtown Minneapolis hotel, but instead would welcome a drop-in visit at Thatcher's GTA office instead. Humphrey then pointed out to the Kennedy people that "presidents come and go, but Thatcher remains".

And so it was that presidential aspirant John F. Kennedy visited with M.W. Thatcher at the GTA board room in the cooperative's St. Paul headquarters just about a month before his victory over Richard M. Nixon (whom Thatcher also received in his cooperative office a short-time later). Humphrey joked that "the GTA board room to many of us was known as 'The Throne Room'."

Humphrey, a durable giant of a politician and unsuccessful presidential candidate in 1968 told the Thatcher Memorial Service congregation assembled in Arlington Hills Lutheran Church, how "agriculture brought me closer to a flamboyant but wise character".

Bill Thatcher, he added "ran the Farmers Union Grain Terminal Association, one of the world's largest grain cooperatives in the world... an accountant by training, he had been a kind of prairie radical, part of the Non-Partisan League in North Dakota... he had an uncanny business sense which he turned not to personal profit, but to helping farmers... "

Humphrey portrayed Thatcher in these words: "He was a demanding, irascible, occasionally petulant friend... devoted and generous, too, to farmers and politicians alike... in Washington (D.C.), with evangelical fervor, Bill Thatcher banged on any door to convert people to his views... a political maverick whose only ideology was helping farmers market their products profitably... he supported anyone who could help – (Democrats) Eugene McCarthy (Minnesota), George McGovern (South Dakota), Quentin Burdick (North Dakota), me, and at the same time he aided very different political characters ... he was even a close friend of (Republican) Robert Taft (Ohio) and supported him in the 1952 presidential primaries against Dwight Eisenhower... because Taft was more friendly toward cooperatives... "

Humphrey added, "Bill Thatcher never felt he had to wear bib overalls or faded denim to prove his identification with his farm members... and he didn't believe in ashes and sackcloth... to the contrary, he felt a certain elegance pleased his members... he drove a big car and dressed like a banker (he, in fact, became a director of a bank), socialized with 'the Establishment', and even held a seat on the (Minneapolis) Grain Exchange – the normal domain of conservative Yankee businessmen whose interest was profit and not (farm) parity... "

Thatcher, who didn't retire until he was 85, died in Dec., 1976, at age 94.

Perhaps there is no greater agricultural figure in Minnesota history than M.W. Thatcher, the guiding genius who took a failed grain marketing concern, and with the loyalty of farmers, transformed the St. Paul facility into one of the world's biggest marketing concerns in north central United States. His story is a legendary accounting of a time of tremendous advancements and progress not only for the giant farm cooperative he headed, but a successful career that reflects an enormously admirable dedication to the family farmer rather than for personal fortune-building which unquestionably he could have done just as well.

Yet he devoted his skills and energy instead to the budding farm cooperative movement in a career which spanned more than a half a century, and didn't end until he retired 20 years later than when the average person ends his career.

Time and time again Bill (he never used his real name "Myron William Thatcher") Thatcher espoused firmly the importance of a mutually beneficial relationship between the emerging regional cooperative and the equally new and young Farmers Union movement. In 1949 Thatcher said, "In this economic war, our greatest associate and ally is the National Farmers Union and its state units... " This was a constant theme he preached continuously to all who would listen, and to those he felt might help him and the cooperative help farmers harvest more dollars out of their produce.

Four years later, in 1953, Thatcher spoke even more strongly about the relationship between the regional cooperative and the Farmers Union. "We must never forget that our greatest asset, our greatest strength, lies with the people who conceived this great (Farmers Union Grain Terminal) Association, who built it, who use it, and who believe in it. We must never forget those people... bound together, just as GTA is, by a great farm organization, the Farmers Union... their common objectives are the same."

Thatcher's strong feeling for farmers and their interests had its birth early in the 20th Century during his first trips as a young, Indiana-born accountant working then with struggling grain cooperatives. In a 1938 autobiographical account written for the Farmers Union *Herald* newspaper, Thatcher wrote:

"The court records (in a 1915 North Dakota litigation brought by the state's Attorney General against Equity Cooperative Exchange) carried mute evidence of the infamy of the private grain trade to destroy the farmers' cooperative grain movement... "

That North Dakota case, Thatcher added, clearly sounded the alarm and alerted "farmers to know they had a real fight to find their place in the private grain markets of the country". As a young accountant, Thatcher was indelibly shocked by what he found while auditing Midwest grain firms – fraud and outright exploitation. Years later he would admonish his farm audiences to "send your hoofs and bushels to your own organization (the cooperatives) and they will be bullets for you... give your hoofs and your bushels to your enemies and they will be used to blow your brains out... "

This kind of message sounded often caused predatory interests to see Bill Thatcher as "a deadly and ruthless opponent" and dangerous to the greedy competitors. Long-time Farmers Union Editor Milt Hakel would write years later that these market interests "would have done anything and spent limitless funds to vilify and destroy him (Thatcher)".

A. W. Ricker, the Farmers Union *Herald* editor described Thatcher as "the best-known and loved – and hated – man in the Northwest". Truly he was known by all – by the farmers-followers who saw in Thatcher the chance to break free in the marketplace dominated then by the grain merchants, and by the agribusinesses threatened by the wave of cooperative gospel being preached and advocated by the rebellious Thatcher who once had been on their payroll. Now, Bill Thatcher wore two widely-recognized faces – a saint in rural areas and a demon to be destroyed because he threatened the all-powerful inner sanctum of the grain exchange. His presence promised hope to the grain growers struggling to survive while signaling that a new marketing frontier had dawned in the marketplace where some grain merchants had amassed fortune after fortune at the expense of the farmer. Now, Thatcher was sending a new, untested market force and competitor into their arena. The status quo of the Minneapolis grain world would be no more.

Thatcher's rise to greatness and eventually to a sort of unmatched agricultural immortality had its beginnings in the wake of the financial collapse of the old Equity Cooperative Exchange, one of the nation's first grain commission firms. Located in St. Paul, Equity succumbed to a combination of factors that eventually led to voluntary receivership in Mar., 1923. M.W. Thatcher was named manager of the bankrupt Equity, the concern that had pioneered in terminal grain marketing and which also had built the first cooperative terminal elevator (on St. Paul's Mississippi River Waterfront) as well as established the first livestock commission cooperative (later known as the Farmers Union Marketing Association).

In his biographical portrait of "M.W. Thatcher, Cooperative Warrior" published in the book, "Great American Cooperators" (American Institute of Cooperation, 1967), author Robert Handschin noted Thatcher's awareness as an accountant auditing grain and milling firms at Minneapolis and Duluth of the fact some grain merchants were "amassing fortunes while grain producers struggled to live". Thatcher was to reverse a substantial flow of grain, and put the farmer-owned cooperatives into the forefront in marketing in a way never before seen.

Handschin said the Equity's foremost problem related to the opposition of the private grain trade to the presence of cooperatives on grain exchanges. There was harassment, boycotts and even legal challenges against cooperatives which also had to battle depressed markets and lower grain prices as well in the early 1920s in the wake of sharp market declines following the economic boom years wrought by World War I.

Many years later, Thatcher would explain the strategy he championed and which was used to rebuild this failed grain-marketing cooperative:

"When the Equity (Equity Cooperative Exchange) fell, we recognized that information and education had to be an important part in the building of any cooperative organization. In addition, we felt, a cooperative had to be identified with a grassroots farm organization that really spoke for its members... "

In 1949, Thatcher, president and general manager of Farmers Union Grain Terminal Association, spoke about some of that early history of the cooperative, and its link with Farmers Union:

"In this economic war, our great associate and ally is the National Farmers Union and its state units. Anybody who is familiar with the history and development of Farmers Union Grain Terminal Association knows that we are part of the family – the Farmers Union family.

"We have had much to do with bringing the Farmers Union into the Northwest, and (I) served as chairman of the organizing committee which brought this organization into Wisconsin, North Dakota, Montana and Minnesota in the 1920s.

"When we set out to organize and bring cooperatives together, we were also determined to build a farm organization to help carry on our work."

So in this statement, Thatcher makes clear the priority he placed on linking the regional grain-marketing cooperative with a farm organization. Just a year earlier in 1948, in another speech, Thatcher had spelled out this relationship just as clearly noting "information and education had to be an important part in the building of any cooperative organization".

83

At the same time Thatcher had the belief that cooperatives alone could not do the economic job for farmers. This kind of thinking was based on his years spent in Washington, D.C., when he was the head and chief lobbyist for the Northwest Farmers Union Legislative Committee. His lobbying stint dovetailed with the early, formative years of FDR's "New Deal" Administration. In his annual report to the stockholders of the GTA in 1940, Thatcher reflected on those historic years:

"The cooperative movement could not save farmers from 25-cent wheat. It took an Agricultural Adjustment Act and (government) loan programs, and marketing agreements, and relief checks and a new farm credit structure, like the Farm Credit Administration, (federal) crop insurance, the Commodity Exchange Act, and a dozen other pieces of legislation – things American agriculture would not have had except for our work in the Farmers Union... "

Thatcher had resigned in 1932 as general manager of the old Farmers Union Terminal Association in order to become the chief lobbyist for the organization in the nation's capital. Years later, according to Farmers Union Editor Milt Hakel, Thatcher told how he invited nine grassroots leaders from the Midwest to come to Washington, D.C., where he "schooled" them over a period of several days. He acquainted them with Washington geography, how to find the Department of Agriculture, the Capitol, The White House, as well as how to visit with Congressmen and others about issues and concerns of farmers.

Nearly 40 years later, Thatcher was to recall his "schooling" session in Washington, D.C. The occasion was a Farmers Union meeting in Barron, WI, in 1973, at a time when Farmers Union bus trips to the nation's capitol had been replaced by airplane, fly-in trips instead. Farmers Union editor Milt Hakel told how tears came to Thatcher's eyes as he listened to a report made by two women who had been among a group of 126 which had recently flown to Washington on a Farmers Union plane on a legislative trip.

Later, Hakel recalled, Thatcher spoke to the audience and explained why he had been moved to tears by the women's fly-in report. It was, he said, because he remembered so well how back in the depression era days, he had to coach farm leaders on how to find their way around Washington, D.C., and tutor them on how to visit with Congressmen. What a contrast that was to the 1970s when farm women filled a plane and set out to do what Thatcher had helped others learn to do more than a generation earlier.

In addition to developing a farm lobby which could speak for itself, the years spent in Washington provided Thatcher with the rare opportunity to be directly involved in the development of historic farm legislation and the enactment of legendary farm programs such as the Agricultural Adjustment Act (AAA) of 1933, the debut of federal crop insurance, emergency credit, the Commodity Exchange Act, and other pioneering measures that emerged during the "New Deal" Era of FDR. Many of these historic programs were the result of Presidential executive order, and not the product of Congressional action.

Those formative years for federal farm programs and related activity undoubtedly figured into Thatcher's career-long support for the Democratic Party and its leaders, and a partisanship he never tried to conceal.

A key step came when Thatcher was appointed by President Roosevelt to an advisory committee on farm tenancy in 1936. This group's activity led to the passage of the Bankhead-Jones Farm Tenant Act of 1937 which provided a means for poor or bankrupt farmers to fight for a chance to continue in farming, as well as a means to gain financial support to maintain a portion of their farming operations. Years later Thatcher reflected on this period of historic farm legislation which anchored much of the "New Deal" activity for rural America:

"If we hadn't had a Farmers Union, you would not have had a Resettlement Administration, or a Farm Security Administration (the predecessor agencies to the U.S. Farmers Home Administration that in the 1990s got yet another name, Rural Economic and Community Development and then yet another new name in 1996, USDA – Rural Development)...

"You would never have had a revitalized Farm Credit System except for Farmers Union... " The Farmers Union also played a key role in the establishment of the Central Bank for Cooperatives during the early FDR years, with Thatcher's influence once more a factor in this major farm credit development.

"OUR GREATEST ASSET... GREATEST STRENGTH... THE PEOPLE... "

When a group of Farmers Union bus-trippers stopped as Thatcher's office in the Twin Cities en route to Washington, D.C., in 1951, he told the group why he and his colleagues made the choice to be closely-allied with this organization: "(Because) it (Farmers Union) is the only organization that started out to get a full fair share of income for farmers, so that they can have the same advantages that the people have in the cities."

Two years later Thatcher in his presidential report at the 1953 GTA convention put it even more strongly: "We must never forget that our greatest asset our greatest strength, lies with the people who conceived this great association, who built it, who use it and who believe in it. We must never forget those people. Those people, those farm families on the land, are bound together, just as GTA is, by a great farm organization, the Farmers Union. The common objectives of the Farmers Union and your Farmers Union Grain Terminal Association are the same."

The farm organization, Thatcher added, "helps give purpose to a cooperative; and, the cooperative helps give permanence to a farm organization".

Another person who credits the success of American cooperatives to the Farmers Union is Joseph Knapp. In his book, "The Rise of American Cooperative Enterprise", Knapp observes:

"Undoubtedly, the Farmers Union can be given much credit for stimulating the early growth of many soundly-conceived and practically-operated cooperative organizations. While The Grange supported cooperative enterprise as a general policy, the Farmers Union made it the cornerstone of its program."

The intimate, beneficially fruitful relationship of Thatcher and Farmers Union dates back to 1925, when young Thatcher sought to build a thriving, growing cooperative on the financial ashes of the old Equity Cooperative Association. In his writings of the "Cooperative Warrior", M.W. Thatcher, Robert Handschin relates: "To revive the Equity, Thatcher personally visited each of over 600 farmers who had $133,000 due from their failed grain pool, driving 27,000 miles over dirt roads to do so... eventually every cent was repaid...

"To put a more solid foundation under the co-op, Thatcher turned to the National Farmers Union which in 1925 gave him and two others (C.C. Talbott and A. W. Ricker) $500 to organize Minnesota, Wisconsin, North Dakota and Montana. The Union could tie together co-op and legislative goals, and carry on co-op education and promotion with 5 percent of co-op savings.

Farmers would gain power through their 'bushels, hooves, and ballots'. Soon scores of new oil co-ops added another department which in 1931 became the Farmers Union Central Exchange... " In 1926 the Equity was replaced by the Farmers Union Terminal Association.

Figure 25 Elevator M

THE CASE OF "ELEVATOR M"

One sure-fire method of destroying a grain-marketing firm is to discredit its reputation by charging short-weights or skimming, or by attacking its integrity in sampling, inspecting, grading, or marketing the grain. Such tactics were encountered by M.W. Thatcher in the early 1930s, just as the grain-marketing cooperative was beginning to become a major factor in the Minneapolis arena.

While there were plenty of enemies of the upstart farm cooperatives in those early, formative years, one major foe was the non-profit trade organization known as the Chamber of Commerce, chartered then by the State of Minnesota as a grain exchange. Ironically, Thatcher earlier had been offered the job as secretary of the exchange when he was a rising young accountant specializing in auditing grain and grain-processing companies.

Existing grain concerns and grain exchanges operated like exclusive trading "clubs" and steadily denied membership to cooperatives. Proof of this kind of animosity is found in the matter of "Elevator M", a facility owned by the Farmers Union Terminal Association in Minneapolis.

The Elevator M episode began in 1931 when allegations of a somewhat vague, generalized and largely technical matter were brought against the cooperative, accompanied with charges that Thatcher and some of his staff members had "violated" regulations pertaining to grain inspection and weighing.

This ploy lasted weeks as the grain exchange officials sought not only to discredit the farm cooperative, but also to raise serious questions about the honesty of Thatcher and his operations. "He (Thatcher) was the target of a wicked effort to destroy him and his grain regional cooperative," recalled Roy F. Hendrickson, long-time executive of the National Federation of Grain Cooperatives. "It came in 1931 at the hands of a major segment of the organized grain trade in Minneapolis at a time when his cooperative was denied a seat on that (Minneapolis) exchange for reasons rooted in fear of the rise in organized farm bargaining power."

The effort to discredit the cooperative was led, Hendrickson said, by the secretary and assistant secretary of the Minneapolis Chamber of Commerce. The formal hearings lasted several weeks in the State Senate chamber in the capitol, with first-term Minnesota Gov. Floyd B. Olson presiding. Where the grain exchange hired a pair of attorneys who had long associated with the Governor in their law practice, Thatcher shrewdly retaliated by hiring Tom Davis, a key Olson campaign supporter to represent him and his cooperative. In the end, Olson was seen as impartial and after a long study of the court record he fully exonerated the Farmers Union Terminal Association and Thatcher. Governor Olson made clear he found no evidence of unscrupulous or illegal dealings in weights, grading, or grain inspection practices.

Then again, in 1944, a cooperative enemy, Ray P. Chase, resurrected some of the Elevator M charges. Chase was a skilled co-op critic, as well as a politician who was able to attain significant offices and appointments such as a commissioner for the Minnesota Railroad and Warehouse Commission, state auditor, and Congressman at large. Chase's attacks vanished after the Minnesota Attorney General ruled that GTA had not violated any state law.

Then in 1946, a group of 11 commercial grain companies brought a suit charging that GTA was not "a true cooperative". This matter would wind up before the Minnesota Supreme court which on Feb. 14, 1947, held that GTA indeed was a true cooperative, that patronage refunds were not profits and that GTA could purchase grain consigned to the cooperative. This matter ended with the Federal District Court dismissing a grain trade appeal on Dec. 31, 1947.

Figure 26 Wheat protien sample can (1927-28 tin can campaign)

THE "BATTLE OF THE TIN CANS"

Cooperative expert Joseph Knapp wrote that as early as 1907 the president of the Texas Farmers Union had charged, "While the Grange and the (Farmers) Alliance sought to cheapen farmers' plows, sugar and clothing, Farmers Union seeks to get the maximum price for his cotton and potatoes."

In the northern farm country, the Farmers Union Terminal Association led the fight to get wheat growers a better price, zeroing in on obtaining a premium for their high-protein content grain. This program came after tests at North Dakota's Agricultural College and at a state laboratory in Great Falls, MT, indicated that hard spring wheat had superior milling qualities because of its inherent higher content of protein. That made this crop more valuable in its end product uses even though prices paid to growers at the time failed to reflect any extra value for this special wheat.

Where most local elevators were not equipped to test wheat for higher protein content and establish a premium price for this kind of product, it was doubtful if many farmers would have trusted them anyhow to report "honest" figures. So, in 1927 the Farmers Union Terminal Association began what some called "The Battle of the Tin Cans".

Actually, it wasn't a battle at all. It was a campaign where growers could send, by common postage, samples of their crop in baking powder tin cans for unbiased, objective testing concerning the weight, grade and protein content of wheat to the Farmers Union Grain Terminal Association's quality laboratory for a fee of 25 cents. So in the 1927-28 crop years thousands of farmers submitted wheat samples and pocketed premiums of as much as 42 cents more per bushel of wheat as a result of their submissions. The household-type baking powder can was later replaced by a commercially-made can which had a small screw-on cap provided by FUTA to grain farmers.

This historic pioneering grain quality experiment proved monumental. It enhanced the value of the crop by hundreds of dollars of most welcome income for the individual struggling growers, while adding millions to the total value of the wheat crop. All told, there were an estimated 50,000 wheat growers who provided such samples, and then harvested premiums ranging from 40 cents to 70 cents a bushel over market price.

This "tin can" market message sounded loudly and clearly across the nation's northern Grain Belt – the new cooperative had proven itself in the most tangible, rewarding way possible. It had pulled off what must be acknowledged as a phenomenal marketing ploy – demonstrating in the most convincing way possible – that "wheat" was not just wheat, and that farmer-growers deserved to be paid more for the real value of a crop richer in protein content, and thus, worth much more than the price it had brought in the same markets previously.

Now for the first time ever growers moved toward a more honest and true price. The "tin can" demonstration also convinced growers of something many had suspected – that most grain merchants had been routinely cheating them of the true value of their crop, and that now, they had a feisty, partisan organization that would stand up to the grain merchants and do battle for them and their right to achieve higher, more equitable prices.

One of those who pocketed a precious first-premium price was Selmer "Sam" Bergland of Roseau Co., Minnesota, the father of Robert Bergland who was a life-long Farmers Union member and Jimmy Carter's U.S. Secretary of Agriculture, 1977-81. "The premium my father got was equal to the market price of the crop itself," Bob Bergland related about the first grain premiums paid to growers in the late 1920s. "That sort of thing made my Dad a true believer (in cooperatives), and me, too."

Then when the senior Bergland saw the overnight success of the new oil cooperative in his county, he became even more of a booster for the new cooperatives which were causing historic, meaningful changes as close as the co-op oil truck that hauled fuel to the Bergland farm. And because of the close relationship with Farmers Union, that grain cooperative benefited as well in the surge of new, welcomed popularity linked with new financial prosperity for what farmers grew and harvested.

1980 Co-op Ad Told of The Tin Can Price Revolution

Many years later, the historic episode of the "tin can" campaign was recalled in a two-page, centerfold advertisement in the Aug. 4, 1980, edition of *Co-op Country News*, the successor publication to the Farmers Union *Herald*, sponsored by Farmers Union Grain Terminal Association of St. Paul. The ad's headlines read: "Fifty-two years ago a small tin can started a wheat price revolution", with a subhead proclaiming, "That tin can battle won a fair price victory for generations of GTA farmers... Our first cooperative protein test gave wheat farmers an extra $4.5 million dollars... "

The text of the ad read in part, "Until 52 years ago, grain farmers were victimized by unfair pricing and buying practices... but in 1927 a pint-sized tin can started a wheat-pricing revolution... across the Great Plains farmers filled those cans with wheat samples... and thousands and thousands were sent by mail for (protein) testing for 25 cents each... over 30 million bushels were moved through co-op marketing channels, and high protein premiums gave wheat growers an extra $4.5 million... like a genie from a bottle, out of those tin cans grew hundreds of country elevators, river and seaport terminals, barges, trucks, and railcars... a muscular, world-wide food marketing organization built, owned and managed by 150,000 GTA farmers... your partnership for the 1980s - GTA, build on it... "

However, this tremendous tribute to the "tin can" wheat campaign that really opened up a new and broader frontier for farm cooperatives was marred by only one significant omission. Sharp-eyed, knowledgeable veteran observers quite familiar with the successful protein-testing experiment spotted immediately a flaw in the artist's close-up reproduction of one of the history-making, fabled baking soda tin cans: The artist had not reproduced a true rendition because two key words were conspicuously – and suspiciously – missing from the address label on the tin can – "Farmers Union".

It was seen by these older-generation partisans as another step toward further divorcing the farm organization from its true heritage as a godfather of the 1980 vintage Grain Terminal Association, just as a few years earlier the "Farmers Union" wordage had been deleted from the GTA itself. This kind of generation "gap" would appear later when the famous GTA moniker itself would succumb to yet another change, and the empire built in part because of the "battle of the tin cans" would abandon all public semblance of its Farmers Union legacy and assume a rather meaningless name reflecting none of its long, important cooperative organization history, or anything pertaining to the commodity on which both cooperatives grew – grain. Instead, the name chosen was "Harvest States".

However, several years later, in a special historical booklet, published by Harvest States Cooperatives, proper historical credit was given. A photo of one of the original "Grain Sample" tin cans appeared in a section titled, "The Beginnings", and clearly showed the "Farmers Union Terminal Association" label on the can. Plus the writers credited the "Battle of the Cans" as "one early skirmish the organization (FUTA) fought and won in 1927 which resulted in a significant change in how farmers were paid for their grain".

FDR PRODS WALLACE INTO ACTION

In the late 1920s when the farm economy was reeling and struggling, Iowa-born President Hoover promised a new cooperative marketing approach – the Farmers National Grain Corp. It started off badly after being established in the fall of 1929, just ahead of the historic stock market crash. Hoover's plan was to do what had never been done – build a nation-wide marketing network. One of those assigned to help make a national marketing system a reality was the cooperative genius, M.W. Thatcher.

Author Robert Handschin, who had known Thatcher since the late 1930s and who was hired by Thatcher in 1943, recalled that by 1937 the national grain marketing program had been abandoned, and the focus had shifted instead to using "regional" cooperatives to achieve better returns for farmers. But, alas, there were droughts amid the economic throes of the Great Depression, and the regional cooperatives lacked funds. Things looked so extremely bleak, that a much concerned M.W. Thatcher went directly to The White House, and brought his concern personally to the attention of The President.

FDR, an astute politician was aware of the troubles gripping rural America, such as the Farm Holiday Movement, for example. He had talked personally by telephone with one of its leaders, John H. Bosch of Kandiyohi Co. after a federal marshal was threatened in Montevideo.

FDR's Secretary of Agriculture (Henry A. Wallace), a farm magazine editor from Iowa and son of a former U.S. Agriculture Secretary (H. C. Wallace), also knew quite well the tension and somber, troubling mood of the mid-continent people, including "farm strikes" in his home state.

Robert Handschin wrote about the historic Thatcher-FDR visit and the historic memo FDR authored, in his profile "M.W. Thatcher, Cooperative Warrior" in this way: "By 1937 co-op grain marketing had to be turned back to the regionals. But after three years of drought and depression, they were without funds. Thatcher went to see President Roosevelt on Dec. 29, 1937. During that visit he inscribed on Thatcher's short memo (that had been written in longhand by Thatcher on The Willard Hotel stationery) these words: 'HW: Try to work out, FDR'. The 'HW' was FDR's Secretary of Agriculture, Henry A. Wallace. And this was done." And the famous FDR note to HW later was reproduced in the Farmers Union *Herald*.

So those few brief words written on the Thatcher's memo by FDR proved historic in a couple of ways: The new financial aid from the Bank for

Cooperatives to help the regional cooperatives anew, and the new loans provided by the Farm Security Administration (later, Farmers Home Administration) so farmers could obtain loans to invest money in shares of stock in local and regional cooperatives. And, over time all of the loans were repaid in full.

In his book, "The Advance of American Cooperative Enterprise: 1920-1945", author Joseph G. Knapp noted the role played by Thatcher in closing down the Farmers National Grain Corp., and reestablishing the regional cooperative marketing organizations. Thatcher's Farmers Union Terminal Association was "the largest of the associations to be freed under the new arrangement" when the Farmers National was dissolved May 31, 1938. On that date it became Farmers Union Grain Terminal Association. Knapp said there were "problems" – lack of financing from the district banks for cooperatives, and lack of funds and services of many closed local elevators which could be used to re-start the regional co–ops.

The new Farm Security Administration loans made to impoverished farmers made it possible for them to buy shares in cooperative elevators to be serviced by the regionals which now had access to Bank for Cooperatives money as well. Without the farmer-loans, though, Knapp wrote, it would have been impossible for the regionals – including the biggest, Farmers Union's GTA – to quickly reestablish themselves as viable grain marketers. All told, there were 177 local elevators in Midwestern and northwestern states restored to service by means of the financial and technical assistance from the relatively new USDA agency, the FSA, Knapp summarized.

Thatcher himself often paid tribute to this USDA agency, calling the Farm Security Administration (originally the Farm Resettlement Administration) "a national godsend to agriculture".

Progress followed almost immediately. A half-year after his White House visit, Thatcher reported on June 1, 1938, that the Farmers Union Grain Terminal Association had 121 local association members. Within two years another 129 local cooperatives became GTA members.

And, the young regional cooperative, that borrowed $30,000 from the Farmers Union Central Exchange to get started, marketed 17 million bushels of grain its first year. Such success, author Handschin wrote, came in part because of "the able help of key men who had worked with Thatcher since the old Equity era".

Thatcher was now on his way. In 1939 he organized nine regional cooperatives into a new "National Federation of Grain Cooperatives". He became its president, an office to which he annually was re-elected for 26 years, or the duration of his grain managing and marketing career. However, Thatcher saw to it that this national organization's bylaws prohibited it from "having the power to engage in business of any kind", or in brief, to become a new, nationwide marketing agency.

Earlier, in a 1936 speech about grain marketing cooperatives, Thatcher expressed concern about the political environment in these words: "Until national cooperative marketing may be freed from quadrennial political swings or may become so financially independent that it can challenge national political panaceas, national cooperative marketing ... will operate without reasonable business projection and with a result known only to the gods ... "

The National Federation of Grain Cooperatives which Thatcher headed from its origin commanded attention of national and world leaders. In 1967, for example, at the Federation's 29th annual spring meeting in Washington, D.C., the featured speaker list read like a Who's Who in American agriculture. In the Co-op Grain Quarterly Spring 1967 issue, an article quoted Orville L. Freeman, the U.S. Secretary of Agriculture, who told the conference, "Muscle in the marketplace... this is an idea thoroughly familiar to your Federation... M.W. Thatcher, a founder and driving force in this Federation since its beginning, first proposed it... "

Freeman, the former three-term governor of Minnesota – the leading state in farm cooperatives, added, "The co-ops in this Federation have had a tremendous effect over the years... increasing the efficiency of grain marketing, reducing industry charges for shipping and handling grain, assuring an honest and competitive market... Yet, it is fair to say, that successful as your cooperatives have been, they have not yet been able to provide the bargaining power that farmers need to significantly influence the price of grain... you are not yet able to negotiate the price for wheat the way, say, steelworkers are able to negotiate their hourly wages... "

The Federation in 1967 had a membership of 26 regional grain marketing cooperatives, owned by 2,500 local cooperatives in 19 states which are in turn owned and controlled by more than a million farmers, according to the publication. Two of these bore the "Farmers Union" name – Thatcher's Farmers Union Grain Terminal Association of St. Paul, MN, and the Farmers Union Cooperative Marketing Association of Kansas City, MO. Two had "Farm Bureau" in their name – Farm Bureau Cooperative Association of Columbus, OH, and Michigan Elevator Exchange (Farm Bureau Services, Inc.) of Lansing, MI.

The publication noted that a recent survey showed Minnesota led the nation in the number of farm cooperatives (1,108), and in cooperative membership (625,875), ranked second in dollar-volume of farm products sold ($783 million), and third in net value of supplies handled by cooperatives ($182 million).

At the same time Thatcher was organizing the national federation of grain cooperatives, the dynamic Thatcher assumed yet another role – legislative representative for the National Farmers Union.

In April, 1940, Thatcher helped engineer a mass meeting of 21,000 Midwest farmers to promote Thatcher's old friend, Henry A. Wallace, for the office he was to achieve later that year – the U.S. Vice Presidency.

The following year a long-held dream of Thatcher's became real – a modern, big (4.5 million bushels) terminal elevator was constructed on the southern shores of Lake Superior in Wisconsin. It was the tallest, most efficient grain-marketing facility ever built – a reputation it was to maintain for many years.

ANTI-COOPERATIVE ATTACK BACKFIRES

More historical progress for the growing regional cooperative came the next year, in 1942, when GTA took control of a durum wheat mill at Rush City, MN This represented the first U.S. Grain cooperative to branch out into milling, and setting the business stage for the next ventures into feed manufacturing and oilseed processing enterprises.

Then, after the May 1, 1943, acquisition by Thatcher's GTA of the 57-year-old St. Anthony and Dakota Elevator Co. with its network of 135 elevators and 38 lumber yards, a new enemy force took shape – the National Tax Equality Association. This anti-cooperative organization representing competitors of the farm cooperatives became alarmed by the phenomenal, and rapid growth of the Farmers Union Grain Terminal Association.

After all, the record spoke for itself. In a matter of a few years, the St. Paul-based farm cooperative had gone from an empty-pockets situation, pleading for financial help at the loftiest levels of government – the U.S. Presidency, to the status of a highly-successful, expanding organization marketing millions of bushels of grain and expanding by leaps and bounds. In less than five years, Thatcher had directed the GTA growth from a membership of 121 local associations on June 1, 1938, to an organization boasting nearly 425 locals, 135 newly-acquired country elevators and 38 lumber yards plus the world's newest, most modern, multi-million-bushel Great Lakes grain terminal.

It was this phenomenal, sudden growth that disturbed the competing businesses, and spawned the National Tax Equality Association in the early 1940s. Millions were spent by this organization in a bitter, prolonged battle over tax laws, government credits, cooperative purposes and missions, and related matters. Again Thatcher was a dominant figure in the counter-argument in the war with the well-endowed tax "equality" association. He played a major role in organizing state and cooperative organizations as well as national committees to respond and rebuke charges made by "the tax group" as well as to create stronger pro-cooperative convictions among those cooperators who might be influenced by the debate, according to his biographer, Robert Handschin.

Former St. Paul newspaper farm editor Alfred D. Stedman rightfully labeled "the co-op tax fight one of Thatcher's brilliant victories" in an article published in the Farmers Union *Herald* when Thatcher retired. Stedman wrote how Ben C. McCabe of Minneapolis, former president of the arch-rival of cooperatives – the National Tax Equality Association, in the end wound up selling his network of 57 country marketing points and elevators to Thatcher's cooperative.

It was a remarkable victory for Thatcher and GTA – winning over the one-time enemy, and gaining purchasing rights to the McCabe network of elevators and feed plants sprinkled across the states of Minnesota, Montana, South Dakota, and North Dakota in 1958.

In the Sept. 22, 1958, Farmers Union *Herald*, McCabe is quoted as saying in a statement: "I believe for all concerned that it was fortunate to make this sale to the Farmers Union Grain Terminal Association, a cooperative. A cooperative has many advantages over private business.

"First is customer ownership. The majority of farmers prefer to do business with a cooperative in which they share the ownership."

The verbal concession by McCabe that co-ops indeed had "many advantages" plus the sale of his network of elevators and feed plants had to represent one of the career highlights for M.W. Thatcher who began his remarkable career struggling with an indebted cooperative that had at best a dark future at the time.

The McCabe transaction increased the number of GTA "line houses", or country marketing points, to 211 facilities boasting a capacity of nearly 18 million bushels. That compares to 163 line houses in Minnesota, the Dakotas and Montana before the McCabe addition. But now it also meant that the number of country elevators shipping grain to GTA totaled 440! And, Thatcher announced that most of the 350 McCabe Company employees would be retained. This result is why the Thatcher-GTA-McCabe transaction drew such high praise from St.

Paul journalist Alfred Stedman. He pointed out how the later generations were not knowledgeable about the serious attack brought by the anti-co-op tax group headed at one time by the successful Minneapolis businessman McCabe:

"Though now not always clearly remembered by the younger farming people of today, that fight (between the co-ops and the tax group) then seemed to mean life or death for the co-ops. With deadly aim, the NTEA blade had struck straight at the co-op's jugular vein – the time-honored legal recognition of a basic difference between cooperatives and other businesses: the former selling and processing members' own products and the latter dealing in and processing the farmers' products for the private profit of others.

"The bullseye target in this fight was the co-ops' usual practice of making partial cash payments to members on delivery of their products and of completing payments as cash or securities as patronage dividends at the end of the year... " The co-op enemies wanted to persuade Congress, the courts and others that these patronage dividends were ordinary corporate profits and should be subject to the corporation income tax rate of 52 percent.

"Instead of being weakened or destroyed, GTA and the other great co-ops survived... it was one of Thatcher's most brilliant victories... "

Then later in 1958 nearly 5,000 persons attended GTA's 20th annual meeting held in St. Paul. It was the largest turnout ever for the cooperative, with farmer-members pocketing $3.4 million for marketing their grain through their own cooperative system. But the real news was made in a special joint meeting of management and the boards of directors of the Grain Terminal Association and the Farmers Union Central Exchange held in conjunction with GTA's annual meeting.

This special session was called "a momentous occasion" by Tom Steichen, general manager of the Central Exchange. What happened was the two cooperatives reached an agreement concerning farm supply business activity. GTA would cease distributing farm fertilizers, while the Central Exchange would confine its feed and seed sales to Wisconsin. The stated objective was "to insure adequate supplies of high quality feeds and fertilizers" to the 300,000 member-owners of the two growing regional cooperatives.

The cooperative action agreement emphasized the fact that GTA's basic role had been in grain marketing, while Central Exchange had the same status in the petroleum field. A sharper focus on these areas of expertise was needed now that GTA had acquired the McCabe Company assets, officials said.

"The Central Exchange has earned $50 million in savings for Upper Midwest farmers because it was basic in petroleum," Steichen said. "Our next objective is to become basic in fertilizer." Thatcher termed the joint cooperative program "a thrilling hour for all of us" in his statement, noting that regional cooperatives were gaining greater recognition in the grain marketing field through the National Federation of Grain Cooperatives which he has headed since its formation nearly 20 years earlier.

THATCHER PROPOSES "FOOD FOR PEACE" IN 1943

In 1943 as World War II waged furiously across both oceans, Thatcher assumed yet another major role – a radio news and editorial commentator broadcasting on a four-state network (Minnesota, Wisconsin, North Dakota and

Montana). With this exposure Thatcher began promoting a "Food for Freedom" policy for the United States, advocating using American farm productivity for a post-war "Food For Peace" effort.

Thatcher often referred to farmers as "soldiers of the soil", painting them as patriots doing what they could to feed the American military forces, and then, after the war ended, helping feed war-devastated Europe. In the case of his appeal for "Food for Freedom", once again, the dynamic Thatcher planted well the seeds of change and progress, although 20 years would pass before President Kennedy started a "Food For Peace" effort, with George McGovern of South Dakota as its director.

Figure 27 M.W. Thatcher at the microphone

THATCHER'S GTA MADE USE OF RADIO, TV

M.W. Thatcher and Farmers Union Grain Terminal Association used the electronic media to spread the message of cooperatives, farmers and farm policy. In the fall of 1968, records show that GTA's "Daily Radio Roundup" prepared by the GTA Public Relations Department in St. Paul broadcast on 22 radio stations in four states (Minnesota, North Dakota, Montana and South Dakota). North Dakota broadcasts aired over stations in Fargo, Bismarck, Minot, Williston, Jamestown, Grand Forks, and Hettinger. Minnesota stations broadcasting GTA reports were located at Willmar, Mankato, Marshall and Worthington. In South Dakota, the stations were in Aberdeen, Sioux Falls, Yankton and Watertown, while GTA aired in Montana over stations at Billings, Baker, Great Falls, Plentywood, Havre, Wolf Point and Sidney.

The daily radio programs ranged from 5 minutes to 15 minutes in length, and including weather, markets, news and news commentary. For example, on Oct. 2, 1968, the GTA lead story pertained to how a bushel of hard winter wheat that the South Dakota farmer sold locally for $1.27 a bushel wound up in Rotterdam, the Netherlands, a few weeks later where it was available for more than twice as much – $3.42.

The broadcast explained that the export arithmetic came from the U.S. Department of Agriculture. It cost 39 cents to transport the South Dakota wheat to the Gulf of Mexico. It cost 11 cents more to haul that same bushel of wheat to Rotterdam, and six cents more to move it on to a processor. Adding on the $1.55 a bushel import duty in the Netherlands put the European price of wheat at $3.42. What was that duty used for? The broadcast said it went to subsidize the export of surplus soft wheat grown mainly in France into the world market, and, of course, competing against American wheat as well.

Sometimes the broadcasts featured interviews with farmers, agribusiness leaders and farm agency and program officials. Reprints of the broadcasts were available to listeners from GTA.

In 1960 Farmers Union Central Exchange advertised on 11 television stations in four states (Minnesota, Montana, North Dakota and South Dakota, , and on radio in five states, including Wisconsin.

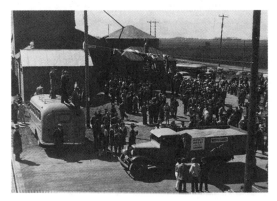

Figure 28 Mercy Wheat kick-off in Climax, MN

"MERCY WHEAT" SENT TO WAR-RAVAGED EUROPE

For years, the old saying goes, "Wheat growers have said, 'Plant your wheat in faith, watch in hope, and reap in charity." However, that axiom turned true after World War II as farmers and their cooperatives provided a hundred million bushels of "mercy wheat" for war-ravaged, hungry Europe.

The national campaign to combat hunger saw Farmers Union organizations and cooperatives in the northern and northwestern states responded immediately when the call came from the U.S. Government for food for the hungry in war-devastated foreign lands of Europe in the spring of 1946.

The "Mercy Wheat" effort had been started in April, 1946, by Fiorello H. LaGuardia, the popular mayor of New York City who had been appointed head of the United Nations Relief and Rehabilitation Agency by FDR. LaGuardia had a regular radio program on a New York station which he used to make his appeal for food aid. One of those who responded was M.W. Thatcher, who actually participated in one of LaGuardia's broadcasts in which donations of wheat were solicited. Thatcher carried the appeal for food aid to Farmers Union Grain Terminal Association, and became widely recognized as a prime mover in the "Mercy Wheat" campaign.

The Farmers Union *Herald* reported the tremendous response to the Government's plea this way in April, 1946: "Even before our Government officially put the program into effect, in the Northwest on April 10, preliminary work resulted in pledges of nearly one million bushels of wheat for immediate delivery in an all-out campaign against mass starvation... "

Within a month, nearly one-fourth or 25 million bushels of wheat had been donated, with 16 million bushels coming from Northwest grain growers alone. Then, before the end of the second month – when the harvest was in full swing – 81 million bushels had been collected.

The "Mercy Wheat" campaign included a "Mercy Wheat Rally" held in Climax, MN. Hundreds turned out to show support for this unique humanitarian effort. (Seventeen years later, in 1963, a "Salute to Wheat Day" was held in that same community, with speakers recalling how farmers had responded in 1946 to the call for "Mercy Wheat").

In the end, the flow of "mercy wheat" from American fields through the cooperative structure was shipped across the seas, and millions were fed. It was one of the brightest moments in the cooperative's history.

THATCHER SOUGHT AN "AGRICULTURAL RELATIONS ACT"

Thatcher also endorsed, and promoted publicly on three occasions in his long career, legislation that would establish a National Agricultural Relations Act, laws that would grant farmers the kind of legality and rights guaranteed organized labor. However, even this powerful dean of agricultural leaders could not garner sufficient support for a collective bargaining program for farmers – a goal he was never to achieve despite attempts beginning in 1941, then again in 1966, and in his farewell address at the 30th annual meeting of his regional cooperative, the Farmers Union Grain Terminal Association, in 1968.

It was one of the few things which Thatcher sought for farmers that he wasn't able to establish in one way or another, even with his excellent rapport with lawmakers and policy-makers at all levels of government.

In 1966, at the 29th annual meeting, Thatcher, then 84 years of age, made this strong appeal for federal legislation for farmers: "American farmers are feeding the world – and they ought to get paid for it." Then he noted criticism of farmers production and marketing costs, before adding, "Farmers will never have the power to sit in on price determination in Washington (D.C.) – where farm prices are made – until they have laws that are as strong as those written for organized labor, organized banking, organized transportation, the organized petroleum industry or any other great group that is protected by federal legislation."

What he was urging here – once more – was the passage of a law similar to federal legislation enacted as National Labor Relations Act. He suggested a panel of "independent representatives of farmers" work through a National Agricultural Relations Act instead of enduring farm programs administered at the whims of politicians, USDA officials and the federal Director of the Budget.

"Slowly but surely the USDA has moved into the marketing field," Thatcher charged. That fact coupled with the power of the Director of the Budget means together "they can slash away at the dwindling margins under which farmers and farmers' institutions are forced to operate".

Thatcher's old dream of such federal legislation, first made a quarter-century earlier, would finally be introduced in Congress two years later, by U.S. Senator

Walter Mondale of Minnesota. However, this "agricultural bargaining bill" as it was called, never materialized as Thatcher had long envisioned, and desperately wanted, and what he thought offered a more permanent solution for farmers instead of periodic and changing federal farm programs.

THATCHER CALLED FARMERS "SOLDIERS OF THE SOIL"

After the end of World War II, things changed on and off the farms of America. The once influential and powerful farm bloc, so influential in the 1920s that it earned the description of farm-minded Congressmen to be "sons of the wild jackass" from New Hampshire Senator Moses, withered both in size and importance.

Attacks became common on the same breed of government farm programs that had helped shed the dark shroud of the depression years only a decade earlier. The powerful grain trade sharpened its knives to whittle away things like USDA's grain storage bins, eliminate the Wheat Agreement, take control of the export business, double-tax farm cooperatives, and kill off commodity price support loan programs that had their debut under Henry A. Wallace.

In such a climate farm-minded politicians as the famed Hubert H. Humphrey of Minnesota as well as Harry S. Truman of Missouri came to the rescue. Aroused Midwest farmers sent Humphrey to the U.S. Senate (ousting one veteran legislator, House Ways and Means Chairman Harold Knutson, a Republican identified by some as "anti-cooperative") and put Truman in the White House. This, however, only slowed for several years the time when price supports would be reduced, and the flow of people off the land would accelerate as if by some unwritten federal farm policy. Thatcher was to fight this change in government policy and posture under a slogan that caught much attention: "Farm Prices Are Made in Washington".

Thatcher's influence was so powerful politically that the late distinguished U.S. Senator and Vice President Hubert H. Humphrey was the perennial banquet speaker at the GTA's annual meeting. In 1967, Humphrey spoke as the incumbent U.S. Vice President to 10,000 at the GTA's annual meeting, a year before he would run for President as the Democratic candidate. However, earlier four other U.S. Senators had addressed the same GTA house of delegates – Eugene McCarthy of Minnesota, George McGovern of South Dakota (and the Democrat who would be the party's unsuccessful presidential candidate in 1972), Milton R. Young and Quentin Burdick of North Dakota, Carl Mundt of South Dakota, Mark Andrews of Oregon, Mike Mansfield of Montana, Gaylord Nelson of Wisconsin, Wayne Morse of Oregon, and Carl Curtis of Nebraska.

In his 1967 annual meeting statement, Thatcher noted the times during his career that "the soldiers of the soil" had responded to national emergencies, from World War I to World War II and the Korean Conflict. Then, after the wartime emergency, the American farmers have had to struggle with scenarios of surpluses and soft markets. Observed Thatcher: "Again and again we have been asked to produce, and the 'Soldiers of the Soil' have done exactly as they were asked to do and came up with an abundance of wheat, feed grains and oilseeds to help take care of the needs of the world... the greatest weaponry this country has for peace and the end of wars is its great capacity and its heart to help people all over the world with food and fiber until they learn how to help themselves. No soldier has saved as many families over the world as has the American farmer... "

PHENOMENAL GROWTH FOR GTA: MORE THAN 600 MEMBERS

Thatcher's record spoke for itself. By 1950, only a dozen or so years since the Farmers Union Grain Terminal Association really swung into operation, it had become the largest grain marketing cooperative in the U.S. It boasted a wide array of services for its members, with annual sales of 100 million bushels – representing a fourth of the total grown in the territory it covered. Savings, or profits, averaged $2 million per year.

In 1950 GTA had memberships in 615 local or (rail) line elevators with a combined capacity totaling 28 million bushels and owned by about 150,000 farmer-producers. The terminal elevators in the Thatcher-built domain could hold another 19 million bushels. So all told GTA represented a grain marketing machine that could at one time hold nearly 50 million bushels.

From 1938 to 1950 the cooperative had netted $22 million, paid to its farmer-members a total of $3 million in cash while building up a net worth of $20 million.

In the *Grain Quarterly*, Winter 1967-68 publication, the scope of Farmers Union Grain Terminal Association was spelled out in a detailed regional cooperative directory. It boasted 560 associations with facilities that included those used to market seven crops – wheat, barley, oats, soybeans, corn, flaxseed and rye:

1. St. Paul – grain marketing (5.3 million). 2. Mankato – soybean plant (4 million). 3. Winona – barley malting plant (1.4 million). 4. Minneapolis – grain marketing (6.6 million). 5. Rush City – durum flour plant (200,000). 6. Superior, WI – grain marketing (18.4 million). 7. Milwaukee, WI – barley malting plant (6.3 million). 8. Shelby, MT – grain marketing (750,000). 9. Lewiston, MT – grain marketing (550,000). 10. Great Falls, MT – grain marketing (630,000). 11. Sioux City, IA. – grain marketing (850,000). 12. Sioux Falls, SD – feed mill. 13. Minneapolis – flaxseed processing. 14. Edgeley, ND – feed mill.

In his summary, Thatcher reported that his GTA had terminal and subterminal space totaling 33.3 million bushels, anchored by the Superior, WI, jumbo facility with its 18.4-million- bushel capacity that is the world's largest inland port elevator. The cooperative now was marketing grain to world markets shipping down the Mississippi River as well as via the St. Lawrence Seaway. The business volume then was averaging more than $350 million yearly, with a record 175.3 million bushels of grain handled in 1967.

The logo used by the regional cooperative was a circle, featuring a plant symbol (several heads of wheat) in the center, the initials "GTA" in capitalized, boldface type at the top, and the slogan "The Co-op Way" at the bottom. Noted the Grain Quarterly writer in the publication, "When GTA, as this regional cooperative is universally known, was organized, cooperative grain marketing had been undergoing a decline due to depression, devastating drought, and need for helpful co-op legislation. It needed dynamic leadership and careful business management. It got both from Mr. Thatcher... today GTA is the nation's largest grain marketing cooperative, is an extensive processor of soybeans, barley, flax and durum wheat, and does a business volume of more than $350 million annually... M.W. Thatcher has guided this cooperative throughout its meteoric rise... "

In addition, the giant regional grain cooperative alone provided a million dollars, or 5 percent of is net savings, to the Farmers Union's educational fund. This was a major support of money that in the future, at the hands of a different generation not as attuned to the cooperative movement, would come to a controversial end. Also, in its dealings with cooperative patrons, GTA and its 615 affiliates or locals automatically deducted annual Farmers Union dues. Again, this widely-accepted procedure at a future time would be halted.

These two things – the automatic 5 percent deducted and dispatched to the Farmers Union (the organization Thatcher many times publicly proclaimed as "our greatest asset and ally") plus the automatic Farmers Union dues checkoff – would end in the years ahead.

A Real Champion for Rural America – M.W. Thatcher

In retrospect, the lengthy Thatcher record spoke quite loudly for itself – unparalleled cooperative progress, the likes of which had never before been seen, and likely would never occur again in the history of American agriculture. Will future historians pose this question: Could, or would, this all have happened without a leader like M.W. Thatcher? Was he a man for the times? Or, a man ahead of his time?

In any case, history is bound to portray M.W. Thatcher as a fiery speaker, articulate and compassionate, a steadfast farm advocate and leader, the son of a former Indiana Granger, who could not only communicate from the wheat field and country grain elevator board room but also to the highest office in the land – The White House – but at the same time produce concrete action and meaningful results where it counted, grassroots rural America.

In two interviews made in June, 1995, and in early 1996, Robert Handschin of Minneapolis recalled some of the rare qualities of his old boss, M.W. Thatcher: "He (Thatcher) was a convincing person, a man who met you on even ground, who was always positive and enthusiastic, always ready and willing to learn...

"When he married his fourth wife (in 1931), his honeymoon trip took them to the one and only world wheat conference... it was held in London... from there the newlyweds went on to Russia because Thatcher wanted to see what was going on there in farming. They stopped in different countries because Thatcher was curious about the condition of farmers and what he might learn from them, and from what was going on...

"Thatcher was there at a time when farmers needed a new organization, new leadership and new hope... there just wasn't the leadership around ... you had Farm Bureau, the Farmers Union, The Grange, they were there, but they weren't doing what needed to be done... they were weak in much of the area... it was a time when farmers were looking for something new...

"Thatcher's story is one of trying new things, and when something doesn't work, trying something else... he had some kind of stick-to-it-ness... " In his article, "Cooperative Warrior", Handschin wrote: "Thatcher brought business ability to his cooperative career as well as leadership skills he developed over the years... He knew how to concentrate on a single issue, how to make it clear to farmers and others, and how to select the best time and means for action... he fully understood how to combine economics and politics if farmers were to protect themselves from those who dominated both...

"His personal warmth to those he served or worked with also characterized his platform speaking, at which he became a master. Friend of Presidents, Senators and Congressmen, he also could make each farmer and his wife in a large audience feel his personal concern for them, or could be thoughtful about small details in the lives of those he met... the confidence he inspired in people made it possible for them to drop their fears and band together in hope for their common good... "

Handschin recalled, "He (Thatcher) always told cooperators that the main goal of a co-op should be the welfare of the family on the farm, not just a good balance sheet or large patronage dividends...

"Among his (Thatcher's) monuments are the cooperative he twice rescued from defeat; the Farmers Union which without his help might not have survived; the Federation of Grain Cooperatives to which he gave vital leadership in its founding years; and the farm and cooperative laws that were his contribution to the 'legislated economy' of which he so frequently spoke to farmers. At mid-century, cooperative grain farmers had both great growth and heavy competition ahead of them, while M.W. Thatcher, already at an age when most managers retire, was yet to achieve some of his greatest successes."

The distinguished farm cooperative leader generally appeared in public looking pretty much like a banker or businessman – bespectacled, always wearing a dark business suit and often a bow tie. But his message usually was that of a farm populist, seeking better things for his clients – the farmer-patrons of the cooperatives, and delivered in such a way he generated warm reactions from the listeners.

"Bill, you tell it just the way us farmers want to hear it," a farmer yelled at Thatcher as he walked from the podium where he had just addressed an auditorium full of farm and cooperative people. Indeed, many long-time observers felt Thatcher appeared just as popular, maybe more so, with farm people at the end of his half-century of leadership as he did in the early, formative days when the future of the cooperative movement was quite uncertain.

It is questionable if any other person, thrust into the same environment in those same years, could have begun to make that half-century anywhere near as productive for agriculture as did the agribusiness accountant from Indiana, Myron William "Bill" Thatcher.

THATCHER'S RECOLLECTIONS OF THE EARLY YEARS

In the elaborate 30th anniversary report of the Farmers Union Grain Terminal Association that was titled, "A Generation of Growth", M.W. Thatcher reviewed the cooperative's early history. "GTA was conceived out of the bitter experiences of farmers' efforts that date back to 1907," Thatcher wrote. "Farmers suffered many unhappy experiences from that date and on into the '30s. There were few federal laws and programs to protect farmers or their cooperatives during those years." He went on to relate how he as a young accountant gained "an intimate understanding of what happened to the first large grain cooperative (The Equity Cooperative Exchange) which was finally forced to close its doors in 1923" as well as his days spent as a part of the Non-Partisan League of North Dakota which provided farmers with a means to press for legislation that would help farmers.

Thatcher recalled the Hoover Administration's efforts to bolster the ailing farm economy, with its Federal Farm Board, the nation's first price support program (wheat was pegged at $1.25 a bushel in Minneapolis), and the use of substantial credit to build cooperatives. Then, when Franklin D. Roosevelt was elected president, new efforts to help farmers swung into high gear.

"From 1932 to the end of 1937 (years when Thatcher was a lobbyist in Washington, D.C.), we were able to develop and see made into laws, programs that were absolutely essential to farmers if they were to build such an institution as GTA."

Thatcher then identified five key programs he helped develop and for which he lobbied hard and long: 1) The Farm Credit Administration with its production credit loans and its lending to cooperatives. 2) The Farm Security Administration (later Farmers Home Administration) which provided funds as the fabled lender of last resort, and enabled more than 100 local cooperative elevators to be organized. 3) The Commodity Exchange Act which provided federal regulation for all commodity exchanges nationally and which required the exchanges to permit cooperative membership without which cooperatives could not have functioned. 4) Federal Crop Insurance. 5) The Commodity Credit Corp., with its ability to provide commodity loans that established a floor, or minimum, price for grain. (Thirty-five years later Thatcher would be one of a group of people invited to a special ceremony by U.S. Secretary of Agriculture Orville L. Freeman to Washington, D.C., where each was honored as "First Farm Act Pioneers" for their efforts in enactment of the historic federal farm programs of the early FDR years in the 1930s.)

It was while he worked for such monumental farm programs in Washington that Thatcher developed his proposal for a plan and program to enable the building of a giant cooperative like he envisioned for GTA. Thatcher related how he took his memorandum hand-written on The Willard Hotel stationery to The White House in his Dec. 29, 1937, visit with FDR.

"I took the program (memo) to The White House... and gained the approval of the President, Mr. Roosevelt," Thatcher wrote in his 30th annual GTA report. "He initialed my memorandum and added instructions to Secretary Wallace to 'try to work out'... The President told me that this was the only time he had given his written approval to the establishment of a business enterprise sponsored by the federal government. I think every GTA patron should be proud to know that a Secretary of Agriculture and a President are a real part of the history of our cooperative. However, the real founders of GTA were thousands of loyal farmers and it is this loyalty and dedication which has enabled this organization to write such an amazing record of progress over these past 30 years... "

At the time of the historic FDR "try to work out" memo, GTA had no money. But what the President did via Wallace was to provide $300,000 for GTA plus another $1.2 million in operating capital. "Of course, these were loans," Thatcher added. "They had to be repaid, and were in less than two years."

At the same time Thatcher reviewed the cooperative's "Generation of Growth", he also urged passage of the National Agricultural Relations Act. "If you are going to win national bargaining power, your cooperatives must be made stronger... with expanded business volume, your ballots and your strong, united voice, this cooperative will be a great force in the coming national decision on farm bargaining power just as you have been a leader on this front for the last 30 years... " Thatcher's plea at his farewell annual meeting wasn't all that new. More than a quarter-century earlier at the GTA's 1941 annual meeting, he had first called for farmers to have legislation akin to that given to organized labor. Then, he reiterated his hope for a National Agricultural Relations Act again in the mid-1960s. It was to remain something never to happen, and one of the few goals set by Thatcher that he did not achieve.

THATCHER'S VIEWS OF A FAMILY AGRICULTURE

"I believe the farm future depends on how successfully the farmers organize politically and economically, just as big business and big labor. And cooperatives provide a means for this economic organization... "
M.W. Thatcher, 1959 Interview, The Minneapolis *Tribune*

Farmers could not have had a stronger supporter than M.W. Thatcher. He often spoke out for preserving the family-type agriculture: "It is a way of life, the backbone of America and must be protected," he said in an interview with the Minneapolis *Tribune* in 1959. "Farmers are entitled to a standard of living on a par with others... " To protect family agriculture, Thatcher advocated "organizing politically and economically" within the cooperative structure, using the same tactics as did other segments like businesses and organized labor.

At the time of Thatcher's death of an apparent heart attack when he was 93, Minnesota Farmers Union president Cy Carpenter described him as "the kind of leader who built the best of America... in his visits with Presidents, business leaders and others he made it clear that he represented the family farmers of this nation... his contributions to farmers placed them in the highest level of respect and acceptance... "

Said B. J. Malusky who succeeded Thatcher as general manager of the super-co-op he built from scratch: "His astute and tireless leadership was instrumental in developing what today (1976) is one of the greatest cooperative grain marketing and processing systems in the nation... Mr. Thatcher devoted all of his energies to serving the farm families that grow our food... that was his mission in life and the legacy he leaves will be honored and respected for generations to come... "

HUMPHREY CALLS THATCHER – A "POLITICAL MAVERICK"

Former U.S. Vice President and U.S. Senator Hubert H. Humphrey, a long-time friend of Thatcher's, paid this tribute: "The farm cooperative movement, which received great leadership from Bill Thatcher, was the most significant development in American agriculture since the Land-Grant College. He was a political maverick, and never let partisanship stand in the way of the interests of family farmers.

"Bill Thatcher supported Democrats or Republicans or anyone else if he thought they helped farm people."

Then Humphrey recalled something he had written in his book about Thatcher: "I said he helped design farm legislation and government policy in a larger measure than any other farm leader. He (Thatcher) was close to former Presidents Roosevelt and Truman. I wrote that Bill Thatcher didn't go to see Presidents; Presidents came to see him."

When Thatcher announced his retirement at a GTA board of directors meeting May 28, 1968, in St. Paul, Humphrey was one of the first to pay tribute to the long-time cooperative leader who was then 85 years old. "A living legend – a man whose voice has always been heard throughout the land wherever there was economic injustice or social indignity," Humphrey was quoted in the Farmers Union *Herald* of June 3, 1968. "The story of the Grain Terminal Association is the story of a man and an idea. The two are so closely intertwined that you cannot think of one without the other. As Emerson once observed, 'a great institution is but the lengthened shadow of one man'. And in the case of the GTA, that man is my long-time friend, Bill Thatcher."

Thatcher learned early in agrarian politics that there had to be some "political muscle" behind the efforts for farmers to work and speak for themselves", said Edwin Christianson, then national vice president of the Farmers Union. "He insisted early that farmers needed a farm organization to strengthen their cooperative efforts and also to make their legislative goals more readily attainable. He also recognized that another dimension of activity was necessary for farm and rural people – taking an individual and group interest in progressive politics."

However history portrays M.W. Thatcher, Christianson said, "We are convinced that the key to his success has been his faith in people – his faith that they could find better ways, that they could devise a more honest marketing system, that they could make government more responsive to the nation's needs.

"The faith and the hopes of Bill Thatcher have been sustained over these 50 years. Not everything has been achieved that he may have dreamed of, but the mechanism is here with which it can be achieved if those of us who continue at the grassroots will retain the spark that he gave to this great organization – the Farmers Union."

Tony Dechant, president of the National Farmers Union, said in a telegram to Thatcher, "Your record of service to agriculture is unequaled anywhere." Dechant said Thatcher in his 50 years of service had "bought and sold more grain than any man living in the world today... " He added that two of Thatcher's most significant contributions were the slogan, "Farm Prices Are Made in Washington", and his 25 years of fighting for legislation to give farmers "bargaining power".

Another telegram sent to the GTA board of directors came from William L. Guy, the governor of North Dakota. "I admire tremendously the zeal and enthusiasm that Bill Thatcher has constantly applied to advancing the best interest of the individual farm family," Guy said.

THATCHER'S GOAL – "WELFARE OF THE FARM FAMILY"

Wrote long-time co-worker Robert Handschin in that June, 1968, Farmers Union *Herald*, "He always told cooperators that the main goal of a co-op should be the welfare of the family on the farm, not just a good balance sheet or large patronage dividends... on his retirement he can look back at age 85 on some 54 years of service to American farmers... his was a turbulent but highly productive life in the service of farmers and their cooperatives which well earned him the title, 'Cooperative Warrior'."

"I can speak of him from first-hand knowledge as I have seen him in action many times," wrote Roy F. Hendrickson, executive secretary of the National Federation of Grain Cooperatives. "Repeatedly I have seen Bill Thatcher go far beyond the call of duty to fight for farm programs, for farm cooperatives, and in defense of achievements endangered by destructive drives initiated by forces opposed to the ascendancy of group bargaining by producers."

That same edition of the Farmers Union *Herald* contained photos of Thatcher with these U.S. Presidents – Harry S. Truman (1950), John F. Kennedy (1960), and Lyndon B. Johnson (1964). Former President Johnson sent a telegram congratulating Thatcher on his retirement after a career which LBJ described as laying the "groundwork in a half-century of cooperative leadership that has brought us closer to the day when equity can be assured for American agriculture".

Kennedy, as a U.S. Senator and Democratic presidential candidate paid a visit to Thatcher's GTA office on Sunday, Oct. 2, 1960, while campaigning in the Twin Cities. He was photographed with Thatcher, sitting at Thatcher's desk, while Thatcher was showing Kennedy some grain checks issued in 1932 when prices were historically the poorest for producers. Earlier Thatcher had visited with Vice President Richard M. Nixon, the Republican presidential (at Nixon's invitation), while he was campaigning in Minneapolis.

After the election of 1960, Thatcher received a telegram from President-elect Kennedy which read in part, according to a report in the Dec. 19 Farmers Union *Herald*: "Please extend my warmest greeting and best wishes to those attending the annual banquet of the Farmers Union Grain Terminal Association... you have every right to be proud of your successful record and your effective role as one of the nation's great cooperative institutions... farm people of America have a vital role to play in fulfilling America's leadership destiny... they must be able to share fairly and equitably in our economy to enable them to fulfill that role... farm income can and must be improved if we are to maintain our nation's economic strength... "

Long-time Minnesota Association of Cooperatives executive director Ed Slettom noted how Bill Thatcher had helped originate that organization just as he had founded the national organization of grain cooperatives. "He saw the need for all cooperatives to be united in an organization that could represent their mutual interests in legislation, education and public relations. Bill Thatcher made an indelible mark on American agriculture and farmer cooperatives. He will be long remembered as an articulate spokesman for the rights of farmers to gain power in the marketplace. He was an outstanding leader, a champion and warrior for farmers and their cooperatives."

GTA member-owners were kept informed of cooperative news in part by a special publication, *GTA Digest*. This publication was issued six times a year by the regional cooperative. It listed all major offices of the organization and their telephone numbers. For instance, there were seven grain marketing offices in five states listed in the May-June 1970 issue, along with 10 feed plants and six processing units. The publication then was edited by Thomas W. Henderson.

In summary, M.W. Thatcher was a strong dominant force in the farm cooperative movement from early in the 20th Century, beginning as a young Chicago accountant disturbed by what was happening to hard-working farmers at the hands of the grain merchants when he moved to the Northwest and Minnesota. For more than 40 years, from the 1920s to his retirement in 1968, he was both the general manager of a growing, highly-successful, pioneering grain-marketing cooperative and an ardent, eloquent and at times fiery speaker and outspoken advocate for family farmers who possessed respect and loyalty from grassroots America to The White House and the halls of Congress. He was at home visiting with farmers, popular with farm people speaking from the podium, congenial in the cooperative board room, or visiting with important politicians and government leaders.

THATCHER RESPONDS TO "RED SMEAR"

A sample of Bill Thatcher's adept verbal skills was demonstrated in the fall of 1950 when New Hampshire Senator Styles Bridges called the "Brannan Plan" (which called for 100 percent of parity prices) "the Stalin plan, inferring that those who supported the price support measure were "communists". Thatcher's response, published in the Sept. 18, 1950, Farmers Union *Herald* suggested that "Bridges is either an unmitigated liar or dupe of schemers".

In a bylined article, Thatcher wrote, "We are not going to elaborate on the Senator's reckless but diabolically clever smear tactics, except to ask, 'What is it that motivates the Senator? He has made a frontal attack on price supports, on parity prices, and on the (Truman Administration) Secretary of Agriculture for advocating such programs.

"It was an easy hurdle for Senator Bridges to jump from calling the parity program a 'Stalin plan' to lumping the 450,000 members of the Farmers Union as red-dominated because they have the courage to support such a program." Noting the quick response by other U.S. Senators to denounce Bridge's' comments, Thatcher added, "The New Hampshire Senator did not compile the distortions, half-truths and outright lies. It is apparent where much of this material originated. It would seem clear that the purpose of the master-minds is to smear every person in the Administration and to put on a price-flexing blitz against the farmers of this nation. If the Senator was aware of this, he has earned a place in the eyes of the public as a unmitigated liar. If he was not aware, then obviously, he is the gullible dupe of scheming economic fifth columnists.. . . "

Then Thatcher closed his editorial with an invitation to the Senator, inviting him to the annual meeting of the Farmers Union Grain Terminal Association in St. Paul, adding, "As our guest, he can see a real farm organization function in the best democratic principles of a people's organization ... he can see for himself how wrong he is ... . "

Later that same year, 1950, M.W. Thatcher was to formally propose a "Food and Freedom" program, focusing on the tremendous abundance produced on America's farms. A by-lined article in the December Farmers Union *Herald* spelled out the kind of program that within a few years led to American international food aid – the P. L. 480 program, and Kennedy's Food For Peace Program.

Thatcher wrote, "The year 1950, midway in the 20th Century, also is a division between an old and out-moded era and the beginning of a new one... there are problems confronting farmers on three levels: 1) His own welfare, with all its problems of how to raise a family, earn a living, and get along with the rest of the world. 2) The success of his cooperatives, and problems of buying and selling that are beyond his immediate community. 3) The problems of his nation and those of the rest of the world.

"Yet, all of these problems are linked together. The world is so small that we live on each other's doorsteps. Time and space do not mean a great deal when measured by atomic bombs and jet propulsion. But through all these tense times, the wild surges and violent outbursts of hungry and oppressed people, there runs a thread of hope. It is food. American farmers are the greatest food producers the world has ever known. Let us use that great abundance to feed a world that hungers for food and freedom."

In 1950, Thatcher outlined this proposal: "In exchange for parity prices at home, the farmers of this nation will agree to donate the surplus – that five to 10 percent carryover that is called 'a surplus' and 'a market glut' – to the people who are hungry and penniless. It will be the gift of American farmers to the hungry people of the world... would it end wars?... we don't know, but it's an idea worth trying... exporting guns and tanks and bullets... makes neither friends nor peace... our plan would be cheaper, and it might work... isn't it worth a trial?... "

Then the farm leader, who at the time was 67 years old, closed his editorial column with a paragraph noting that it was "the Christmas season... . in time of abundance we are admonished to fill our hearts with thanksgiving, and share our bounty with others... what greater good could America offer to the world than to help feed a world hungry for food and freedom?"

Such words and ideas illustrate the depth of feeling and concern that were possessed by this legendary farm leader. In short, it is convincing that there was none like Bill Thatcher before, and, indeed, none like Bill Thatcher since.

A 30-YEAR COOPERATIVE SUMMARY

"The cooperatives represent the greatest and most efficient marketing, processing and merchandising machine ever constructed by farmers anywhere in the world... it will serve them well for generations to come, if they diligently build, patronize and protect it... "

A Chronology of Thirty Years of Cooperative Achievement, published by Farmers Union Grain Terminal Ass'n., 1968.

When M.W. Thatcher retired after 30 years as general manager of Farmers Union Grain Terminal Association in May, 1968, a small booklet was published which listed historical highlights achieved by the regional cooperative during that period.

The 10-page pamphlet was titled, "A Chronology of Thirty Years of Cooperative Achievement by Cooperating Grain Farmers and their Local Associations". It provides for one not familiar with GTA and its emergence into a super-cooperative a history in brief which traces major events, and developments from its humble beginning on the heels of the Great Depression to its imposing presence as one of the world's great marketers of grain three decades later.

"The benefits of working together should never be forgotten when measuring the value of our cooperatives... " concludes this booklet. "They have meant much to the farm families, and they can mean as much or more in the future if new generations on the land want to make cooperation work for them.

"The cooperatives are completely owned by farmers, and represent the greatest and most efficient marketing, processing and merchandising machine ever constructed by farmers anywhere in the world. It will serve them well for generations to come if they diligently build, patronize and protect it."

At the close of cooperative business for the 1967-68 fiscal year, the record of success spoke for itself:

Nearly 3 billion bushels of grain had been marketed "The Co-op Way" (wordage featured in the GTA logo) during the 30-year period beginning in 1938.

Farmers saved themselves $73 million, with $50 million of that reinvested in their regional cooperative facilities and the balance distributed in cash to the farmer member-owners.

The world's largest "marketing, processing and merchandising machine" they built in 30 years now could obtain farmer-owned grain at 600 locations in Upper Midwest states through a network of 400 local, member associations (many of which had branch operations as well) and 200 "line stations" as well.

Statistics about grain marketing, the booklet suggested, "doesn't tell the whole story". By cooperating, farmers also gained much more than limited savings of a few cents per bushel. At the same time they have gained strength to be an important influence in getting and keeping legislated benefits for agriculture – important things like commodity price supports, commodity loans, wheat certificates and such.

Here are a few of the 30-year highlights of Farmers Union GTA during Bill Thatcher's inimitable reign:

1936 – Congress passes the Commodity Exchange Act, guaranteeing that farmers' cooperatives could not be excluded from membership in grain exchanges. Farmers Union Grain Terminal Association is incorporated on July 15.

1937 – M.W. Thatcher leaves Farmers National Grain Corporation and returns to St. Paul to prepare GTA for operations.

1938 – GTA begins operations June 1 with $30,000 borrowed from Farmers Union Central Exchange, plus $1.2 million from Farm Credit Administration's Bank for Cooperatives (with direct approval of FDR). The new cooperative has 121 local associations and only one terminal elevator (located in St. Paul) with two branch offices (Duluth and Great Falls, MT).

1939 – Loans from Farm Security Administration (later known as Farmers Home Administration) help farmers finance 129 more local elevators. GTA now has 250 members, handles 17 million bushels with $144,000 net savings its first marketing year.

1940 – GTA leases Elevator "M" in Minneapolis. First managers' "school" held (in Fargo).

1941 – A big building year – the large Superior port terminal constructed; sub-terminals built in Montana at Shelby and Lewiston. GTA Digest debuts Oct. 15.

1943 – A cooperative enemy appears. It is called National Tax Equality Association. Thatcher proposes "National Agricultural Relations Act" for post-World War II protection of farmers. GTA establishes line elevator division after acquiring three firms, begins own radio program aired Sundays and daily five times per week.

1946 – Of the 80 million bushels of "Mercy Wheat" provided nationally for war-ravaged Europe, 16 million bushels came from GTA and its farmer-members. This humanitarian feat produced national headlines for the cooperative's contributions. Wartime ceiling prices on flax ended in October, and prices almost doubled (hit $7, with 76 cents per bushel refund).

1947 – GTA sells wheat for $4.06 ½ a bushel – highest ever. Post flax-boom permits record savings ($4.6 million). Minnesota Supreme Court makes unanimous, historic ruling upholding GTA's right to buy members' grain.

1948 – GTA buys Sioux City terminal, establishes Traffic Department, and finishes its 10th year with $3.5 million savings and net worth of $17 million.

1950 – First farmer tours of GTA facilities are held, with tour totals destined to average 10,000 persons each year.

1951 – Co-ops team up in Washington, D.C., to block punitive tax measure advocated by anti-co-op group, National Tax Equality Association. GTA provides $10,000 for wheat rust research effort.

1952 – St. Paul elevator hit by big Mississippi River flood. GTA handles one-billionth bushel of grain. Farmers boost production in response to Korean War.

1954 – Superior terminal expands once more, adding 1.4 million bushels to its capacity. M.W. Thatcher launches "GTA Family Farm Survey".

1955 – St. Paul elevator modernizes, adds two truck dumps and automatic boxcar unloader.

1956 – GTA again leads fight to gain "yes" vote for wheat production controls and price protection. Eisenhower vetoes 90 percent of parity wheat price supports.

1958 – A big year for GTA. It buys McCabe elevators and feed plants (McCabe had been a leader of anti-co-op group), adds 10 million bushels of storage at Superior terminal, and Jack Barry becomes first GTA person to be elected to Minneapolis Grain Exchange Board of Directors.

1959 – S. S. Asia fills with GTA grain headed to Denmark in historic first shipment via long-awaited St. Lawrence Seaway. Vice President Nixon helps M.W. Thatcher put down yet another serious co-op tax threat.

1960 – Vo-ag scholarship program sponsored by GTA begins, as does 50-year and 75-year recognitions for local cooperatives. Presidential candidate John F. Kennedy visits M.W. Thatcher in his office at GTA.

1961 – The 2-billionth bushel of grain is handled by GTA. Agriculture Secretary Ezra Benson leaves office and $7 billion in surpluses as Minnesotan Orville Freeman becomes President Kennedy's farm chief. Freeman introduces Feed Grain Program and raises price supports.

1963 – GTA ends 25th year, with 200 line elevators, 450 member-affiliates and 2.25 billion bushels of grain marketed. Buys St. Louis river terminal with two other regional co-ops.

1965 – Another big Mississippi River flood hits St. Paul elevator. GTA buys Froedtert Malt Corp. of Milwaukee.

1967 – GTA and seven other regional cooperatives began construction of 5-million-bushel export facility on Mississippi north of New Orleans. It begins processing sunflower seed at Minnesota Linseed and Honeymead facilities, and reports net savings of $4.2 million on 175 million bushels handled.

1968 – B. J. "Barney" Malusky succeeds M.W. Thatcher as general manager May 30. Lowell Hargens, who began his career as a GTA accountant in Great Falls, MT, in 1940, is named assistant manager.

Figure 29 Original Farmers Union / Equity St. Paul River Terminal

Figure 30 GTA's Superior, Wis. Terminals

Figure 31 1940's Farm Equipment

Figure 32 Roseau Farmers Union Co-op Oil c. 1930's

Figure 33 Yeild 30 bu./acre, harvest 50-80 acres/day

112

Figure 34 Farmers Union GTA Office, So. St. Paul, MN

Figure 35 Old Equity Exchange Building

5. BUILDING A GIANT SUPPLY CO-OP...

"We must live with change to live at all."
Thomas H. Steichen, General Manager, Cenex, 1971.

"The catalogue of cooperative accomplishments is large. You brought credit where it seemed unobtainable. You brought light and power to rural Americans when it appeared out of reach. You brought modern telephone service to tens of thousands... And lowered the price of gasoline and fertilizer to many thousands more... Your goal has been opportunity for all and not just success for some. You have achieved so much because you cared most about people."
Lyndon Baines Johnson, President, 1964.

"Some of the pioneers have said that the story of the growth of the Central Exchange sounds like a fairy tale... "
Emil A. Syftestad, Gen. Mgr, Farmers Union Central Exchange, 1955.

"I voted against changing the name ("Farmers Union Central Exchange" to "Cenex")... I thought it had served us well over the years... It was unfortunate... it was close to a 50-50 vote... "
Joseph E. Larson, ex-Cenex President, member of its board of directors, 1960-1990.

Looking back at history, the early 1930s absolutely were not the best of times for starting a new cooperative venture. The tentacles of "the great depression" were beginning to grip hard at the economic throat of America in the wake of the Wall Street stock market crash of 1929. Confidence in the Hoover Administration farm board programs had vanished. And farm leaders were worried about the fledgling cooperatives they had just started to build, particularly considering the cooperative track record up to that time was not all that good.

Things looked so bleak that a joint special meeting was called on the thirteenth day of Feb., 1932, by three cooperative organizations – the Farmers Union Terminal Association, the Farmers Union Livestock Commission Company and the one-year-old Farmers Union Central Exchange. Out of that session came one important message instructing management of the three cooperatives to be extremely cautious, and to move slowly.

A resolution adopted that day by the representatives of the three Farmers Union cooperatives read in part: "We hereby recommend and instruct the management of the Farmers Union activities not to engage in any program of expansion, and to discontinue any such activities as may be dissipating the funds of the activities...

"We recommend that every effort be made to conserve and keep intact the resources of the business activities... "

The bottom line of this in-house statement clearly was emphasis on "going a little slower, building solidly, and, thereby maintaining the confidence of the (cooperatives') members. "That attitude has been maintained throughout our first 25 years of operations," noted Farmers Union Central Exchange General Manager Emil A. Syftestad in the cooperative's silver anniversary report.

It could well be that the tough, economic times in which it appeared on the agricultural scene in Minnesota was precisely the best opportunity in which to start what eventually would become one of the nation's major supply cooperative success stories.

The year was 1931 when the Farmers Union Central Exchange, the organization known today as Cenex, officially was created. Earlier it had been only a small department of the Farmers Union Terminal Association which was managed by M.W. Thatcher.

The Exchange really had its origin about four years earlier, in 1927, when the grain cooperative's board of directors consented to a new, fully-owned cooperative subsidiary. The new Exchange was set up to be a wholesale supplier of the many farm supplies needed in daily operations by farmers – things like binder twine, coal, salt, flour, feed, tractor fuel, lubricating oil, and such, purchased in carload lots, with tires added in 1929 or 1930. And when real horsepower faded in use and tractors began to invade agriculture in a big, definite take-over, the needs of farmers changed – with quick growth in demand for tractor fuels, oil, grease and the like – products not needed only shortly before.

This unexpected spurt in demand created by the invasion of the tractor into farming is thought to have caused interest in the decision to make the Exchange a new, separate cooperative, and end its relationship with the parent grain cooperative, the Farmers Union Terminal Association.

Thus began an important aspect of the cooperative movement in rural America, the development of local "oil companies". Most of these were organized after the major fuel firms balked at making rural deliveries to farmers. Also, it was felt these firms also were charging excessively for such products. By the end of 1929 there were about 20 of these local oil cooperatives in four states, Minnesota, Montana, North Dakota and Wisconsin purchasing their petroleum needs from the new Farmers Union Central Exchange.

EARLY EFFORTS TO ORGANIZE AN EXCHANGE

Interest in cooperative farm-related business activity had existed for many years as evidenced by the fact that in 1921 in Wright Co., Minnesota, members of Farmers Union's Dewey Local raised seed money to establish a Farmers Union store. And that a year later five Farmers Union locals in Jackson Co. jointly ordered carload shipments of flour, sugar, fruit and even machinery. And in 1923 a Farmers Union Exchange is organized in Jackson after $6,000 was raised.

So the idea of such a farmers' exchange was not exactly new or novel in the late 1920s when the regional farm supply cooperative was started by Farmers Union Terminal Association in St. Paul.

The first recorded farm cooperative grain elevator in the United States had been built in Dane Co., Wisconsin, in 1857 – or nine years before a cooperative grain elevator was established in Minnesota (at Watson in 1886).

Then in 1903 farmers organized the "Minnesota Farmers Exchange" in an attempt to try and correct some of the unfair and illegal practices common then in grain trade commerce. But when this new Moorhead group tried two years later to gain membership on the grain exchange in Minneapolis, it wasn't successful.

The Moorhead clan tried again in 1908, organizing the Equity Cooperative Exchange. This action led to attacks by the private grain traders, then a costly court battle in 1915, and finally bankruptcy and receivership in the early 1920s. Out of that shaky beginning was to emerge one of the northwest's giant grain cooperatives that would one day market a fourth of all grain in the Minneapolis-Duluth marketing centers.

However, such a dubious beginning didn't bode well for another farm cooperative business which would use "Exchange" in its name. At the same time the old Equity Exchange was reeling on the financial ropes, interest in exchanges existed in other business pursuits, including making such an "exchange" a subsidiary of a going grain cooperative. Such a step required a special kind of leadership. And there was such a person available – M.W. Thatcher had become general manager of the bankrupt Equity Cooperative Exchange of St. Paul in 1923. And Equity would become the base for a cooperative organization that would make history – the Farmers Union Terminal Association – only three years later.

Thatcher was known, Syftestad pointed out in the 25th annual Exchange report "for his experience in cooperative work, and one who recognized the importance of having a strong, active Farmers Union membership organization to assist the farmers in obtaining proper legislation to protect them in the market place". The Terminal Association provided a $35,000 line of credit to the new Exchange, and a single share of stock valued at $25 was sold to 21 local cooperatives as well.

It became evident that farmers could not raise enough money to provide for all of the needed bulk tanks, delivery trucks and equipment to establish Farmers Union Oil Cooperatives in every locality. Consequently, a plan was developed to set up regional retail bulk plant petroleum stations to better serve the rural customer-patrons, with the Exchange providing the financing for this network of oil centers.

In April, 1932, when M.W. Thatcher resigned to head up the new Farmers National Grain Corporation in Washington, D.C., Emil A. Syftestad became general manager of both the Farmers Union Terminal Association, which was now a part of the National Grain Corporation, and the Exchange. Gross sales by the Exchange in 1932 were $1.7 million. These sales included a new item marketed by the cooperative for the first time – Brunswick tires for cars and trucks.

Syftestad was to become known for his adherence to a "cooperative doctrine of growth without debt", a policy which helped the Exchange redeem and retire capital stock despite the ups and downs of the farm economy.

In 1934 the Exchange's board authorized the construction of its first oil blending plant, warehouse and offices after getting a Bank for Cooperatives loan. And C.C. Talbott, a widely-respected North Dakota Farmers Union leader and organizer who was president of the Exchange, suggested the possibility that someday this young cooperative would own its own refinery and oil wells at the cooperative's annual meeting that year.

A "MODERN" MACHINE – THE "CO-OP TRACTOR"

Figure 36 The Co-op Tractor - From left to right are Alex Lind, Ralph Ingerson and E. A. Syftestad

In 1935 a new cooperative venture began – farm machinery. Some farmers felt that farm machinery manufacturers were lagging behind in the design, manufacture and distribution of more modern and efficient machines compared to the progress evidenced in the automobile industry. The critics thought tractor manufacturers were too inclined to keep on producing what they had always produced and were too slow in making changes. The out-moded, steel-lugged, slow-moving farm tractor was a good example what the current manufacturers felt belonged on the farm.

When they learned of a new type tractor designed by an automotive engineer, Farmers Union Central Exchange immediately began exploring its development. Demonstrations of the new tractors were held in several states, and the two models introduced became known as Farmers Union Co-op Tractors No. 2 and Co-op No. 3. These revolutionary machines had five forward speeds, electric lights and starters, and zipped down the road at speeds up to 35-40 miles per hour on rubber tires. These machines were versatile, and could handle heavy field work chores as well as light, faster operations.

The tractors were first built in Michigan, then West Virginia and Indiana. The emergency during World War II interrupted production which also provided major tractor manufacturers ample time to try and copy some of the popular Co-op Tractor's features. Later these Co-op Tractors were assembled at the Central

Exchange's headquarters in St. Paul beginning in 1948. After World War II, things changed, though, and in 1952 the Co-op Tractor enterprise was sold to Cockshutt Farm Equipment, Ltd., of Canada. Cockshutt had been the source of various farm machinery items sold by the Exchange either under the Canadian name or the Co-op label, and widely advertised in the cooperative publications sent to patrons and members.

There was a joint effort with other regional cooperatives to develop a manufacturing plant in Arthurdale, VA, and the concept of a full line of farm machinery and equipment being produced there. A plant was built and dedicated, but became a victim of the World War II emergency when a moratorium on manufacturing was put into effect.

A long-time Central Exchange employee, Ralph Ingerson, was called "creator of the Co-op Tractor" in an obituary published in the Farmers Union *Herald* in April, 1950. This article told how Ingerson, who grew up on a farm near Maple Plain, MN, had been hired in Feb., 1929, as field organizer of the original Farmers Union Exchange. The Exchange then was a small subsidiary operation of the old Farmers Union Terminal Association.

Ingerson's chores at first were supervision of the very first Co-op Oil cooperative associations in Montana and North Dakota. Then, he was moved to the Exchange headquarters in St. Paul and put in charge of developing a wholesale cooperative oil business.

Then in 1934 and 1935, when the Exchange made its debut in the business of distributing hardware and farm machinery, and the development of the Co-op Tractor occurred, Ingerson was made supervisor of these enterprises.

During the 17 years he worked for the Exchange, he helped organize and establish a new group of regional cooperative wholesalers that became the National Farm Machinery Cooperative. It was the organization that owned and operated farm machinery factories, including the facilities that produced the Co-op Tractor.

"Although the venture in the farm machinery field did not prove to be highly successful, it did prove to the cooperatives that there must be a distributive system built up large enough in sufficient quantity ... to maintain an efficient operation," Syftestad said.

The truth of what the veteran cooperative agribusiness leader was saying also could be found in the Exchange's foray into yet another field, the retail grocery business.

GROCERY MARKETING – A "WORTHWHILE" EXPERIMENT

In 1942 the Farmers Union Central Exchange ventured into another new enterprise – the grocery business. There was no reason to believe that cooperative grocery stores wouldn't become as successful as earlier cooperative business breakthroughs into the world of rural retailing.

Consumers Cooperative Wholesale of Superior, WI, operated a successful grocery store enterprise, and was a source of groceries for Farmers Union outlets. One popular item quite common was a special that became known as "Farmers Union coffee".

The Central Exchange did establish a grocery department, with retail and distribution centered in a newly-acquired warehouse operation located in Williston, N.D. for the benefit of patrons-customers mainly in northwestern North Dakota and northeastern Montana This grocery business venture all began when the Central Exchange purchased the Northwest Cooperative Society and its inventory and facilities in Williston, ND

After three years, it was reported that "the grocery operations to date have been about on a break-even basis". Then in 1947 the report on grocery operations observed in part that "margins have been insufficient to meet costs".

In 1948 the story was similar. The report summarized: "The handling of groceries at the one warehouse has been an experiment. It has not been very good. The volume is small and the margins narrow."

The 1949 report told how "we have liquidated our grocery inventory at below costs and below markets... it was a worthwhile experiment... "

A post script on the grocery effort: One of the items sold was Farmers Union brand coffee. It gained popularity through its availability at Farmers Union Oil Companies where it was served free to customers and the public.

SYFTESTAD WARNS OF NEW ANTI-CO-OP FOES

In a speech made at the annual convention of the North Dakota Farmers Union in the fall of 1944, Emil A. Syftestad reflected on the growth of Farmers Union Central Exchange, and how it survived difficult times and attacks by money interests that opposed the cooperative's entry into the wholesale farm supply business.

Then, in a published text in the Farmers Union *Herald*, he called attention to a new threat and issued this warning to patrons: "I see some influences which today are attempting to curtail, if not destroy, the cooperative movement in this country in much the same manner as they attempted it about 30 years ago.

"The stakes are higher today in terms of the number of cooperatives, members, dollars, and understanding. The enemies are greater in number, more widespread and amply financed... I urge the directors and managers of every cooperative to become alert to the danger before them, and cooperate in defeating this menace which confronts us... "

Syftestad recalled how "pioneer cooperators" had fought vested interests a generation earlier. "Go back a few years to refresh your memory of the struggle cooperatives had, and the obstacles they had to overcome, and the type of opposition which they had to endure," he told the delegates in 1944. Looking back at history, he added, "should be of interest to those not in the movement until the past 10 or 15 years because it is a background which they will need to be of the greatest value in assisting the pioneers in planning a counter-attack... it will again arouse their fighting spirit in defense of the cooperative movement of which they have been a part for 30 to 35 years... "

When the movement was in "its early days", Syftestad related, "many difficulties were encountered... the private or independent grain elevator operators resented the fact that farmers who produced the grain should ever have any thoughts of marketing their own grain through a cooperatively-owned grain elevator... this resentment spread to other private businessmen in some localities...

farmers did not establish elevators for the sake of merely having a grain elevator... they had learned that they were not receiving as much for their grain as they should... through the operation of their own local cooperative grain elevators, farmers have saved themselves millions of dollars which otherwise would have gone into private channels... "

Later, Syftestad added, "Farmers also recognized there could be some economy in the purchasing of their farm supplies, particularly in major supplies such as gasoline and other fuels... they learned of a wide spread or margin between the cost of the fuel delivered locally and what the farmer paid for it... the savings through cooperative oil companies encouraged farmers to expand into serving themselves with many other items of merchandise commonly used on their farms such as tires, batteries, auto and truck accessories, Co-op farm tractors, farm tools, fertilizer, flour, wire products... "

The growth of cooperatives which in 1944 for the Central Exchange represented 350 local retail outlets serving 150,000 farm families in five states was seen as disturbing and alarming to some business interests, Syftestad said. "They fear that these ' super-co-operatives' are going to take over all business," he stated.

Consequently, Syftestad added, "A cooperative writer has stated that "the cooperative movement is the United States and Canada faces the severest trial and the greatest peril of its existence... there is hateful opposition, backed by unlimited wealth and powerful influences that seek to destroy it... " The threat in 1944 was similar to what had happened 20 years earlier when grain traders in Minneapolis conspired against buying grain from Equity Cooperative Exchange and the St. Paul Grain Exchange, according to a report issued by the Federal Trade Commission. The Commission had investigated the Minnesota market, and determined that the Minneapolis Chamber of Commerce had been behind a boycott where its agents or others refused to buy grain from cooperative sources, and also did other illegal activities in an attempt to discredit the cooperatives.

The Commission also found that during the 1914-17 period, when efforts were made to discredit Equity by spreading false reports that questioned its honesty in its grain dealings, the cooperative was being cheated by the grain buyers who resold Equity grain for immediate profits as large as 14 cents a bushel.

Syftestad then warned about the current anti-co-op attack, begun by a group called the National Tax Equality Association. Its president was Ben C. McCabe, a Minneapolis man who owned a string of grain elevators.

This organization had employed people to stir up criticism against cooperatives among local business interests, Syftestad reported. "They are attempting to build up a prejudice against the cooperatives method of doing business," he added. "This is one front on which every farm organization and every cooperative must stand united, or they all go down in defeat."

"CENEX – WHERE THE CUSTOMER IS THE COMPANY"

The slogan for the Farmers Union Central Exchange in 1950 was "50 Million in '50", a theme adopted at the cooperative's 19th annual meeting which meant a target of sales surpassing $50 million that year. This goal was announced before a record 4,000 delegates, stockholders, managers, members and visitors at the Exchange's annual meeting held in the Lowry Hotel in St. Paul.

Net savings for 1949 totaled $2.8 million, down from 1948, General Manager E. A. Syftestad reported. The year's performance was caused by smaller gross dollar sales and lower net margins – the same problem plaguing many local and regional cooperatives the previous year.

In his annual meeting address, Syftestad urged greater support of the cooperative: "Today there are about 150,000 farmers who are members and patrons of cooperatives affiliated with Farmers Union Central Exchange. If these 150,000 farmers fully understood how effective they could be by purchasing all of their major farm supplies from their own cooperatives, local and regional, such as the local cooperatives and the Central Exchange, and did make such purchases, they would double and probably triple the volume of the local cooperatives and the Exchange, which would result in tremendous savings at both levels much of which today are going to other interests never to return to the farmers in any form."

While there is no way of knowing just how effective Syftestad's words might have been, the record shows that phenomenal growth of the Exchange was ahead.

THE SYFTESTAD STORY, AND A "WORD-WAR" WITH THATCHER

Emil A. Syftestad was a native of New Rockford, ND, wore three vocational hats shortly after joining the Farmers Union Terminal Association in the early 1930s. First, he had been hired by M.W. Thatcher when the Farmers Union Central Exchange was founded. Then, after Thatcher headed to Washington, D.C., to be chairman of the new Farmers National Grain Corp., Syftestad succeeded Thatcher as general manager of the Farmers Union grain marketing cooperative. But Syftestad also became business manager of the Farmers Union *Herald*. Any one of those jobs in itself seemed a full-time position.

When Thatcher returned in 1938 to start the Farmers Union Grain Terminal Association, friction between the two men developed. That's because Thatcher envisioned one cooperative management in charge of both grain marketing and farm supply agribusinesses. When his idea was vetoed by C.C. Talbott, the highly-respected North Dakota Farmers Union leader, Thatcher tried another ill-fated tactic – an eight-page letter charging a variety of short-comings by Syftestad written in 1940. Syftestad fired back with an 18-page letter sent to cooperative officials, directors and managers – the same people who had received Thatcher's hotly-worded letter. And Syftestad made it clear he would "do battle" if necessary to defend himself.

"If I had first written you a letter such as you directed at me dated April 8th, I believe that there would be a roof missing at 1923 University Avenue (Thatcher's office)... ", Syftestad wrote in his reply letter. He added that it took two days for him to cool down after reading Thatcher's letter.

Thatcher had brought up the debt owed to GTA by the Central Exchange, and other critical matters. But Syftestad responded, "Who kept the movement alive while you (Thatcher) were in Washington (D.C.) From April, 1932, to April, 1937? It was the state Farmers Union organization, the Junior movement, the Farmers Union Education Service, the Farmers Union *Herald*, and respective state newspapers, the Livestock Commission Company, the Terminal Association, the Exchange, all the cooperative directors and the Northwest Legislative Committee.

"Then, there are the many thousands of individuals at every crossroads who have given much of their time, money, and patronage... to build the cooperatives and the Farmers Union movement. To them must go the greatest thanks of all... it could not have been done without them... "

Despite the "word-war" letters, and the resultant tension between the two cooperative leaders, both cooperatives received a combination of leadership and support that enabled both to grow far beyond what the founders had envisioned. But E. A. Syftestad did not live to see it happen. He died suddenly of a heart attack in July, 1957.

So ended the career of Emil Syftestad which had begun behind a lone desk across which came the first petroleum product transactions for Farmers Union members. Syftestad had seen that single desk become first a separate department, and then grow into a separate cooperative entity destined for both growth and great success. The few dollars saved for farmers in those early years turned into a total of $11.3 million paid in cash as patronage refunds to farmers during the last 14 years under Syftestad's leadership.

His successor was Thomas H. Steichen, who began working for Central Exchange in 1935, lured to St. Paul by Syftestad from New Richmond, WI, where he had been manager of a Farmers Union Oil Company. Within two years Steichen would be a principal in the historic joint cooperative action program developed with M.W. Thatcher. Under that agreement, the two growing regional cooperatives would spell out areas of business activity in which each would operate.

Steichen would call the agreement between the Central Exchange and Thatcher's GTA "a momentous occasion", explaining that his cooperative would now do in farm fertilizer what it had done for farmers in petroleum. He pointed out that the Central Exchange had earned $50 million in savings for Upper Midwest farmers through its expertise in petroleum marketing.

M.W. Thatcher praised the agreement as "a thrilling hour" for farmers, noting how his grain-marketing cooperative had grown. Now, he explained, the cooperatives had agreed on sales areas and activities to avoid costly duplication of services while guaranteeing farmers would continue to receive high-quality feeds, seeds, and fertilizers efficiently along with superior marketing of their grain. The addition of the McCabe Company elevators and feed plants earlier in 1958 by GTA emphasized a need to more clearly identify areas and services each of the two cooperatives would provide, Thatcher reported.

A joint statement about the cooperative action agreement stated that the Farmers Union Grain Terminal Association would cease distributing fertilizers while the Central Exchange would confine its distribution of feeds and seeds to Wisconsin. "Cooperation between the two regionals has been growing in recent years," the statement read. "Their aim is to further improve the relationship and thus contribute to the performances by each organization of a more efficient job for their 300,000 farmer-owners and patrons... "

So ended apparently, and publicly as well, evidence of the earlier strained relationship that was known to exist between the two cooperatives at the loftiest levels of management.

STEICHEN PRAISES FARMERS UNION AS "GREATEST HOPE"

Strong support for Farmers Union came in a speech Thomas Steichen delivered at the 65th annual convention of the National Farmers Union in Oklahoma City, OK, in Mar., 1967.

"The Farmers Union program, both nationally and at the state level, offers the greatest hope our farmers have to achieve the necessary bargaining power to gain some measure of economic justice for rural America," Steichen told the delegates.

"This is why the (Farmers Union) Central Exchange from its very beginning devoted so much time, effort and money working with and supporting the state Farmers Union organizations in our Midwest states... recognizing that cooperation is a two-way street, Central Exchange support of the Farmers Union is a matter of daily operating policy... "

Then Steichen noted, "Evidence of that Central Exchange support is the fact that to date total education funds paid by Central Exchange now amount to $5.7 million... these funds are for the purpose of promoting and encouraging cooperative organization and education by state Farmers Unions... this has undoubtedly helped to strengthen the effectiveness of Farmers Union in our area... this arrangement permits the state Farmers Union organizations to share in our cooperative success... "

He said Central Exchange represented nearly 350,000 farm families who own more than 850 cooperatives affiliated with the regional cooperative in an area from the Upper Midwest and the Great Lakes to the Pacific Ocean.

"We believe your (Farmers Union) efforts to strengthen the family farm have more than justified farmers' and cooperatives' support of Farmers Union," Steichen added. "Your efforts to protect farmers from an unregulated free market have saved hundreds of millions of dollars of farm income in our area. Some have estimated that the programs you have sponsored and supported have saved from 30 to 50 percent of the farm income in recent years... how can you measure the effect this has had on our rural economy, on our co-ops or on the entire country?"

CENTRAL EXCHANGE GROWTH CONTINUES

By the early 1970s, its leaders reported, Central Exchange was more than 100 times larger than it had been at its outset back 40 years earlier. It was marketing 20,000 different farm supply items. Its sales then topped $200 million annually.

Where it had taken 32 years for the cooperative to surpass the $100-million-mark, sales then advanced to more than $200 million only eight years later.

And now, in the wake of such phenomenal growth, some people began to wonder, "Just how big should a co-op be?" Some had a ready answer for this question by saying the cooperative "should be big enough to challenge and compete price-wise in the market arena, and big enough to deliver the supply items modern agriculture required. Big enough to respond, yet small enough to hear the (co-op's) owner, alone in the field."

In 1963 nine states were part of the Exchange family, Total savings topped $23 million, and $37 million of Exchange stock investments had been retired through 1956 in 23 annual patronage payouts. It was a remarkable record. However, the news of the Central Exchange's success was dampened by the Aug. 30, 1963, death of Thomas H. Steichen. At age 60, he was struck down by the same thing which felled his illustrious predecessor, E. A. Syftestad – a sudden heart attack. So ended Steichen's 28 years of history-making service to the Farmers Union Central Exchange – a period of truly phenomenal growth where the cooperative was a pioneer and leader in many new agribusiness fields.

For 15 years Steichen, a native of Watertown, SD, served as the CEO of one of the nation's premier regional supply cooperatives. Under his leadership, Central Exchange grew steadily from $75.8 million in sales in 1957 to $203 million in 1971. His career in cooperatives began in 1935 when he was named manager of the Farmers Union Cooperative Oil Co. in New Richmond, WI. He joined Central Exchange a year later when this co-op was only four years old. In 1966 the Central Exchange ranked fourth in sales and 471st on the list of the nation's top 500 companies. One of those who took note of the growth of Central Exchange was Edwin Christianson, president of the Minnesota Farmers Union and almost always a speaker at its annual meetings (just as Central Exchange officials were religiously always speakers at the Farmers Union annual convention). Observed Christianson at the 33rd annual meeting of the Farmers Union Central Exchange held in St. Paul in Mar., 1964:

"Each year, as we meet here at the annual meeting, it is possible to take note of some new landmark in the history of service to farmers by this great cooperative institution. This past year, history has been made in reaching a volume of more than $100 million in annual sales. What a tremendous figure... it took some time to reach this level... (it) reached $5 million in its eighth year of operation, in 1939... and it took another four years to reach $10 million... $20 million by 1946... seven years later, in 1953, the volume expanded to $50 million... Then in another four years (in 1957), the volume reached $75 million, and now, six years later, here we are with a $100 million volume... "

Christianson told the delegates that "such an achievement was possible, first of all, because people had the wisdom to establish Farmers Union Central Exchange, but even more important, they had the determination and the staying power to stick with their own institution and build it to what it is today... " He noted a "savings" of $10 million which he said "improved the balance of trade in our own communities", keeping those dollars at home rather than into the coffers of distant corporate interests.

He went on to list accomplishments for farmers he attributed to the Farmers Union through its many battles to gain cooperatively, and legislatively. "When we reach the situation where we have today, with 60 percent of our wheat, 40 percent of our soybeans and 26 percent of our corn going into export markets, none of us can afford to take the attitude that we need to be concerned only with what goes on within our own fence lines. We farmers are in the export business... "

In July, 1967, at a regional Central Exchange meeting, Christianson talked of great changes confronting agriculture, including for the business cooperative, fewer farmer-customers. "Looking back several years, who could have foreseen that the production and handling of chemicals would play such a major role in our farm production and the farm service industry?

Or, that our interdependence with world trade would be such that a good crop of sunflowers in Russia would have an immediate effect on the price of our soybeans in Minnesota?"

He noted the development of a network of "Soil Service Centers" by the regional cooperative as an example of how new technology could be brought personally to farmers in a way just as important as the price or content of fertilizer. "Your competitors can match price and contents," Christianson said. "But they do not offer a service that is in tune with the farmer's needs, or that is as close to the farmer-customer himself... "

IT WAS A TIME OF CHANGE

So it was a time of great change. And perhaps unexpectedly in many farm circles, it was a time of change within the Farmers Union Central Exchange itself. Thus emerged a new red-and-white logo bearing one word – "CENEX" – as the main advertising identification for the cooperative. It would no longer brandish boldly any of the four words used from its 1931 origin, "Farmers Union Central Exchange". The new lone word to be used now in bold, capital letters, was "CENEX". No more "farmers" or "union".

And with the debut of the new name came a new slogan printed beneath CENEX, "Where the Customer is the Company".

Left in the wake of this modern logo "progress" was the familiar red, green and gold Farmers Union Co-op seal used for many years. "But the principles of the pioneers who had seeded the effort were certainly not abandoned," was a statement printed in the Cenex account of this change. Despite such words of encouragement that the Farmers Union connection and emphasis would be preserved, in reality the stage was set for a different business bent, away from the farm, and, away from Farmers Union. It was all in the matter of doing more business, the change-makers seemed to be saying.

The development of the new logo came under the Cenex regime of John McKay, who also became "president" rather than general manager of the old FUCE. McKay's background contrasted with that of his illustrious predecessors – Thatcher, Syftestad and Steichen who had accounting backgrounds. McKay was a public relations and marketing director.

In the cooperative's 50-year publication, "To Gather Together – CENEX – The First Fifty Years", the authors write, "Change. It fertilized cooperative growth. In its earliest years, cooperation had been an economic option that helped farmers survive. In the 1940s, change had repositioned cooperatives to ever larger circles of influence and risk. Throughout the 1950s and 1960s, change carried with the ring of transition as all levels of farmer-cooperation brought themselves into modernity. As the 1970s came to the union of farmers, renewed efforts were placed on self-sufficiency and preparation for the future. The world had become interdependent and interconnected. Arab oil-sellers and Russian wheat buyers were to be included in the strategy of American cooperatives."

Part of this "change" would be the removal and diminished appearance of the two words representing the long-standing "Farmers Union" relationship with this cooperative. As far as the public and customers were concerned, that historic and meaningful terminology – "Farmers" and "Union" had passed into the pages of history.

One of those who voted against changing the name of the Farmers Union Central Exchange was Joseph E. Larson of Climax. He joined Farmers Union in

1946 when he bought his grain and potato farm in Polk Co., and was active in the Neby local which met in a two-room country school house and staged a minstrel show yearly to raise money. In time Larson became the local president, and in Mar., 1960, advanced to the board of directors of the Farmers Union Central Exchange, filling a post held since 1938 by E. H. Reitmeier of Crookston. Eventually he became chairman of the Cenex board of directors.

IT WAS "A CLOSE VOTE" TO CHANGE NAME

For 30 years Larson would sit on the Exchange's board. And as "things changed", he recalled, the change included a different name. "They didn't want 'Farmers Union' any more," Larson said. "The name was too long, they said, and some felt, too, that it had reached the point where it wasn't good business to have it part 'Farmers Union' and part big business. But I voted against it, but like anything else, if you don't have the votes, you lose... I thought it had served us very well for many years... It was unfortunate". The vote to eliminate "Farmers Union" from the name of the Exchange was close to a 50-50 deadlock, Larson remembered.

Not all of the Exchange's cooperatives followed the name change policy, though, including the Farmers Union Oil Co. that operates out of Larson's home town of Climax.

In North Dakota, a special resolution was adopted, according to Farmers Union President E. W. Smith in a letter mailed Mar. 30, 1973, to Farmers Union Oil Co. of Alamo, ND. "We were concerned last year at the trend taken by the board and management of Farmers Union Central Exchange in that it appeared we were losing the tie to Farmers Union in putting up the CENEX logo in place of the Co-op shield. The shield has a real meaning to our people.

"We said the next step would be to remove the name, and in a couple of years, the 'Farmers Union' would never again be mentioned. Now it has already happened. Telephones are being answered now say 'Cenex Service Center'... "

The last annual report bearing the Farmers Union Central Exchange, Inc., name was published in 1981. CENEX first appeared on the annual report the following year in bright red and black letters aside "1982 Annual Report" printed at the bottom. The Farmers Union Central Exchange, Inc., wording, however, appeared at the top of the annual report cover.

The same wording appeared on consecutive reports through 1984. Then in 1985 the annual report cover did not mention "Farmers Union" or "Central Exchange". Instead, the cover identification read, "CENEX 1985 Annual Report, For 55 Years the Country's Best". However, on Page 3, the first printed page, the report begins with these words, "CENEX, Farmers Union Central Exchange, Inc., is one of the nation's leading agricultural supply cooperatives, serving as a wholesale supplier to 1,600 retail cooperatives. Its market includes 13 states from Wisconsin to the Pacific Northwest... it is ranked 237 on the 1985 list of Fortune 500 companies... "

The report listed the main business activity as "to supply essential agricultural (production) inputs, such as refined fuels, lubricants, fertilizer, feed, seed and chemicals... also passenger and tractor tires, batteries, auto accessories, fencing products, livestock equipment, and a variety of other products... " It also pointed out that Cenex's cooperative farm management service, called "CENTROL" and begun in 1980, "is now the nation's largest", and that its PUMP 24 electronic fuel-dispensing system permitted customers in 250 communities to purchase motor fuels around the clock.

In summary, total sales have generally increased over the years, topping one billion dollars for the first time in 1979 (the cooperative's 48th year of operation), and then the two-billion-dollar mark in 1993 (the cooperative's 63rd anniversary year).

TOTAL CENEX SALES IN RECENT YEARS:

1980 – $1.2 billion. 1986 – $1.093 billion. 1992 – $1.779 billion.
1981 – $1.325 billion. 1987 – $1.138 billion. 1993 – $2.048 billion.
1982 – $1.47 billion. 1988 – $1.259 billion. 1994 – $2.184 billion.
1983 – $1.40 billion. 1989 – $1.365 billion. 1995 – $2.244 billion.
1984 – $1.487 billion. 1990 – $1.567 billion. 1996 – $2.683 billion.
1985 – $1.365 billion. 1991 – $1.567 billion.

Where Central Exchange had total sales of only $61,000 in its first year (1931), serving a tiny network of 21 affiliates in four states, in its 60th year Cenex sold a then record $1.6 billion in sales through 1,925 locals in 15 states and achieve a net income of $44.1 million (returning $17 million in cash to its member-owners).

In that 60th annual report, Noel K. Estenson, Cenex president and chief executive officer, noted the successful joint venture with Land O'Lakes, Ag Services begun in 1987, and a "recommitment" to what he called "our core businesses of petroleum" service and products. This meant, he added, emphasis on "strategies for securing your regional's (cooperative's) future as a financially sound business in its rural marketplace". As this is being written, Cenex has 2,200 employees.

The "partnership" of Cenex/Land O'Lakes into the agricultural services arena marked the linking of two of the nation's leading agricultural supply cooperatives. "We are drawing on the very best products, services and people our two cooperatives have to offer, in a new spirit of cooperation... ", commented Cenex President Darrell Moseson at the time of the consolidation.

"That spirit will produce positive benefits for our member-owners across our 15-state trade territory."

Prior to the joint venture, both cooperatives had their separate but similar agricultural service divisions supplying individual farmers and local community cooperatives with petroleum, fertilizer, farm chemicals, feed, seed, and many other agricultural products and services. There was obvious duplication of services, and competition for customers.

The yearly savings in yearly operations, it was estimated, would be $10 million. Under the plan, Land O'Lakes would concentrate its ag supply activities in the feed and seed areas, leasing and managing the feed and seed assets of Cenex. In turn, Cenex would specialize in petroleum and energy products, and expand its base by leasing Land O'Lakes petroleum assets and managing the combined petroleum operation. Together, the two co-ops would manage the merged fertilizer and agricultural chemical divisions.

Ralph Hofstad, president of Land O'Lakes, described the joint-venture agreement as "an important step in needed cooperative consolidation... which will lead to even greater cooperation between the organizations." Land O'Lakes Board Chairman Claire Sandnesss said, "It is an ideal way for farmer-owned organizations to streamline operations and become more efficient."

The consolidation was approved by a 90 percent margin in voting which followed a series of local meetings outlining the proposed joint venture.

In 1996 the two regional cooperatives boasted combined sales annually of $3.6 billion, and 140 years of combined experience out of headquarters in the Twin Cities. Through its agronomy marketing this division had the largest dealer network in North America, owned nearly 40 percent of CF Industries, and yearly was selling more than 4 million tons of fertilizer. It was a major marketer of crop protection products, distributing more than 1,100 products through 25 "Express Centers" strategically located mainly in the Upper Midwest area, and employing a corps of crop specialists.

On its own, Cenex is a wholesaler/reseller marking more than 1.7 billion gallons of refined fuels, and representing the largest cooperative refiner in the nation with a 42,500-barrel per day refinery at Laurel, MT, and a 74.5 percent owner in the National Cooperative Refinery Association at its 80,000 barrels per day refinery at McPherson, Kans. It also had a network of Cenex Convenience Stores, car and truck stops, and Cenex Fast Lube centers.

Land O'Lakes at that time was a major supplier of animal and poultry feeds, with nearly 250 manufacturing plants. It operated a research farm ("The Answer Farm"), and was a basic seed producer marketing alfalfa, corn, forage seeds, soybeans, sunflowers, and turf seeds.

Marking the 60th year of Cenex, Noel Estenson said, "It is a time to reflect on our rich heritage, and it's also a time to rededicate ourselves to the future... a strong cooperative system is the inheritance left to us by its founders, and it's the gift we will leave to the younger producers of tomorrow... "

In that same 1990 report, Cenex board chairman Elroy Webster of Nicollet, MN, pointed out "much has changed over the last six decades, but one thing hasn't – the ties that built this system and hold it together, remain strong and will endure... "

THE FARMERS UNION 'LINK' IS ATTACKED

There were critics of the closely blended identities of "Farmers Union" and the two major regional cooperatives based in the Twin Cities – the Central Exchange and the Grain Terminal Association. Edwin Christianson mentioned this adverse sentiment concerning "Farmers Union" wordage in his July, 1967, speech at a Central Exchange meeting.

Christianson told how an unidentified official of Farmland Industries, Inc. (which was the new name adopted by the former Consumers Cooperative Association of Kansas City which also later got a new logo that replaced the old, familiar red, white and blue "double-circle Co-op" used for generations), had criticized both Farmers Union and the Farm Bureau for their "partnership concept of farmers and their farm organizations and cooperatives" in a recently-published magazine article.

The Farmland Industries spokesman had alleged that "the Farmers Union organization and cooperatives operate with a split personality – that the farm organization had handicapped more than it helped", Christianson reported. "He even questioned if our Farmers Union cooperatives are true cooperatives because of the relationship with the farm organization," he added.

"This (Farmland Industries) spokesman seems to fail to recognize that farm cooperatives and farm families survive together, or not at all... His organization once carried the name 'cooperative' in its identity but no longer does so... It is moving away from farmers into a 'big business' concept... The next logical step in such a trend is to reorganize as a private, for-profit type organization... "

Christianson then stated, "This attack... upon Farmers Union... (represents) a serious threat to us and the well-being of family farmers... cooperatives must take a role in improving farm income through legislated farm programs if the family farm is to survive... certainly, there is no place in agriculture for farm cooperatives if there no longer is a family farm type of agriculture... "

A LOOK BACK AT THE ROOTS OF CENEX

In 1950, just a year shy of the twentieth anniversary annual meeting for the FU Central Exchange, Credit Manager C. J. Mitchell authored "Seeds of Cooperation" which was published in the Farmers Union *Herald*. In this article Mitchell, who had been one of the first field representatives of the Exchange, wrote that he felt it was important to recount for new members and others some of the early history and challenges encountered by the cooperative during its first 19 years.

"Comparing the Exchange of today with that of 19 years ago, it does not seem possible," Mitchell wrote. "It seems like a tale in a story book, but the facts are there...

"We started as a brokerage in a rented building with only a few employees and a few pieces of office furniture. Many of the contracts we had with suppliers were vicious... it now has grown to where it can supply to affiliated companies most of the supplies needed on the farm.

"It owns refineries, production and gathering lines for petroleum, its own blending plant for oil, it has heavy investments in feed plants for the patrons who want Co-op feed. It is in a position to furnish fertilizer. It has built up a farm machinery program even to the extent of assembling its own Co-op Tractor. That tractor belongs entirely to the Exchange, and no other cooperatives are part of this program. The Exchange also is in position to provide all farm tools, hardware items, steel sheets, grain bins, electrical appliances, and, in fact, almost any item the farmer needs on his farm can be purchased through his own organization... "

Mitchell recalled "in the early days the Exchange had encountered obstacles created by competitors determined to discourage the farmers and their cooperative movement" through attractive price discounts. The cooperative had to ask for financial help only twice, Mitchell related:

First, when the Exchange was just getting started in the early 1930s. It borrowed $50,000 from the Farmers Union Terminal Association. (Its net worth in 1950 when Mitchell authored his history was $15.8 million).

Secondly, the Exchange in 1943 obtained $250,000 from its affiliates when its oil refinery at Laurel, MT, was purchased. It subsequently repaid that indebtedness. (In July, 1991, Cenex announced it was building a $30 million refined fuel delivery pipeline linking Minot to Fargo. The cooperative had built the first leg of that pipeline from Laurel to Glendive, MT, in 1954. Then it acquired the Oil Basin Pipeline and extended the Laurel pipeline to Minot in 1961. The Minot-Fargo line will stretch over 250 miles.)

Mitchell wrote, "To me it seems only right and proper that after 19 years of successful endeavor that we should inform people who have come into this organization in later years of some of the early history and trials which the Central Exchange went through to build up an institution that would be of value to the farmer and safeguard his interests."

The success of the Exchange, he said, "is due primarily to a move that began in 1925 when a group of determined farmers met in Des Moines, IA., and formed what became known as 'the Corn Belt Federation of Cooperatives', and to an organization committee made up of C.C. Talbott (of North Dakota), M.W. Thatcher (general manager of the Farmers Union Grain Terminal Association) and A. W. Ricker (editor of the Farmers Union *Herald*)... "

"This group and a few determined organizers stayed on the job building Farmers Union membership and encouraging and teaching cooperation. This was the job of everyone in those days... "

TOP EXCHANGE/CENEX MANAGERS/PRESIDENTS/CEOS

Here are the men who managed the Farmers Union Central Exchange / Cenex since the formation of the supply cooperative:

1931 – M.W. Thatcher (who also continued to manage the Farmers Union Terminal Association).

1932 – Emil A. Syftestad (who also managed the grain terminal association from 1932 to 1938).

1957 – Thomas H. Steichen (who had worked for the Exchange since 1935).

1972 – John McKay (who had worked in public relations previously).

1974 – Jerome Tvedt.

1980 – Darrell Moseson.

1986 – Noel K. Estenson.

Figure 37 Combining Then

Figure 38 Combining Now

Figure 39 MN Gov. Karl Rolvaag, l., Thomas Steichen, a
Farmers Union Central Exchange General Manager

Figure 40 New York Mills FU Headquarters 1940's

6 ...POST-WAR CHALLENGES...

"I farmed through the depression years of the 1930s and I know what happened when we farmers lost our income. I saw banks and business people go bankrupt. These are experiences seared in my soul... "
Roy E. Wiseth, candidate for State Representative.

ROY E. WISETH, FU PRESIDENT, MINNESOTA POLITICIAN

Figure 41 Roy E. Wiseth

Brief Biography
Born: Aug. 12, 1900, Hatton, N.D.
Died: Aug. 11, 1969, Wiseth farm, Goodridge, MN.
Employment: Life-long farmer, Goodridge, MN
Community service: Active in many civic organizations; state Farmers Union president, 1949-50;
State Senator, 1954-62; DFL candidate for Congress; president, Northern Cooperative Association of Wadena; board of directors, Thief River Falls Farmers Cooperative Grain and Seed Association; member, Goodridge School board; member, First Lutheran Church Council; Board member, Minnesota Farmers Home Administration; executive committee, DFL.
Family: Married Ida Theresa Peterson, Nov. 25, 1923; six children.

DEATH COMES IN A MINNESOTA FARM FIELD

Tragically, the limp, crushed body of the father was to be discovered by one of his young sons. It was found lying in the field where he had been at the controls of the farm tractor pulling a plow just as the veteran farmer had done countless days before. But on this day, no one really knows what really happened. One can only speculate exactly how farm leader and former State Senator Roy E. Wiseth of Goodridge lost his life.

The son, Kenneth Wiseth, took the limp, injured body of his father and rushed to the nearest emergency center, Northwestern Hospital in Thief River Falls. It was sadly to no avail. And his father was pronounced dead on arrival, just one day before his 69th birthday, and 10 days after a once-in-a-lifetime trip to Scandinavia – his wife Theresa's homeland.

Officially in the coroner's statistics, the death of Roy E. Wiseth goes into the records as a farm accident. That his life had ended in a tractor mishap plowing in a field on the family farm five miles northeast of Goodridge.

However, that report didn't begin to reflect the full scope of what had happened that fateful day. It was much more than just another farm tractor accident. Or, that this tragedy had brought to an abrupt end the career of a man who had been elected state president at the eighth annual convention of Minnesota Farmers Union nearly 20 years earlier. Or, that now dead, was a farm leader, a former state senator and a one-time candidate for Congress, and a leading citizen long active in civic affairs, in building farm, cooperative, and church organizations.

Nor did the death reports indicate how a young Roy Wiseth had fallen deeply in love with Theresa Peterson, a young lady born in Sweden and who, like so many others had done in the late 1800s and in the early 20th Century, had just recently emigrated to Minnesota. Young Miss Peterson and her family found their new home in the Iron Range mining country of northern Minnesota. When she met her husband-to-be who came as a young man to the iron mines to work (as he later would say "for wages"), Ida Theresa Peterson spoke only Swedish. However, despite this lingual shortcoming, a romance blossomed.

LESSONS OF LIFE CAME EARLY

The lessons of life for Roy Wiseth had been learned well over the years in rural Goodridge, on the farm, working "for wages" in the mines, and during tough economic times of the 1930s.

Roy Wiseth recalled some of these things in his campaign literature when he was seeking to be elected State Representative from Marshall Co., Minnesota. "I came to Marshall Co. with my parents in 1904. My father owned a small farm and country store in Moylan Township. As a young man I worked for five years on the Iron Range (where he met his wife, Ida Peterson) where I came directly into contact with the problems that a family has making ends meet while working for wages.

"I own and operate a stock and grain farmer in Moylan Township where I have farmed since 1925. I farmed through the depression years of the 1930s and I know what happened when we farmers lost our income. I saw banks and business people go bankrupt. These experiences are seared in my soul and no high-pressure lobbyist can change them, or me. The farm program helped bring us back... we do need laws to protect us... "

His political campaign message were based on lessons learned from his personal history in northern Minnesota, a heritage he shared with countless others of his generation – a generation which sustained bleak and stressful years, and times that left indelible, life-long impressions. This kind of message to voters would eventually help him win important leadership positions, in the State Legislature, as well as in the influential Minnesota Farmers Union.

A DISTINGUISHED VISITOR WITH AN OFFER

Thanks to a long career, Wiseth was to earn praise and political endorsements from two prominent Minnesotans, Hubert H. Humphrey, then a U.S. Senator, and Orville L. Freeman, then the Minnesota Governor. And a favorite story he often told is how one spring day in the early 1950s two visitors strode across the plowed field where Roy Wiseth was doing field work. One of the two men was a young man destined to become U.S. Senator, then Vice President, and eventually, a Presidential candidate.

This impromptu field visitor who walked across Wiseth's plowed field that day was Walter Mondale. He and his colleague were convinced Roy Wiseth should run for the U.S. Congress, and told him so that spring day.

"The U.S. Congress needs men of vision and foresight, men who will work to make America and the world a better place in which to live for us and coming generations. (State) Senator Wiseth is just such a man, and will be a valuable addition to our hard-working DFL Delegation in Washington," Senator Humphrey stated in a Wiseth for Congress political brochure.

"Senator Wiseth has made a distinguished record in the Minnesota Senate," stated Governor Freeman in the same pocket-sized, printed document. "He has worked conscientiously and diligently to further the welfare of Minnesota's citizens. And he merits the support of voters of the Ninth Congressional District for the office of Congressman."

Such high praise was justified for Roy E. Wiseth, although it came near the end of his impressive career. And while he never won the Congressional seat Walter Mondale thought he deserved, Wiseth did serve a pair of four-year terms in the State Senate, beginning when he was first elected in 1954, and ending eight years later.

STRONG SENTIMENT FOR FAMILY FARMERS

Roy Wiseth boasted a strong feeling for the fabled "family farm". "It is the keystone of the economy for much of northwestern Minnesota," he wrote in a campaign brochure. "Under the present farm policies the farmer's income has declined drastically, costs of administering the program have soared, and consumers' costs have gone up. A program of realistic price supports coupled with acreage allotments will return prosperity to rural areas... "

He spoke out for helping "small business and working people", noting that "as the son of a country store operator" he would oppose "letting corporate monopoly destroy small business – the cornerstone of democracy". And his campaign leaflet contained a photo of nine grandchildren and an appeal for federal aid to education, "to invest in America's future, its greatest asset, its youth... "

Roy Wiseth's grandson Vance was a national Farmers Union delegate in 1974, when he was in his early 20's. Asked what he felt was the most important issue confronting the convention delegates, young Wiseth answered: "It is the age-old story of fair prices. I think this convention reflects that farmers are done with people making promises they don't keep. Farmers are through with that uncertainty. If they have to take action and reduce their production in order to get the price they want, farmers are going to do it. It may be disastrous to the consumer to do this, but they will then realize how important farmers are, and that we want to stay in business and make a profit like everyone else."

Roy Wiseth's dedication to Farmers Union was something which was passed on to two other generations, to his son Robert and his wife Ethel. Robert Wiseth was long active as president of the East Marshall Co. Local and a member of the state executive board of directors. And grandson Vance also has been active nearly life-long, and sought the office of state vice president in 1986. In a campaign brochure, Vance Wiseth, then 32 years old, stressed the fact that he was a "Third Generation of the Wiseth Family Active in Minnesota Farmers Union".

The younger Wiseth was not to win the Farmers Union vice presidency, and was to encounter personal misfortune a month after his marriage. A news account published in the Grygla *Eagle* on Nov. 2, 1978, told how the young farmer had lost his right arm in a farm harvest time combine accident. Somehow, the machine was put into operation by a co-worker who was unaware Vance Wiseth was adjusting the combine at the time.

The young farmer, who with his wife Lori (then expecting their first child) operated a 480-acre farm between Goodridge and Grygla and rented another 780 acres, was hospitalized. After two weeks in a Fargo, ND, hospital came every-other-day trips for treatment at hospitals in both Thief River Falls and Grand Forks. Later came weekly therapy at Grand Forks.

Shortly after the accident, neighbors and friends came to the Wiseth farm in a caravan of tractors, trucks and farm equipment at 8 a.m., according to the Grygla newspaper. This group of about 40 men and women spent the day doing field work, farmstead chores and finished the Wiseth's newly-built home. The Farmers Union Oil Co. of Thief River Falls dispatched a fuel truck to the farm and donated fuel to refill the tanks of the farm tractors involved in the "Good Neighbor Day" activity.

FARMERS UNION TRADITION CONTINUES

Nearly 20 years later, Vance's mother, Ethel Wiseth reported her son is "doing just fine" and that he took over the family farm operation following her husband Robert's death. In 1997, Vance Wiseth also operated a hay business enterprise, both as an alfalfa producer and as a marketer using his own semi-truck and trailer equipped to transport the big hay bales (both round and square) enclosed in plastic.

His father, Robert Wiseth, who was a dedicated Farmers Union leader locally as well as statewide, had become involved in the late 1940s. He attained the presidency of the Little River Local, and then East Marshall Co. president. Then he moved up to the state executive committee, and eventually its chairman. At the same time he was active locally, as president of the Faith Lutheran Church Council, and president of the Farmers Cooperative Grain and Seed Co.

His career was cut short by cancer in 1981, and one of those who spoke in eulogy was state Farmers Union president Cy Carpenter.

Ethel Wiseth, a talented amateur artist and art contest winner, was given a special recognition at the 1972 state convention. One of her acclaimed paintings was an oil portrait of the legendary Minnesota Farmers Union state president, Edwin Christianson. It had been a part of a "Farm Family Living Project" display set up at the state Farmers Union convention by Robert and Ethel Wiseth.

That painting eventually was to be placed in the parlor of the Christianson home in St. Paul, his wife Nora wrote to the Wiseths in Jan., 1973. "It is a beautiful piece of art... I never knew you were such a great artist," Nora Christianson said in her letter. "Ed and I shall always remember you folks and all of our good friends in Farmers Union and those thoughts will always be cherished... it is so sad when a person has a severe stroke as Ed had and has to quit all the work he enjoyed so much... "

A second letter came from Lura Reimnitz, the long-time state Farmers Union education director: "Last night I took your painting to Mr. Christianson, and he recognized it alright," Lura Reimnitz wrote Dec. 14, 1972. "He sat and looked and looked at it, and he would say, 'Oh Gosh!. . . Oh Gosh!' and shake his head, look at it again, and say again, 'Oh Gosh!'. So you can rest assured that your portrait was very warmly received."

In her letter, Lura Reimnitz complimented the Wiseths on their "Farm Family Living Project" display, saying she thought it ranked "tops" with any ever displayed. The Wiseths were given a Farmers Union "Farm Family Living Project" plaque at the 1972 state convention. Lura ended her letter with this comment:

"My association with the Wiseth family has been very, very memorable from back in the mid-1940s when I stopped at the home of Roy and Theresa Wiseth when I was working on organization in Marshall Co. It is an association that both (Lura's husband) Oskar and I will remember fondly."

NEEDED (FINANCIAL) HELP CAME FROM CO-OPS

Roy Wiseth won the Farmers Union state presidency in a three-way contest in Nov., 1949. He earned 6,184 votes – 269 more than did Einar Kuivinen, the incumbent, and nearly twice as many as the third presidential hopeful, state staff member Russel Schwandt.

In his campaign speech, Wiseth had informed the delegates that his first priority as president would be "to find some method to borrow (money) from (the Farmers Union) regionals".

The state organization was struggling at the time, rift with political feedback caused by open support by some Farmers Union members for the somewhat controversial U.S. Presidential candidate Henry A. Wallace in the 1948 elections. Evidence of this was the fact that Tony Dechant had been moved into Minnesota by the National Farmers Union as state secretary.

Roy Wiseth was to keep his campaign promise – not only did he gain financial support from the regional cooperatives, it was put to a novel, good use in an obvious way: $13,000 from Farmers Union Grain Terminal Association was used to purchase a bus. GTA had asked, though, that one use of that bus should be trips for members to tour regional cooperatives and their farmer-owned facilities.

In all, Wiseth's regime received three loans totaling $32,000 from GTA, Farmers Union Central Exchange and Great Plains Lumber & Supply. Coincidentally, the three loans almost equaled the $31,659 indebtedness Tony Dechant had reported at the organization's eighth state convention.

WISETH SEES CHANGE, DETERMINATION

In a news column published in May, 1950, in the Farmers Union publication, Roy Wiseth recalled asking delegates to elect him "for the opportunity to break the bonds that have kept us in a state of inertia". In the column, Wiseth explained he felt progress lagged in Farmers Union "that was the result of an ineffective program of coordination of all the resources available" in the organization.

Since then, Wiseth said he saw encouraging signs – the money from Farmers Union Grain Terminal Association to purchase a bus, and funds to operate it and use it to expand membership, the contacts with cooperative fieldmen who were "helping build good relations" among co-ops and Farmers Union, and the determination of members to seek changes.

Wiseth described the Farmers Union "as on the March... behind this great movement are some of America's greatest farm leaders plus the pent-up determination of farm people everywhere to have a real voice in their own and their children's future, that they shall have a fair share of the wealth they create... for these people who may never know a 40-hour week, who willingly toil from sun-up to dusk, want justice spelled out in equal opportunity... these things are all bundled up in one package – parity price for their products... "

Then he added in a concluding paragraph : "The Minnesota farmer today, with the experience of the early 1930s in his background, has turned his back to the wall. He is picking up the only weapon that is offered to him – the Farmers Union. When the fight is over, men in high places will know that the farmer will not meekly accept tax evictions, farm foreclosures and peonage as his reward for producing abundantly."

WISETH URGES SUPPORT FOR "BRANNAN PLAN"

Roy Wiseth came away from the National Farmers Union convention in Denver in 1950 inspired. In the *Glimpse* of April that year he wrote, "It was an inspiration to meet farm people from many states and discuss the many problems facing agriculture wherever farm people live, and to hear U.S. Secretary of Agriculture Charles Brannan in his honest, straight-forward manner speak for the Brannan Farm Program."

But then Wiseth added, "He (Brannan) said in humble sincerity that the Brannan Plan wasn't his plan, but the very best plan that they had been able to devise... And he said that if someone doesn't like the Brannan Plan, then they should present another plan and that he would be glad to consider it."

The arch-opponent of the Brannan Plan was the Farm Bureau. Wiseth told how Harold Cooley of North Carolina, chairman of the House Agriculture Committee, had reportedly told a Farm Bureau group to "Either come up with something better than the Brannan Plan, or stop standing in the way."

However, in the end, the Brannan Plan did not become law. And, with the election of Dwight Eisenhower and a new Agriculture Secretary, Ezra Taft Benson, any hopes of getting such legislation vanished.

WISETH URGES END OF FARM BUREAU-EXTENSION TIE

A trip to Washington, D.C., was made in the spring of 1950 by Roy Wiseth to testify regarding a Senate bill that called for the separation of Farm Bureau and the Extension Service. The legislation would end federal funding in states where the long relationship of the farm organization and Extension had not been ended.

Minnesota Senator Hubert Humphrey was among the supporters of the proposal, which was to happen. "I have great respect and admiration for the Farm Bureau... every farmer should be free to join the organization of his choice... ," Humphrey said. "However, every farmer will agree that no one of the farm organizations should have the free services of county agents paid for by the U.S. Government, to solicit membership or enlist support for the program of that one organization... "

WISETH SPEAKS OUT AGAINST COMMUNISM

Wiseth was one of the first Farmers Union leaders to speak out against the charge that Farmers Union was "Communistic". That ill-founded assertion originated in a speech made by Senator Styles Bridges in Sept., 1950. In short order, Wiseth and his staff worked up a rebuttal pamphlet denouncing Bridges' statement as a "lie", and quoting Midwest Senators like Humphrey of Minnesota, Young of North Dakota and Mundt of South Dakota in sharp rebuttal of their East Coast colleague in the U.S. Senate.

The last section of this pamphlet was written by Roy Wiseth. In part, he wrote: "Our despite for Communism is only equaled by our despite for the methods used by Senator Bridges and his political associates. The answer to this kind of rot is to go out and build a bigger and stronger democratic farm organization. Farmers Union is that kind of an organization.

"To hate Communism is not enough... what can you and I do as farm people to stop its growth? First, we must understand why Communism flourishes in a country where the masses are underprivileged... underprivilege is caused by a system of distribution that concentrates the wealth and power that wealth gives in the hands of a few... your cooperatives will do the job of distribution and return the excess margins to the people who will spend it for other wants and needs... cooperatives need fair legislation to exist and expand... this is our answer to Communism or any type of totalitarianism. – Roy E. Wiseth, president, Minnesota Farmers Union, Willmar, Minnesota."

In the July, 1950, edition of *Glimpse*, President Wiseth wrote glowingly about prospects for Farmers Union, alluding to his feeling that the days of "whispering about 'isms' and other bugaboos in dark corners" are gone. "Gone," he added, "are the letters and questions on theoretical ideologies far outside the sphere and realm of farmers' interests and understanding."

Wiseth said in his travels across Minnesota, "I sense increased respect for Farmers Union. Farmers have been telling me about the regionals they have visited on the Farmers Union bus tours. They are astonished about the size, the large number of employees, the volume of business, and the variety of services they perform.

"Others speak about the local co-op annual meeting where some fieldman from the Farmers Union Livestock Association, Central Exchange, GTA or Farmers Union insurance, has spoken of the great need of farmers banding themselves together by joining Farmers Union, the one farm organization that will never settle for less than equality for farm people whether you call it 100 percent of parity or some other name."

Noting the recent growth in Farmers Union insurance, Wiseth predicted, "The groundwork is laid for an insurance co-op in Minnesota that will in a few years take its place alongside our big three regional cooperatives".

What Roy Wiseth didn't know was that insurance was an entirely different kind of enterprise, and a risky, highly competitive business. And that in the years ahead, financial troubles awaited, both for Minnesota and the insurance-parent, National Farmers Union Insurance Co.

That same issue of *Glimpse* included a report by Art Tisthammer, Farmers Union Insurance director in which he announced the hiring of 31 new licensed agents, which virtually doubled the number at work in Minnesota. A list of 60 agents was printed, including the name of Bob Bergland of Roseau, the member-agent who would 27 years later become the U.S. Secretary of Agriculture.

Tisthammer pointed out that 2 percent of all gross renewal premiums in the state is passed on to Minnesota Farmers Union, and that the check in July, 1950, was "well over a hundred dollars".

Figure 42 MFU Bus Group

SOMETHING NEW – BUS-TRIPS TO WASHINGTON, D.C.

One of the innovations which made its debut during Roy Wiseth's presidency was the Farmers Union bus-trip to the nation's capitol. However, the Washington, D.C., trip was not just a "legislative" trip but the first "membership award" trip of its kind. Later, buses transported Farmers Union members on lobbying trips not only to Washington, but to the state capitol as well. Then, less than two years later, a second bus was purchased.

In a 1996 interview, Ethel Wiseth recalled how in those early years the Farmers Union Locals had "fun, dances and things like fund-raisers... it was not unusual to have 300 to 400 people at one of our fund-raising dances in the Goodridge school gymnasium... we had fun putting on budget fund-raisers like that... it's slowed up now, we don't have as many meetings ... it's different... " Ethel Wiseth said she and her family feel Farmers Union is and has been an important part of their lives. "It always worked to get us better prices, to help us achieve better farming. Our whole family has always been Farmers Union-oriented, starting with the youth camps for our kids... it's always been important in our lives... "

However, Roy Wiseth was not to be re-elected to a second one-year term. At the ninth annual Minnesota Farmers Union convention, he would lose to his vice president, the popular Edwin Christianson of Gully in what the Farmers Union *Herald* edition of Nov. 20, 1950, described without further explanation as "a spirited contest". William Nystrom of Worthington had an easier election – no opposition for state vice president since Wendell Miller declined to seek re-election.

And again, some suspected that the election outcome had somehow once more been influenced by the cooperative giant, M.W. Thatcher.

7 THE CHRISTIANSON YEARS

"Sometimes people get discouraged and feel hopeless in the face of the trends which are taking place... progress comes, but it comes gradually, not all at one time... we can, and do, have an impact – if we work at it... "
Edwin Christianson, President, 26th MFU Convention, 1967.

A BRIEF BIOGRAPHY OF EDWIN CHRISTIANSON:

Born: 1918, Gully, MN
Died: Jan. 1, 1983, St. Paul, MN
Education: Graduate, Crookston High School.
Employment: Lumberjack, farmer, cooperative elevator manager, 1932-71.
Community Service: Incorporator, member, Board of Directors, Minnesota Farmers Union, 1942-71; State vice-president, 1949; state president, 1950-72; vice president, National Farmers Union, 1967-71.
Honors: Member, National Agricultural Advisory Committee, U.S. Department of Agriculture; agricultural advisor to various Minnesota governors.

Figure 43 Edwin and Nora Christianson

THE DAY THE HAYMAKER BECAME MANAGER

It's a story told over and over again by Nora Christianson about how her husband Edwin happened to become a country elevator manager. Edwin Christianson was a young, strapping farmer of 24 years making hay that fateful day in 1932. Then here came a visitor – right out into the hayfield where Christianson was working. The intruder was the president of the board of directors of the local elevator. He came right up to where Ed Christianson was working with his brothers.

Then the visitor went directly to the point of his visit, asking the young farmer a question he didn't have the slightest idea was coming: "Would he take over managing the grain elevator?." It didn't take Ed Christianson long to respond. "No," he told the visitor.

Undaunted, the elevator president came up with a second question: "Would you (Ed) rather pitch hay instead (of managing the Gully elevator)?" Again, Ed Christianson's answer was short and to the point. It was an unqualified, "Yes." In short, he was comfortable where he was, and preferred the hayfield to the elevator manager's office.

However, the same man was persistent. He was to return to the Christianson farm a second time shortly later, and again asked the same question. This time the tall, young farmer produced the answer the visitor had wanted to hear in his first visit. The young man's response was affirmative. Yes, now Edwin Christianson said he would take the job he had turned down before.

Not known then, of course, was the fact that eventually the young haymaker turned elevator manager would become an important figure in two movements which would sweep across his native state – farm cooperative development, and a swift, unparalleled growth in Farmers Union.

Nor that in the early 1950s this same man, born of Norwegian immigrant parents who homesteaded in northwest Minnesota, would become such an important figure that he was considered a candidate for governor of Minnesota by the state's powerful Democratic-Farmer-Labor party. And the man who eventually won the DFL nomination that year was Orville L. Freeman, who served three terms as governor and then eight years as the Kennedy-Johnson administrations' U.S. Secretary of Agriculture.

The Minneapolis *Star* reported that "a campaign to draft Edwin Christianson, president of the Minnesota Farmers Union, as the DFL candidate for governor next year (1952) was under way". The news article described Christianson as a good choice because of his popularity in the farm organization, his record as a successful farm cooperative businessman and because he was known to be "a right-winger" within the Farmers Union and one who could create unity within the political party plus he loomed as a vote-getter in southern Minnesota where the DFL lacked solid support.

However, in 1932, Ed Christianson was a young farmer with no real track record other than the fact that he had worked in lumber camps to help pay off farm debts during the early depression years. His baptism into the agribusiness arena in which over time he would win deserved, widespread acclaim and respect was brief and somewhat awkward. Ed soon discovered he would work alongside the manager he was to replace at the small elevator concern for only one week before he was put completely in charge, and took over the financially-troubled elevator as its new manager.

He also found out something else, something pretty significant – the overall elevator outlook was as shrouded in dark depression era financial attire as the rural countryside and its economy. The bare facts were that this new, youthful manager and refugee from the hayfield now was in charge of an indebted

($12,200) and troubled diminutive cooperative which had only an unimpressive sum of $85 in the bank.

Further, that $85 bank account was only $5 larger than the monthly salary Ed Christianson was to be paid. That is, in fact, if there was that much cash left. Also, he inherited any charges outstanding and accounts payable. So clearly, it was a matter of paying the bills, generating enough business to leave proceeds sufficient hopefully to pay his salary of $80 a month – if he could somehow make ends meet.

To add to the uncertainty of his first non-farm job, there was the relentless pinch of the depression. Nora Christianson recalled, "There wasn't too much business then, as times were difficult. Ed had lots of time at first because there just wasn't much business going on."

This proved beneficial. He wisely used that time when things were slow to read up on things, including the three major farm organizations around then – the Farm Bureau, The Grange and Farmers Union. He studied them closely, to see what the differences were and how they might figure into things.

"As far as he could see, the Farmers Union was the only one really trying to help the farmer," his wife was to write years later. She thinks this conclusion led to a contact with the Farmers Union Terminal Association, and most likely a visit with an agricultural leader who would subsequently play an important role in his new profession, M.W. Thatcher.

Certainly, there was no thought then that within 18 years, the untested, young mid-20's Ed Christianson would take the bankrupt country elevator and turn it into a $300,000 debt-free property, plus expand the cooperative business to include a lumber and machinery business, a bulk oil plant, a service station, co-op store, trucking affiliate and a local credit union. Nor that at the same time Ed Christianson's "cooperative Midas' touch" would see him help organize Farmers Union locals in several northern counties and at the same time become the first president of the Polk Co. unit.

Nor that in less than 20 years the reputation of this remarkable man would grow to the point he was seriously considered a likely candidate for governor.

THATCHER, CHRISTIANSON – SOMETHING IN COMMON

Bill Thatcher of GTA and young Mr. Christianson had something very much in common: both early in their careers wound up managing financially-troubled grain elevators. The bleak circumstances involved for both men at such an early stage of their life's work were similar – terrible indebtedness, and little prospect to move things back into the black side of the ledger.

And both men during their careers exhibited a rare, almost inimitable brand of loyalty to, and support for, Farmers Union, in their businesses and otherwise, which somehow seemed to wane in the years after their presence was no longer possible. But at their prime, there was an unquestionable dedication to the concept of a solid "partnership" of cooperatives and the Farmers Union working together. And, both organizations under the influence of these two leaders, Thatcher at GTA and Christianson at the little community of Gully in northwest Minnesota, showed rare and real progress.

147

However, at the outset with a firm terribly short in finances, Ed Christianson had to make solid business contacts somehow. He had to find a source of badly-needed capital to merchandise grain. Eventually he was to gain support from a nearby bank to supplement needed funding to finance his operation that hinged on buying from farmers and then marketing their grain.

One of his early successes came when he transformed the local, troubled elevator into a cooperative. He organized by selling stock at $10 a share to farmers. And when the new co-op was up and running, and the important first patronage earnings paid to farmer-members shy of financial resources during the depression, things began to brighten up considerably.

Nora Christianson recalled how happy one local farmer was when he received his share of the earnings – only in the amount of $125, but "enough to pay taxes" on his entire farm, he happily pointed out to anyone who would listen, including Ed Christianson who had made such earnings possible through his management and the cooperative he founded.

"Ed bought grains of all kinds, and during a rainy season, he would run the elevators all night so the grain wouldn't start to go out of condition, get hot, burn or spoil," his wife recalled. Many a night her home-cooked supper would turned cold as she waited patiently for her husband's elevator chores to finally end for that day. Sometimes it was well after midnight when he'd finally come to their home.

Nora Christianson lent a helping hand, too. She recalled an instance when a government inspector showed up, and checked seed samples at the elevator when her husband was out in the country on a seed-buying trip. In addition to buying grass seeds, Ed Christianson also bought wool and even cord wood to help the farmers, Nora recalled.

ECONOMIC LESSONS HE NEVER FORGOT

The young co-op manager learned as he worked. There was no bookkeeping system at first, so he set up his own, and kept books himself. Then his wife began helping him at night at first, and then later during daytime hours as well. Of course, she never was paid a wage. Then, when business improved enough, a full-time bookkeeper was hired.

All grain under the control of Ed Christianson was shipped by rail and marketed through GTA. This activity and contact with farmers provided long-remembered economic lessons never forgotten by the Gully elevator manager who had inherited debt when he first took over, and knew the value of a dollar.

Ed Christianson told of a depression era event which reflected graphically how low grain market prices can get at times. A local farmer hauled to his elevator a double wagon box of barley he wanted to sell to buy the twine needed for the farmer to complete his harvest. Twine then was sold in a bag which contained six balls of twine. However, in sizing up the transaction, Ed realized there wasn't enough cash from the barley sale to even pay the full cost of the bag of twine. His solution was simple – just remove one ball of twine. It was truly a memorable lesson in grain market economics that likely neither person ever forgot.

Another unforgettable episode for Ed Christianson that occurred during the depression when an older farmer came into the elevator one winter day and asked if Ed would grind some wheat for him in his feed mill. He recognized immediately that this meant a little extra work, because he understood what was going on. So he cleaned out the mill as best he could. Ed knew that wheat wasn't going to be used to feed livestock. Instead, that home-grown wheat would wind up in the oven and eventually on the kitchen table of the old farmer.

It was such experiences which seemed to stick with Edwin Christianson, and to form the resolve and his deep-seated interest in working harder to help farm people. "These things inspired Ed to try and see if he could do something more for farmers," Nora explained. "He began having meetings in different parts of the county (Polk) and talked about three things – the cooperatives, Farmers Union, and organizing (Farmers Union) locals."

After his successful debut with the co-op grain elevator, the next cooperative was a lumber yard. Again, stock was sold. Supplies came from Midland Cooperative's lumber facility in the Duluth vicinity.

Because of the lumber co-op's good reception, Christianson embarked into the cooperative appliance field – but at the same time made plans to provide parts and repair service in addition to just making retail sales. Then he arranged for the purchase of a local Standard Oil service station, and set up a cooperative structure for that business. Next came the acquisition of a local restaurant. Once more, cooperative stock shares were sold, and a "cooperative store" established.

In addition, a local livestock shipping co-op emerged, with the purchase of a modern semi-truck-trailer rig.

When the local bank closed its doors, Ed Christianson organized a co-op credit union. More than just checking services were provided. Members also could borrow money. And after the success of the new Gully Cooperative Credit Union became known, then came telephone calls long-distance "from the East" from firms and organizations which wanted to lure Ed Christianson from the farm country of northern Minnesota. But those job offers were never to be seriously pursued. And Ed Christianson stayed in what some called "the most 'cooperative' town in the United States – Gully, MN (Some critics of cooperatives, however, were known to refer to the tiny Minnesota town in a less complimentary way. They called Gully "Little Moscow".)

LONG, BUSY DAYS AT THE GULLY CO-OP

Here's a look at industrious Ed Christianson at work in the co-op grain elevator, as recalled by wife many years later:

"During threshing (season) he was at work at seven o'clock, or sometimes earlier, and he bought grain until two (a.m.) in the morning or later. The grain trucks filled the roads coming into our small town, both from the North and from the South, backed up about three-fourths of a mile. They had to take turns to enter the elevator and unload there."

In time his success in buying and marketing grass seed was mirrored by his activity acquiring and selling grain.

His leadership did not go unnoticed locally. Ed was elected first president of the Polk Co. Farmers Union organization of locals in that county. "Membership in Polk Co. and also in the state began growing," Nora remembered about that era. At the state level, Ed was a member of the original Minnesota Farmers Union state board of directors after the organization was re-chartered in 1942. He was in his mid-30s then, and destined to become state vice president before his fortieth birthday.

Looking back, Nora Christianson recalls that "Ed's second success in life (his first was establishing a network of successful co-ops in his hometown) was winning the state presidency of the Minnesota Farmers Union (in 1950)." She said members were "few in numbers" when he took office, and then climbed to more than 40,000 in just seven years.

Sudden growth came quickly after Christianson was elected. Organization Department Director Russel Schwandt reported in the May, 1950, edition of Glimpse that membership had exceeded 1949 by a thousand and that 30 new Farmers Union locals had been organized in 17 counties since Jan. 1.

Five counties – Jackson, Redwood, Yellow Medicine, Polk and Kandiyohi – then were boasting "well over 500 families" each enrolled in Farmers Union. And Schwandt predicted that many Minnesota counties would be able to report over 1,000 family-members by Jan. 1, 1951.

This was the start of membership growth that would be the hallmark of the Christianson years. His wife Nora credited several factors for the record growth in Farmers Union: Her husband's dedication and interest which took him to the podium of countless meetings night after night and day after day; a cadre of "good fieldmen", or staff, membership organizers; and the support of leaders like GTA's M.W. Thatcher and his organization as well.

"We held lots of local meetings," Nora remembered. "A couple or more in a week, with always large crowds. I remember once we had a meeting up north close to the Canadian border and we didn't get home until 6 a.m. Then Ed had a 10 o'clock meeting downtown (in Gully) that morning. We took turns driving on the way home, with each of us driving three times. And we stopped, and drank plenty of coffee, of course. But we made it!"

Nora recalled that the state Farmers Union didn't at first have its own newspaper for members. "But GTA was kind enough to let us (MFU) have some space in their paper. And when Ed was looking for a writer and editor, GTA helped put him in contact with Milt Hakel, and he hired him." Hakel's reputation grew as he took advantage of the phenomenal growth of Farmers Union in Minnesota. Eventually, Milt Hakel would advance to Washington, D.C., as the National Farmers Union editor 20-some years later.

Over time Ed Christianson earned a national reputation. Former National Farmers Union James Patton used to say "that if the organization needed any new ideas in building Farmers Union, all they had to do was come to Minnesota", Nora Christianson said. "And Ed always had a lot of ideas. Whatever he went into, he succeeded. He was a very bright person and was always looking ahead to see what should be done to make things better for people. Many of the things he talked about for a long time eventually did happen. And he was always a willing person ready to help anyone. Even farmers came to him and asked his advice about things, like should they buy a certain farm, raise livestock, or what kind of crops to grow."

She added, "Another reason he became such a leader was the fact he thought every person should have the right to learn what was going on in this world. He wanted Farmers Union to have its own newspaper so farmers could find out how the people they elected to office voted on programs that were helping the farmer."

"ADVENTURES IN GOOD CITIZENSHIP"

Nora Christianson pointed out how her husband had helped develop the system where Farmers Union members participated in lobbying trips to the state capitol in St. Paul and to the nation's capitol in Washington, D.C.

"He saw to it that the Farmers Union used buses to transport farmers into the Twin Cities to see for themselves how their cooperatives worked for them. And some of these people and farmers had never been to the cities before.

"Then, later, these same Farmers Union buses took many more of our (membership award-winners) members to Washington, D.C., to visit with members of Congress, agricultural leaders and officials there. Next, Ed helped organize the very first 'Ladies Fly-In' (to Washington, D.C.). This contrasted to early days when he made his first trips to Washington driving his own car. Like the time he and Russel Schwandt traveled together to testify on various programs.

"I also remember one trip when Farmers Union was promoting the school milk program. We took with us a long petition that favored this program which had been signed by hundreds of farmers and their wives. It all was pretty impressive."

The importance of traveling to Washington, D.C., dates far back in time for Farmers Union, Edwin Christianson related in conjunction with the first "Ladies Fly-In" he and his wife Nora organized in the spring of 1966. He told of the long relationship of Farmers Union lobbying activity in Washington, D.C., in a brief message printed in a commemorative brochure prepared after that first lobbying "Ladies' Fly-In" trip to Washington:

"It is now 60 years since Charles S. Barrett, the first president of the National Farmers Union, came to Washington to speak for farmers. It is 17 years since National Farmers Union began bringing farmers by the busload to the nation's capitol to speak for themselves.

"It is one year since Farmers Union originated the technique of bringing farmers to Washington by the plane-load, that is, by 'Fly-Ins'. Now, you have had the distinction of being Farmers Union pioneers again — by being members of the first 'Ladies Fly-In' in history."

The commemorative booklet was mostly pictures taken by long-time Farmers Union member and staff worker-editorial cartoonist Archie Baumann and by Alice Masanz. Their photos chronicled the visit to The White House, with fellow-Minnesotan Orville L. Freeman (then the U.S. Secretary of Agriculture), the breakfast with the new U.S. Senator from Minnesota (and future U.S. Vice President and Presidential candidate) Walter Mondale and his wife Joan, coffees and briefings with Minnesota's Congressional Delegation, sight-seeing activity in Washington, D.C., and, of course, photos on the plane which represented the first airplane flight for a high percentage of the participants.

In this booklet, Ed Christianson concluded his remarks to his plane-load of members with this thought: "We hope this book will be a treasured memento of this once-in-a-lifetime, kind of adventure in good citizenship."

ONLY ONE REGRET: "WE DIDN'T DO ENOUGH"

Nora Christianson was asked by the author: Were there any regrets held by her husband about anything he hadn't achieved? His widow responded noting that one thing was his disappointment that he and others could not convince Congressional leaders how essential it was to get farmers consistently a fair price for their products.

Truly, he felt more should have been done to prevent the economic blows he and others had experienced during the depression years of the 1930s. And, one could imagine his anguish at what took place in American agriculture during the "farm crisis" of the 1980s.

Another concern harbored by Christianson pertained to a nagging issue that had persisted during his long career – corporate farming. He had seen vertical integration move into farming, into poultry first, and then grain-fed cattle. Later, in the 1990s long after his death, that same disturbing trend would emerge in pork production as well.

Also at numerous meetings Christianson had spoken out critically against the acquisition of farms and agricultural land by large corporations, and stressed instead his long-standing preference for the family-type farms.

However, despite his strong personal feelings for "family agriculture", Nora Christianson recalled that her late husband also realistically recognized there would continue to be change, and that "sometimes the family farm would have to be a little larger in order to take care of the family if there were sons who wanted to be involved with their dad in the family farm operation".

Today, Nora Christianson added, she feels it is highly likely that Ed Christianson would be disappointed because of some of the things that have happened and are happening in Minnesota and elsewhere today – the scenario where big corporations have invaded rural areas, and bought up the land, and then compete head-on with their non-farm fortunes against the family farm operations.

When asked why Ed Christianson did not seek the state Farmers Union presidency earlier, Nora commented, "He was happy with his work with all the cooperatives he had been building, and really wasn't sure if wanted to be state president." A couple of members from the southern part of Minnesota came to Gully just to ask him if he would accept the nomination as state president if they came out and nominated him, she recalled. "Ed thought about it for while, but nothing happened. Still, they didn't seem to want to give up the thought of him being state president. So, finally, he did agree to be a candidate. And he won the election, two to one."

(Farmers Union historian F. B. Daniel also feels that "Ed Christianson's interest in, and dedication to, the cooperative development he had underway in Gully was the reason why he didn't think seriously about or seek the state presidency" earlier. Then, at repeated requests from others, Mr. Christianson did ultimately agree to be a candidate, Daniel believes.)

Frequently Ed Christianson was asked to serve on state and national boards and committees. He was a member of Agriculture Secretary Freeman's dairy advisory committee, a member of the Minnesota Governor's Council on Children and Youth and the Governor's Council on Aging, a member of state highway and tax study commissions. In National Farmers Union, in addition to being the national vice president, he also served as a member, vice chairman and chairman of the national organization's executive committee.

During his long leadership career, Ed Christianson worked closely with the late Hubert Humphrey and Orville L. Freeman of Minnesota, as well as with Farmers Union's national president, James Patton. Within the state, he worked hard in the various election campaigns waged by Humphrey and Freeman. Both were known to contact the Gully cooperative leader whenever they sought counsel or advice concerning Farmers Union's position on legislation or farm programs.

SPECIAL RECEPTION FOR NEW, NATIONAL VICE PRESIDENT

A surprise welcoming party awaited Edwin Christianson when he returned home from the National Farmers Union convention in Mar., 1966, as the newly-elected national vice president.

About 100 co-workers, friends and Minnesota Gov. Karl Rolvaag greeted him as the 58-year-old Christianson returned to his office in St. Paul. A banner proclaimed, "Welcome Home Mr. V.P." and a red carpet had been laid across the foyer leading to his private office at the state headquarters of Minnesota Farmers Union on University Avenue.

Onlookers, according to the Farmers Union *Herald* dated Mar. 28, 1966, thought Christianson was, "Obviously stunned by the warmth of the reception, Christianson was escorted on the red carpet ... as the crowd applauded his entrance and as a trumpet blared the Minnesota Songs... "

When he had the chance to speak, Christianson announced that he would do two things:

He would continue to serve as president of Minnesota Farmers Union.

He would fulfill his new duties as national vice president out of his Minnesota office in St. Paul. And he promised "to vigorously carry on the ideals and principles that have made the National Farmers Union the progressive and dynamic organization it is... and I will work hand-in-hand with (new NFU) president Tony Dechant on our Farmers Union program to advance the cause of family farmers... "

Christianson succeeded Glenn J. Talbott, 65, the president of North Dakota Farmers Union and the son of Farmers Union pioneer organizer C.C. Talbott who was the first president of the North Dakota organization when it was chartered in 1927. Talbott had been national vice president for five years.

CHRISTIANSON HELPS INITIATE "GREEN THUMB"

In Gully, he was extremely active in community affairs – on the city council (including a time when it financed and built a Community Building after an earlier facility burned to the ground), a member of the Lund Lutheran Church Council, the Gully school board and the Polk Co. Educational Reorganization Action Committee.

While president of the Minnesota Farmers Union he helped launch several programs, including "National Farmers Union Green Thumb". This program provides for employment and jobs for older, low-income citizens where they work in local community projects. This program meant a great deal to these folks in terms of employment and having jobs they could handle, as well as income earned especially in rural areas where jobs are scarce for senior citizens.

Nora Christianson recalled how at a meeting in Wadena several National Farmers Union Green Thumb participants told how they were able to buy a pair of new glasses, new shoes, new clothes and such, thanks to their new jobs. "They were so happy to be able to do such things," she added.

Other new projects that had their debut during the Christianson Era of Farmers Union included "Green Light" and KIDS, Inc.

A LOOK AT NORA CHRISTIANSON

Nora Christianson is a former elementary country school teacher in Clearwater and Polk counties where she had about 40 students and taught all eight grades. She enjoyed a most successful teaching career, including the setting of a School District 195 record – the most eighth grade graduates (nine) from one class of a country school.

She confided that her eighth grade class record would have totaled 11, but two pupils moved out of the community at mid-term!.

During World War II she was a member of the Red Cross Committee charged with making sure Red Cross qualified nurses were available locally. Toward this end she helped train young women in basic nursing health skills. Also, she worked on the wartime emergency rationing board helping farmers figure out the gallons of gasoline needed for field work and other chores as well as helping local people figure out rationing food items like meat, butter, sugar and such. One of her jobs was to monitor retail grocery store prices and to file periodic reports.

Locally, she taught Sunday School and was superintendent for many years, was president of the women's group and served on various committees. After moving to the Twin Cities, she shared driving duties that took her husband and farm leader Ed Christianson to countless meetings. She recalled the country "mid-winter conferences" which featured a two-a-day schedule – one session at 1 p.m., and the second at 8 p.m. Lura Reimnitz and Nora handled the meeting registration chores and passed out literature to those in attendance. In summertime there was generally a Farmers Union picnic every Sunday. She recalls that "these trips produced many, long friendships".

HIS OWN WORDS: "THE PRESIDENT'S CORNER"

Edwin Christianson's "The President's Corner" column appeared on the front page of the weekly Farmers Union publication, *Minnesota Agriculture*. It featured his photograph, and a terse headline topic which indicated what he had to say that week to his members. There is little doubt that this feature was well-read, too, because it provided a real flavor of top-level thinking in Farmers Union. And it reflected the importance he attached to communicating to his members the news of the organization, the issues about which he and others were concerned, and other information of interest to farm people.

Here are some samples of Ed Christianson's reports to members:

Corporate farming – "Shameful Maneuvering", May 13, 1971:

"But rural people are not stupid enough to be fooled by voting records on legislative bills which never pass. The only thing that will count with many voters will be a recorded vote on a satisfactory bill which did pass... " (Christianson was disappointed because of failure to get meaningful action to enact a tougher anti-corporate farm bill by the Minnesota Legislature.

Corporate farming – "Not Too Late", April 29, 1971:

"With little more than three weeks left in the 1971 (State Legislative) session, you can expect some legislators who don't want to act on certain legislation, to say that it is too late to get something done... they may well try to use this as an excuse... the truth is that it is not too late to act, if the majority faction has any desire to do something... "

(Christianson wanted action on a Senate bill which would strengthen Minnesota's laws covering corporations in agriculture and specifically eliminate an exemption for corporate "farms" smaller than 160 acres.)

Farm program – "Needed: Higher Price Floors", July 22, 1971:

"The most recent federal crop reports certainly should add urgency to the efforts of Farmers Union to obtain stronger farm price protection floors... (the USDA) just increasing the (cropland diversion) set-aside is not enough in our view... support floors and (program) payments need to be increased... " (Christianson was concerned about bumper crops and lower market prices causing reduced incomes for farmers.)

Trade "Wars" – "The Common (Market) Problem, Aug. 12, 1971:

"Farmers here and in the (European) Common Market are realistic enough to know that no one in farming wins in a world price war... (that) no one in farming really wins through subsidized dumping of farm products... much of the problem in Europe as in the United States is that farmers and their organizations have too little to say in farm policy ... " (Christianson was bothered by "protectionist" trade policies both in the United States and in the European Economic Community.)

Bumper crops – "The 1972 Feed Grains Program", Sept. 23, 1971:

"The unfortunate thing was that when USDA encouraged expanded (crop) production for 1971, officials were not willing to give the producer assurance that his price would not be depressed because of added volume... " (Christianson was worried about the impact of an impending bumper crop encouraged by the USDA driving down prices.)

More: "In His Own Words..."

One can sample the flavor and fervor of Edwin Christianson in copies of his speeches which fortunately have been preserved in a file cabinet in the state office of Farmers Union.

Here are a few quotations selected randomly from scores of speeches delivered by Christianson during his long reign as state president:

1952, 11th State Convention – State of the organization:

"In making my second annual report... We have greater income, sizable savings, and a building fund reserve of $7,000... Farmers Union has done a great job for American agriculture, but our accomplishments will not become good enough until no farmer will take below 100 % of parity prices for farm products...

"Not good enough until all farmers have rural telephones, have the electrical power they need, and all farmers have adequate credit resources on reasonable terms... "

Christianson ended his 1952 address with a hope for "world peace", and proposing to "Extend our hand of fellowship to all ... to join us in this crusade" to gain "full parity for farmers... "

1955, 14th State Convention – Agricultural Policy:

"In my opinion, the farm program of (U.S.) Agriculture Secretary Ezra Taft Benson is one of bankruptcy of faith in America and faith in progress... the flexible price support system is conceived in hopelessness, characterized by disparity and doomed to disaster... to put it in one sentence, the Benson theory is that you can divide more and more poverty among fewer and fewer farmers and somehow come up with a higher average income for each farmer... our farm prices today (figured at a percentage of parity) are lower than they have been at any time since 1934... "

1959, 18th State Convention – "The Farmer's Share":

"With farm purchasing power at a 20-year low, with the farmer's share of the food dollar lower than at any time since the depression of 1932-33, and with surplus and price disaster threatening our grain-livestock economy... it is obvious that agriculture needs a better and more effective farm program... "

1959, Minnesota Association of Cooperatives meeting – About Co-ops:

"There are many outstanding farm cooperatives in the nation, and we take pride in the fact that cooperatives bearing 'the Farmers Union' name are among these... we are fortunate here in Minnesota to be home base for two of the truly top-ranking regional cooperatives – the Farmers Union Grain Terminal Association and the Farmers Union Central Exchange... they have contributed much to the stability of local cooperatives and their ability to be of greatest service to their patrons... we in Farmers Union feel that farm cooperatives are just on the threshold of their greatest period of usefulness to farmers... "

1962, 21st State Convention – "Local initiative":

"If there is any lesson that should have been learned over the 100-year history of USDA and 60 years of Farmers Union, it is that there is not likely to be any real progress without grassroots action – without local initiative. The federal government or the state government can propose; they can take the lead. But they cannot be effective if there is apathy or complacency at grassroots... time after time we have seen that it takes a partnership of federal initiative and local initiative before a program will move ahead rapidly... we have seen this in rural electrification, in soil conservation, in rural housing and commodity programs, and elsewhere... we have seen that grassroots participation can make the difference... "

1963, "Salute to Wheat Day", Climax – The "Price" of Wheat:

"There is something especially unusual and significant about the 17 years which have passed since the 'Mercy Wheat Rally' in 1946 and our 'Salute to Wheat Rally' here today. These 17 years from 1946 to 1963 include the only period in the last 100 years that we have had $2 (a bushel) wheat... the only other years, ten of them in this country, that we have had $2 wheat before were in times influenced by war demand... this is something to think about... "

1967, FU Central Exchange annual meeting – On Cooperatives:

"Farmers Union was founded very wisely on a triangular base of 'education', 'cooperation'. and 'legislation', and certainly it must be recognized that these tools have proved useful and successful when we used them with a purpose and a determination... I would not attempt to say which of the three sides of the Farmers Union triangle has been the most important, but it is quite clear to me that except for the contributions of the farm cooperatives, not much would have been possible... "

July, 1967, Farmers Union Central Exchange local meeting – Cooperatives:

"Cooperatives must take a role in improving farm income through legislated farm programs if the family farm is to survive – certainly there is no place in agriculture for farm cooperatives if there is no longer a family farm type of agriculture... "

1967, 26th State Convention – "The Real Issue – Control of Agriculture":

"If agriculture is deprived of programs in which farmers can maintain their integrity and economic well-being, then agriculture will be controlled by someone else and farmers will have little independence left... you have a Farmers Union organization which is working where it counts for the betterment of farm families, through legislation, through cooperatives, and through every other means which promises any help... there are those trying to kill the farm programs, to short-change funding for agriculture, education, conservation, and the 'war on poverty'... ."

CHRISTIANSON CRITICIZES "THE COMMUNIST SMEAR"

"To be successful in taking over this country, the Communists would have to destroy the Family Farm System, and that is the very opposite of what Farmers Union is doing... "
Edwin Christianson, MFU President, Mar., 1957.

The Farmers Union program places itself squarely on record against Communism and other foreign "ism's", Edwin Christianson declared in a Mar. 4, 1957, speech focusing on what he and others at the time referred to as "the communistic smear" campaign. He chose his words carefully to relate his impression as to what happened, and added, "Mark this well, it was... a conspiracy... "

The "smear campaign" began, Christianson said, after Farmers Union began to openly challenge the Farm Bureau's leadership claim of representing farmers in various states, including Minnesota. In 1953, for example, there were 21,288 memberships in Minnesota Farmers Union compared to 49,509 Minnesota Farm Bureau memberships. The significant thing not shown in these figures is the Farmers Union's rapid growth, more than doubling its membership in three years. Nationally, Farmers Union membership in that same three-year period (1950-53) increased by 26 percent while Farm Bureau membership increased by less than 10 percent.

At the same time, he reported, unfounded charges attempting to link Farmers Union with Communism were made, including some published statements which came directly from Farm Bureau sources. The "red smear" tactics grew out of the intense national attention and publicity being given in the 1950s to "un-American" activity and membership in the Communist Party of America.

In the fall of 1950 the Utah Farm Bureau published a statement referring to "the Communist-dominated Farmers Union". A lawsuit seeking damages for libel in U.S. District Court (Utah Central Division) was brought by the National Farmers Union, with a jury decision holding that the published statement was untrue and therefore, libelous. Not only did the jury clear the Farmers Union of the Communist charge, but it awarded Farmers Union $25,000 as damages. A Farm Bureau's appeal for a new trial was denied, with the U.S. Circuit Court agreeing with the decision from the District Court.

In Sept., 1952, the Utah Farm Bureau paid $25,000 damages plus $2,804 interest to the National Farmers Union as settlement of the libel case.

Part of the Farm Bureau's libelous charge was based on a report which Senator Styles Bridges made (under Congressional immunity) and placed in the Congressional Record on Sept. 7, 1950. In that report, the Senator said in part that he had found "no evidence of any attempt... of Communist infiltration into either the Farm Bureau or The (National) Grange... " Then he added, "The Farmers Union so consistently espouses Communist causes, parrots Communist propaganda, and refuses to denounce Communists or their activities as to preclude the possibility of accident or coincidence".

Bridges went on to mention these Farmers Union leaders by name – NFU President James Patton, M.W. Thatcher of the Farmers Union Grain Terminal Association and his assistant, Robert Handschin.

Later, these same statements appeared in a newspaper paid advertisement headlined, "Communist Invasion of Agriculture". At the bottom of the two-column ad was an invitation to the reader to order copies of Senator Bridges' report for 1.5 cents each from the Minnesota Farm Bureau, 420 Commerce Bldg., St. Paul, 1, MN

Also circulated was a poster advertisement by "Wide Awake Anti-Communism Crusade", Box 448, Des Moines, Iowa, which offered for sale ($3 a copy) a 300-page loose-leaf paper bound publication titled, "Concerning Communism and The Farmers Union".

Midwest U.S. Senators responded in the Senate to the Bridges' statement. Sen. Hubert H. Humphrey protested loudly "against any aspersions being cast upon Farmers Union in my state and I would resist and deny any statement or any insinuation that the farmers who belong to Farmers Union by the thousands... are participants in the Communist movement... to call them Communists is to me simply something that is incredible... "

Senator Young of North Dakota pointed out in his statement that FBI records indicate only 70 known Communists in that state. That fact, he added, would indicate that Farmers Union probably has fewer Communists in its membership than probably most other big organizations in the United States. Senator Mundt of South Dakota said he checked House Un-American Activities Committee records and found no Communist connection with Farmers Union leaders in that state where the FBI identified 30 Communists.

Only one Farmers Union member (a man from Ohio) was ever found to be a card-carrying Communist, Christianson reported in his speech in 1957. The Communist conspiracy, he added, seemed more likely found elsewhere, not in non-farm and non-rural areas, but centered among upper "middle-class" persons – doctors, lawyers, teachers, actors, writers, white-collar workers, and others even including the clergy. In general, people who didn't work with their hands because few of the Communist leaders had actually ever worked, Christianson charged.

In short, farming and Communism were an ill-fated match, even though there were well-documented accounts of "Reds" trying to gain support in farm circles during the traumatic times of the Great Depression of the 1930s, preying upon the hard times and the vulnerability of people then to subscribe to new doctrines for relief.

Christianson: "Under Communism There Would Be No Farmers Union"

"Under Communism, you could not own your own farm," Christianson said. "You could not own your own home. You could not belong to a labor union which would seek better pay, better working conditions and shorter hours. Your children would be taught only what the Communists wanted them to learn. You could not belong to organizations of your choice...

"In fact, if Communism took over our country, one of the first things it would do would be to try to wipe out the Farmers Union... Communists have never been able to take over a country which had an agriculture built upon family-owned, family-operated farms... "

Christianson told how a common tactic is for the "smear organization to import hit-and-run smear artists, speakers, or columnists and others" to plant the seeds of distrust and suspicion.

"The only remedy is for farmers to state the truth for the record, and go on with their work," he added.

Christianson then related how one speaker responded when asked if Farmers Union was a Communist organization: "He brings out a pamphlet and says, 'Let me give you the answer from this House Un-American Activities Committee pamphlet' and then gives the impression that Farmers Union is a Communist-controlled organization without saying he was quoting a statement made by an ex-Communist (Benjamin Gitlow). And he doesn't point out that this was only that one person's opinion stated to the Committee."

Christianson concluded his speech relating, "Communism made its greatest gains following the days of 30-cent wheat, 15-cent corn and nickel hogs. That was the time our great American system of government was in danger – not during the period when we had farm prices supported at 90 percent of parity...

"As for us in Farmers Union, we believe a healthy agriculture will mean a stronger nation – stronger in an economic way and stronger against any threats from within or beyond our national boundaries... "

CONCERN ABOUT GI FARM TRAINEE PROGRAM

Testimony before a Congressional veterans subcommittee in June, 1969, Edwin Christianson reported only 23 former GIs from Minnesota were enrolled in a lone GI farm trainee school in the state where the potential was for 5,000 farm trainee students.

"It is discriminatory and unfair that young men who interrupt their educational or farming careers to serve in the military forces cannot use their GI educational eligibility for farm training," Christianson told the Senate subcommittee in Washington, D.C.

The only GI farm trainee school was located in Alexandria, with 23 ex-GIs enrolled. Eleven other Minnesota schools that had applied for classes but were denied because of the rule that students must attend a minimum of 12 class hours per week. This requirement is virtually impossible for full-time young farmers, Christianson pointed out. The schools interested were located at Middle River, Willmar, Montevideo, Madison, Worthington, Jordan, Blue Earth, Waterville, St. James, Hayfield and Detroit Lakes.

There existed a potential for another 30 schools that could be farm trainee sites, with an estimated 5,280 young ex-GIs eligible if a change in class hour requirements could be made, Christianson added. He also urged support for a bill that would reduce the yearly class hour total required from 200 hours to 100 hours of farm training schooling.

Financial assistance for the farm trainees ranged from $105 per month to a former GI with no dependents to $145 for a man with two dependents. Monthly totals increased $7 for each dependent in excess of two dependents.

Christianson told the committee that where only a handful of eligible veterans had enrolled by spring, 1969, there were 30,000 ex-GIs who had taken advantage of GI farm training in Minnesota following World War II and the Korean Conflict. Of those an estimated 80 percent who competed training remained in agriculture, he said.

THEN, NO MORE WORDS, MESSAGES... OR, LEADERSHIP

Such moving words of leadership, of hope, of encouragement and concern for farm people then ended that Wednesday, Nov. 17, 1971, when the dynamic, compassionate Farmers Union spokesman and leader was attacked by what turned out to be a disabling stroke suffered just prior to the state convention. The result was paralysis and an inability to speak.

It was a cruel fate that brought to an immediate end his lengthy and brilliant career speaking out for and working for – and with – the farm people he knew and loved.

The tragic happening occurred at a place Edwin Christianson frequented countless times in his career – the busy international airport in the Twin Cities. He was routinely returning from a two-day meeting of the board of directors of the National Farmers Union in Denver, CO. Christianson wore his usual two "hats" at that meeting – one as the Minnesota state president for nearly 21 years, and the other as national vice president, then completing his sixth year in that office and unquestionably a future candidate for the National Farmers Union presidency.

From the Midway Hospital in St. Paul came the firm recommendation from the Christianson Family that the upcoming Minnesota Farmers Union's 30th convention proceed as planned. The state executive board met via telephone conference call, and then later agreed. The meeting would be held as scheduled with the man then seeking the state vice-presidency, Cy Carpenter, serving as acting president.

A six-month hospitalization was to follow the onset of Ed Christianson's disabling stroke. Then, for the next 11-plus years, his wife, Nora Christianson, was destined to be his chief caretaker almost continually – except for times when he had to be rushed to the hospital because of a reoccurring heart problem.

Then, on New Year's Day, 1983, the long years of suffering endured by the disabled Great Gully Cooperator, Edwin H. Christianson, and the anxious years of compassionate concern and care by his beloved wife, family, and friends, came to an end. Behind he left his wife Nora, a daughter, Lenore Fitzsimmons of Willmar, four sisters, a brother and five grandchildren.

He also left a legacy of distinguished cooperation, and a generation of cooperative-minded friends, associates, and the five Gully cooperatives he launched during the depression years of the 1930s.

Two services were held for Edwin Christianson. The first at St. Timothy Lutheran Church in St. Paul where he and his family had lived since 1951, and the second at Lund Lutheran Church in his hometown of Gully the following day. Burial was in the Gully Lutheran Cemetery, not far from the five farms he owned, nor the offices of the quintet of co-ops he fathered.

THE LEGACY OF THE GREAT GULLY COOPERATOR

What epithet for such a man? How best to sum up such a full, productive life of nearly 75 years for Ed Christianson? How do you convey his greatness or his many contributions in mere words or a paragraph, or even a chapter in some book?

One answer to such a question can be found on a plaque in his hometown of Gully, where his path to leadership and success began. There, one finds a memorial to this man. This bronzed plague features just five words that attempt to sum up the life and career of Edwin Christianson – "Farmer, Cooperative Builder and Leader". They seem to appropriately represent the areas in which he gained widespread recognition, respect, and praise.

This memorial brings into focus for those who stop by the farm community in northwest Minnesota how Ed Christianson for 40-some years had a fruitful career of devoted service to family farmers, to rural community life, and to mankind in general. The plaque reads in part, "Beginning in 1932, when he was only 24 years old, he began to organize cooperatives in Gully, then, throughout Minnesota and beyond... " It notes Ed Christianson's life-long bent that these local, cooperative businesses "should work for the good of the people".

And the bronze, engraved plaque proclaims that under Ed Christianson's leadership, "the cooperatives of Gully grew like mushrooms after a rain". It continues by noting his many years of Farmers Union leadership with these words: "An incorporator of the Minnesota Farmers Union in 1942; president, 1950-1972; national vice president, 1967-1971."

At the bottom, the plaque concludes: "Inscribed June 1, 1984, by his friends and fellow townsfolk in remembrance of a man who shared his talents and energy with his community, his state, his nation, and the world." It was signed by, "The Gully Cooperators".

From his widow, however, the greatest legacy of Edwin Christianson is portrayed in somewhat different terminology. Nora Christianson believes the most significant aspects of her husband's remarkable career was his steadfast, life-long emphasis on two things which he exhibited from the very start – from way back during the depression when he took over the manager's reins of a financially-troubled, ailing, small, country grain elevator:

The importance of farm cooperatives to help make things better for farm people and their communities. The importance of having Farmers Union as a base from which to build cooperatives.

To Ed Christianson these were the important, essential elements for progress. He had demonstrated this time and time again during a life when he gave unselfishly so much of himself, worked long and tirelessly, and shared freely his beliefs and ideals with so many. There seemed never any question where the Gully cooperator put his highest priorities – in fighting for farm people, and their cooperatives. Working together, he felt, there was always a better chance for better future times.

As he told his delegates at their 26th state Farmers Union convention at the Prom Center in St. Paul on Dec. 4, 1967: "Sometimes people get discouraged and feel hopeless in the face of the trends which are taking place... To be sure, there is no such thing as immediate parity, no easy answers, no overnight solutions to our problems... Progress comes, but it comes gradually, and not all at one time... We can, and do, have an impact – if we work at it... "

And, Edwin Christianson did just that – he worked hard, sometimes night and day, and spent his life seeking, and at times finding, solutions. His relentless pursuit of answers and solutions, indeed, might possibly be his greatest gift and legacy left to the generations.

One of his Gully cooperative associates was Harry Hanson, a member of the board of directors and chairman of the oil company. In an interview, Hanson recalled vividly how well organized Christianson was at board meetings. "He had everything figured out ahead of the meeting – why we needed to do something, how it would work out, and all. There wasn't much for us to do but to go ahead and vote to do it," Hanson recalled.

Hanson's sister, Sanna Brovold, managed the Gully cooperative store also organized by Christianson, "the Great Gully Cooperator". However, this enterprise eventually was sold to a private concern.

There was a down-side, though. After Christianson moved up the Farmers Union leadership ranks and left Gully, Hanson remembered, "We had a dickens of a job replacing him. There were no managers who could measure up to him."

And in 1991, three of the Christianson-fostered cooperatives merged. The elevator, oil company and lumber yard consolidated operations and became "the Gully Tri-Coop". And the Gully co-op credit union founded in the Great Depression became an affiliate of Headwaters Federal Credit Union of Bemidji in 1991.

SUMMARY OF EDWIN CHRISTIANSON'S MFU YEARS

1942 – Fergus Falls (Oct. 15-17), member of state board of directors.

1943 – Alexandria (Nov. 4-6), state board member.

1944 – Detroit Lakes (Nov. 9-11), state board member.

1945 – Willmar (Nov. 1-3), state board member.

1946 – Willmar (Nov. 21-23), state vice president.

1947 – Willmar (Nov. 28-30), state vice president.

1948 – Willmar (Nov. 11-13), state vice president.

1949 – St. Cloud (Nov. 8-10), no state office.

1950 – Willmar (Nov. 9-11), state president.

1951 – St. Paul (Nov. 11-14), state president.*

*The state convention has been held in St. Paul or Minneapolis ever since 1951. Edwin Christianson served as president of Minnesota Farmers Union from

Nov., 1950, until disabled by a stroke in Nov., 1971, just days before the annual state convention. State membership officially stood at 12,115 when he became president in 1950. It surpassed 40,000 in the late 1950s, peaking at just under 42,000 member-families.

In 1967 he was elected vice president of National Farmers Union.

THE GALLERY

Figure 44 The Latest: 1920's

Figure 45 25,000 Farmers Jam St. Paul, Minn. Auditorium April 27th, 1940

Figure 46 MFU Milk Petition. Front: Edwin Christianson, MFU President, Minnesota U. S. Senators. Eugene McCarthy & Walter Mondale

Figure 47 NFU Board Meets with President Harry Truman

Figure 48 1946 MFU Board & Office Members & Co-op Managers. Back: Cassavant, Lyngen, Nystrom, Kuivinen, Dwight Wilson; Center: Lura Reimnitz & Oskar, Junice Dalen, Mitchell, Tom Croll; Front: M. W. Thatcher, Christianson, Franklin Clough, Marv Evanson, Emil Syftestad

Figure 49 1947 MFU State Executive Committee. Back: Wm Nystrom, Wendell Miller, Roy Lindall; Front: Wm Cassavant, Einar Kuvinen & Freda Eisert

Figure 50 L to R: Wis. Sen. Gaylord Nelson, NFU Pres. Jim Patton,
Orville Freeman, Mn Gov. & U.S. Sec. of Agriculture,
ND FU Pres. Glenn Talbott, MFU Pres. Ed Christianson

Figure 51 Left to Right: Frank Livingston, Jim Jesson -Pres. Waseca Co. FU,
Louis Huper -Pres., Faribault Co. FU, Nels Wongen, Freeborn Co. FU Pres.

Figure 52 L - R: MFU Pres. Ed Christianson, Orv Freeman, Wm Nystrom

Figure 53 Left: Paul Anderson, Dir., MFU Green Thumb,
Right: Wendy Anderson, MN U.S. Sen., Gov. & State Sen.,

Figure 54 From Left: Glen Coutts, FUCE Bd. Member;
Dennis Forsell, Dir., Green View Field Operations

Figure 55 Tony Dechant, NFU Pres., L., Ed Christianson, MFU Pres., R.

Figure 56 L to R: Ray Grasdalen, MFU Field Staff;
Fred Gronninger, early organizer & GTA Fieldman

Figure 57 Axel & The Co-op Shoppers

Figure 58 Restored Co-op Tractor

Figure 59 L to R: Orville Freeman, MN Gov., U.S. Sec. of Ag;
Wilton Gustafson, long-time MFU staff & bus driver

Figure 60 L to R: Norma Hanson, MFU Vice Pres., Exec. Comm., Co. Pres.;
Harry Hanson, long-time Gully Co-op Bd.,

172

Figure 61 L to R: Hubert Humphrey, MN U.S. Sen. & Vice Pres.
with MFU Pres. Ed Christianson

Figure 62 L to R: Milton Holtan; Leo Klinnert, MFU Bd., FU Ins. Agent

Figure 63 MFU Fly-In Group in Washington, D.C.

Figure 64 Ladies Fly-In Group Departing for Washington, D.C.

Figure 65 MFU Legislative Action in Washington, D.C.

Figure 66 Pres. Lyndon Johnson, a surprise guest at Minneapolis NFU Convention

Figure 67 Frank & Ethel Krukemeier, long-time MFU staff

Figure 68 L to R: Joe Larson, Climax, MN., Chm., FUCE Bd. Ch, NFU Ins. Bd.;
Glenn Long, Manager, MFU Marketing Assoc.

Figure 69 Early group of Green Thumb workers reporting in.

Figure 70 MFU "old-timers" from 1945-1950. Photo taken 1993 L to R, Front: Mrs. Ann (Verner) Anderson, Irene Paulson, MFU Sec., Gwen (Paulson) Birch, Mr. & Mrs. James Youngdale. Back:Ione (Dahlen) Kleven, Edu. Dir., June (Smogaard) Carlson, Olaf Haugo, GTA Bd., Junice (Dahlen) Sondergaard, Edu. Dir., Verner Anderson, Ruth Lorentzen, Garland Birch, Regina (Mrs. Henry) Herfindahl, Alice (Van Dyke) Burlingame, Frances (Wolfe) Struck

Figure 71 L to R: George Mann, GTA Bd., MN St. Legislator;
Ray Novak, Pres., NFU Ins.

Figure 72 L to R: Alec Olson, U.S. Rep., MN State Sen. & Lt. Gov;
Gene Paul, NFO Pres., MFU Member

Figure 73 L to R: Orv Calhoun, Gully Co-op Mrg; Archie Bauman, cartoonist, MFU Sec.; Verner Anderson, life-long activist & organizer; Leo Klinnert, St. Bd. Member

Figure 74 L to R: Howard Peterson, GTA Staff, Co FU Pres.; Bob Rickert, MFU Sec., Bd. Member

Figure 75   L to R: Larry Roe, Dir., Green View;
MN Gov. & Congressman Al Quie

Figure 76 Ale & Emma Rousu, Sugar Bush Local

Figure 77 MFU Foundation Bd meeting, 1997. L to R: Shari Ciccarelli, Sec.,
F.B. Daniel, Dennis Sjodin, Cy Carpenter, Orion Kyllo, Leona Jordahl,
Russ Ruud, Elmer Deutschmann (seated), Curt Wegner.
Not pictured: Brenda Velde, Harold Windingstad, Marion Fogarty

Figure 78 Typical Torchbearer Ceremony

Figure 79 L to R: Dave Roe, MN AFL-CIO Pres,
Gordon Scherbing, CEO & Pres., FU Mktg.

Figure 80 Ferne Simonson, Sec. to Ed Christianson; Emil Syftestad

Figure 81 Honeymead, Mankato, MN

Figure 82 Refinery in Laurel, MT.

Figure 83 M.W. Thatcher

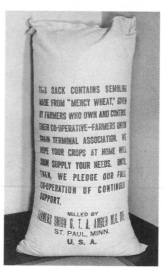

Figure 84 Mercy Wheat

Figure 85 Gully, MN Main Street on Meeting Day

Figure 86 Gully Co-op in the 1990's

Figure 87 MFU Managers at McPherson Refinery

Figure 88 MFU Campaign Yard Signs

Figure 89 L to R: Jerry Tvedt, Pres., FUCE;
Nels Wongen, Pres. Freeborn Co. FU

Figure 90 MFU at the Minnesota State Fair

Figure 91 Bill Walker, MN Comm. of Agriculture, MFU member.

Figure 92 Ed Weiland, MFU Mktg Assoc. Gen. Mrg.

Figure 93 Robert Wiseth, Co. Pres., MFU Exec. Comm.

Figure 94 Devon Woodland, NFO President

Figure 95 Farmers Union Marketing & Processing Association Board, 1996. Seated L to R: Harold Eklund, Paul Symens, Harold Nearhood, William Johnson, Mathew Birgen, Palmer Pederson, Standing, L to R: David Eblen, Dwight Bassingthwaite, Marvin Jensen, Dennis Rosen, Robert Carlson, Dennis Sjodin, Frank Daniels

8... NEW PRESIDENT, NEW HOME

"It (Farmers Union) is more than a farm organization... much more... "
Cy Carpenter, President, Minnesota, National Farmers Union

"The Spirit of '76 is not new to farmers. When this country declared its independence 200 years ago, farmers were in the front line. And they still are. They are in the front line for change, for progress, and a better way of life... "
Cy Carpenter, President's Address, State Convention, Nov. 1975.

"The answer to imperfect farm programs is not to abolish them... To do so is to eliminate a farm economy that has fed the United States and much of the world abundantly and reasonably for decades... "
National Farmers Union President Cy Carpenter in Commentary
Column Published in Minneapolis *Star Tribune*, Feb. 26, 1985.

Figure 96 Cy & Frances Carpenter

SHORT BIOGRAPHY OF CY CARPENTER:

Born: Dec. 21, 1922 Family Farm, Sauk Centre, Stearns Co., MN
Education: country school; graduated, Sauk Centre High School, 1940.
Employment: Family farming and trucking partnership, 1941-56; fieldman, Farmers Union Marketing & Processing Association, 1956-65; state legislative director, 1965-69; state Farmers Union Secretary, 1969-71; state Farmers Union Vice President, 1971; MFU President, 1972-1984. NFU President, 1984-88.
Married: Frances Stauning, Oct., 1958; two children, Richard and Kris.

Community Service: Chairman, K.I.D.S., Inc.; president, Green View, Inc.; Office of Technology Assessment; member, board of directors and president, Green Thumb, Inc.; member, Advisory Committee, University of Minnesota Institute of Agriculture; member, President Carter's National Council on Aging; chair, Minnesota Governor's Council on Aging; frequent spokesman and participant at various state, regional, national and international meetings, conferences, workshops, conventions, etc.

Farmers Union – As president, he was chief executive officer of the organization, and presided over the meetings of the full state board of directors (comprised of the 82 organized Farmers Union counties in Minnesota) and the annual state convention; a member since 1945; state staff, 1965-71; state secretary, 1969-71; acting president, 1971; state president, 1972-1984; national president, 1984-88.

Honors: John Carroll Award for Outstanding Service to Minnesota Vocational Technical Education; Honorary State Farmer degree and Honorary American Farmer Degree – F.F.A.; National Farmers Union Award for Distinguished Service to International Agriculture Award (1984); Minnesota AFL-CIO special recognition; National Soil Conservation award.

AN ARDENT ADVOCATE OF FAMILY AGRICULTURE

"He often was ahead of his time – years ahead," said long-time co-worker F. B. Daniel about Cy Carpenter. "That's why, regretfully, a number of his very good ideas went begging – both during his years as president of Minnesota Farmers Union and National Farmers Union. Too often what he wanted to do for farmers was exactly what should have been done, but for some reason or another just wasn't done"

Daniel continued, "Cy is a tremendous thinker, and usually far ahead of the crowd. Like his idea of a group legal service for farmers. Farmers are gun-shy of lawyers, and consequently, were reluctant to accept his plan to make legal experts more available locally to Farmers Union members."

The "Group Legal Services" plan had its debut in Jan., 1983, just as the real economic crunch of what became known as the "farm crisis" was occurring. Seven districts were established in Minnesota, with field representatives assigned for each district. These representatives were Robert Transtrom, Roseau; Clyde Allebach, Menahga; Roger Vogt, Mora; Fred Tauer, Sleepy Eye; and Earl Butenhoff, Faribault.

Under the plan, all Farmers Union members, and others like managers and directors of affiliated cooperatives were eligible to obtain legal help at a substantial savings under contracts negotiated by the organization. "As farming becomes more and more complex, legal help is becoming essential to good agricultural management," Carpenter explained in announcing the plan.

A special legal handbook was prepared and distributed free to every county organization. It included suggestions on many subjects, from estate planning and property law to farm liability and futures market contracting, as well as bankruptcy and mineral rights.

Legal advice columns were published in Minnesota Agriculture, and members were reminded by the newspaper that as members they could obtain a half-hour of legal advice free from any designated attorney in a county listing of those under a Farmers Union Group Legal Services contract. There was at least one attorney listed per county, with names and phone numbers published periodically in the newspaper. In addition, legal seminars were held, including one session in Willmar which was attended by 50 members of the Kandiyohi Co. Farmers Union. The speaker was William Bernard of Bernard & Johnson of Willmar, and his topic for the one-evening program was estate planning.

A FARMERS UNION DAIRY CO-OP

Carpenter also played a major role in Minnesota's affiliation with the Farmers Union Milk Marketing Cooperative based in Madison, WI. "What was happening was southern (U.S.) dairy interests were trying to shift the focus of milk pricing away from the dairy price support program to Class I bottling sales," Carpenter recalled. "That would have been a windfall for them, but would have sharply depressed prices in Minnesota and Wisconsin where most milk goes for manufacturing cheeses, butter and powder. Further compounding the problem was that federal regulations prevented general farm organizations like Farmers Union from representing dairy farmers in the federal milk market order system where these battles are waged."

As a result, Carpenter added, "It became clear to us that if family dairy farming was to have a future, we needed to form a dairy cooperative to fight our battles." So, the Farmers Union milk co-op was organized in 1971 in Wisconsin, and expanded into Minnesota four years later. Eventually, it marketed milk for more than 6,200 dairy farm families in five states, including Iowa, Illinois and Michigan.

Daniel then noted another Carpenter innovation – developing "allies" for farm interests by working with non-farm groups like organized labor, the clergy, government agencies, and others. "He saw the need to develop a better relationship and an exchange of trust with working people because he felt that the situation of farmers and working people were similar," Daniel added. "He also was interested in working with groups like the International Federation of Agricultural Producers. He had a real knack for developing alliances with other groups. And that's how he was able to 'get the votes' to get things done, in Farmers Union, in the State Legislature, in Congress and elsewhere.

HE CAME FROM FARM COUNTRY

While his background in the farming country of the north central United States was a fairly normal legacy Cy Carpenter of Minnesota shared with thousands of others of his generation, what this former farm boy did with his life is quite extraordinary.

Born in a family of six sons and four daughters to second generation Americans, Carpenter early in life began farming in a partnership with his brothers after graduation from Sauk Centre High School just ahead of the outbreak of World War II. Three of his brothers were serving their country in military service then, and the family's farming and trucking business was expanding.

There was no hint in those days that this rather humble beginning he shared with countless others of his generation, would be the foundation for decades of

193

leadership and family farm advocacy which Cy Carpenter would provide for farmers and American agriculture. Nor that his career would involve thousands of farm families in the battle for a better life for farm people and an improved climate for rural America.

Nor was there a glimmer of the kind of indelible imprints that would mark the path taken by the former Minnesota farm boy in his later years championing the family farm system.

Carpenter's first contact with Farmers Union came as a 23-year-old farmer in 1945. It was a time of transition. The GIs had returned home to the farm after World War II. "It became evident that there was going to be real disruption and trouble in the farm price arena and in farming in total, that the days of unlimited demand were gone," Carpenter recalled. He remembers going to a couple of Farmers Union meetings, and coming away convinced "that this was the kind of organization we had to have, that you had to participate", and "an organization which recognized that you couldn't do it alone".

Such a sentiment led Cy Carpenter to become a part of the Farmers Union. "I joined to have a farm organization to work through and with," he related.

Forty-some years later, he was 62 years old, and a mere 20 years younger than the Farmers Union itself when he was elected the organization's ninth national president in 1984. Carpenter would follow six predecessors that included a pair each from Kansas, Oklahoma, and Colorado. However, Carpenter was elected in New Orleans in a contest with Charles Nash of Ohio Farmers Union (Carpenter won by a vote of 150,000 to 107,000) and became the first from his state and region to attain the national presidency. (Later, in 1988 Leland Swenson, another native Minnesotan and former farm boy who became president of the South Dakota Farmers Union, would succeed Carpenter. It was the first time two consecutive National Farmers Union presidents were natives of the same state.)

The election of Carpenter and the selection of his successor, Willis Eken of Twin Valley, brought this editorial observation in the April 14, 1984, Minneapolis *Star* and *Tribune*: "Many are asking why Eken would give up a powerful legislative position (leader of the House DFL. majority) to step into big shoes (worn well for nearly 20 years by Cy Carpenter, recently elected president of the National Farmers Union) to take on big troubles (financially strapped farmers are having a tough time even paying membership dues)... "

The Minneapolis *Star* and *Tribune* quoted The Farmer magazine which observed, "Agriculture doesn't have a lot of heroes, but Minnesota has provided a lot of great leaders at the state and federal level" and then concluded its editorial with this thought: "We hope to see continuing strong leadership from Eken in St. Paul and from Carpenter in Washington."

In an interview with The Arkansas *Gazette* shortly after he was elected National Farmers Union president, Cy Carpenter spoke out critically against the Reagan Administration and its farm programs. In a word, Carpenter said in 1984, the current farm program "is a disaster to family farm agriculture and the rural community". And he called for the resignation (something that would happen in Jan., 1986) of John R. Block, Reagan's Agriculture Secretary and an Illinois Farm Bureau member.

One of the things needed, Carpenter told The *Gazette*, "is recognition that young and beginning farmers have special needs" – and consequently, he feels that they should receive different levels of farm credit and lower interest rates. At the same time, he proposed a new monetary system where agriculture would be given preferential credit treatment. Carpenter explained, "Agriculture should not have to compete for money on the same level with other things like vacations, color televisions, and so on.

In an interview with St. Paul *Pioneer Press* Agribusiness Writer Lee Egerstrom, a professional honored for his work by Minnesota Farmers Union, Cy Carpenter noted the traditional role of the organization as "an advocate for rural life and family farms". But he added, "There's a fine line you have to walk. A broad-based farm organization needs to be an advocate of the family farm and what's good for rural communities. But you can't get so far out front on economic matters that you hurt the businesses those same farmers own."

Minnesota leads the nation in cooperative enterprises, with 722 cooperatives tracing their origins to the Farmers Union, Farm Bureau, The Grange and other organizations, Egerstrom pointed out in his Mar. 12, 1984, article published before Carpenter became national president.

He wrote how Carpenter is receptive to working with other groups, including competing groups like the influential Farm Bureau. "We can't afford to be wasting time just nit-picking each other to death," Carpenter said.

A PERSONAL LOOK AT CY CARPENTER

"Carpenter is a product of his times – times that changed the economic, social and political fabric of Minnesota," Egerstrom reported. "The farm revolution may have been won in Carpenter's lifetime, but the evolutionary struggle continues and is couched in the rhetoric of the past. Carpenter, a short, intense man, still turns to eye-watering tales of the plight of farmers, not humor, to score points in his speeches and casual conversations."

Cy Carpenter did not have an imposing presence – he boasted not the large physical stature of a broad-shouldered Norwegian farm leader standing tall. Nor did he possess a booming, evangelistic voice that would reverberate loudly and echo in the convention hall where hundreds of partisans listened, ready to applaud lustily at the appropriate time. Yet, he could match verbal punches and mental wits with the best of them, in his speeches championing the family farmer in a fight-for-life struggle against the money-merchants and greedy corporate giants sopping up the economy of rural America and threatening the survival of struggling family-type farmers.

No, Cy Carpenter was of slight (5 feet, 7 inches, 155 pounds) build. And his firm articulation emanated from a rather normal, and not outstanding, set of vocal chords – although his remarks and declarations came across clearly and with intended, unmistakable blend of dedication, sincerity, and appropriate emphasis as he sounded a steady and constant call for better programs, activities and such things for farm people, and as he attacked those who stood in the paths blocking progress and profits for his hard-working legions within the ranks of the Farmers Union.

Make no mistake. Cy Carpenter's efforts more often than not would benefit all farmers, not just those within the membership of the farm organization for which he worked for nearly 40 years. What he – and Farmers Union – sought so often was unselfish: the promise of benefits for all independent, family-type farmers like those of his early days, and like those of his brothers still on the farm, and not really car-marked just for Farmers Union members only.

The dedication to agriculture possessed by Cy Carpenter cut a much broader swath. It was painted with a bigger, fuller brush.

His career as local, state and national leader had its roots in the farming country of his boyhood home, and his early years helping on the farm and also helping operate a family trucking enterprise that augmented the income from a diversified livestock farm. As a boy, Carpenter was one of the family's chore hands caring for crops, livestock, poultry, dairy cattle and such back before the days of true horsepower ended on the farms of the Midwest.

Then, when he left the farm in the mid-1950s to work for Farmers Union Marketing & Processing Association, Carpenter was made more aware of the challenges of livestock marketing. His duties operating as a marketing specialist out of Thief River Falls in northwest Minnesota included physical challenges as well as the demanding and frustrating fluctuations of the livestock market. One day as he wrestled with sheep grading them for prospective buyers, Carpenter was working hard as usual. But this time there was a new element, something else disturbingly more than the routine of sore muscles and sweat. For the first time, and unfortunately not the last, his sheep-sorting duties were accompanied by sudden, sharp chest pains. This discomfort was, he reported later, a sort of "warning". Fifteen years later, he would again sustain the kind of breath-stopping chest pains that would lead to a triple heart by-pass. By that time Cy Carpenter would be at the zenith of his Farmers Union career – the national president.

A FATEFUL TIME OF TRANSITION

After six years as a marketing specialist (he had other chores, like working on Farmers Union membership), Cy Carpenter was ready for a new challenge. At the same time state President Ed Christianson of Gully was just as eager – again – to utilize a wider range of skills possessed by his younger colleague and co-worker whom he knew to be quite interested and to have been active in politics. So Christianson asked Carpenter to move to the Twin Cities and be in charge of the Farmers Union legislative, or lobbying activities. Again, Carpenter planned to limit his tenure at that assignment, willing to handle the legislative chores for only a couple of years. But as it turned out, he was to be a legislative family farm advocate for six-plus years.

A few years later, Carpenter's talents again led to significant promotions – to state secretary, and then a nomination to be the vice president of Minnesota Farmers Union and to become the running mate with the incumbent state president and national vice president, Ed Christianson. Carpenter was now clearly moving up to greater responsibilities, and in line to succeed Christianson, the legendary cooperative leader. But unfortunately, it was not to be a traditional advancement and succession. A cruel fate would tragically intervene at an untimely juncture.

It would happen in Nov., 1971, on an ordinary, routine assignment, returning from a regular meeting of the board of directors of the National Farmers Union at the organization's headquarters in Denver, CO, a trip Ed Christianson had made countless times. But this time it was tragically different – Mr. Christianson would suffer a severe, disabling stroke that would cause a sudden and premature close to his illustrious career. The cooperative giant from Gully would never recover, would never again be able to stand tall and speak for farmers, to make wise, meaningful business decisions in their behalf, to build on the cooperative foundations he and his colleagues had established.

Word came at the Farmers Union office on Wednesday, Nov. 17, about 11 o'clock in the morning of the attack that brought down Ed Christianson. Cy Carpenter, Milt Hakel and Lura Reimnitz rushed to Midway Hospital where Mr. Christianson was hospitalized. After visiting with Nora Christianson for a time, Carpenter and his editor, Milt Hakel, returned to the office. Calls were made to state vice president William Nystrom and members of the state executive committee of Farmers Union. It was decided to wait until the following day to reach a decision concerning the upcoming state convention only 10 days away.

On Nov. 18, 1971, via telephone conference call, the decision was made to go ahead with the convention, especially since that course of action was recommended by Mrs. Christianson and her family. The motion to proceed was made by Hubert Hanson and seconded by John Waisanen, with all on line agreeing with Hanson that the organization's 26th state convention be held as scheduled even though it appeared its distinguished president of nearly 21 years couldn't possibly be present.

The timing of the crippling stroke that struck down Edwin Christianson was such it occurred just prior to the upcoming state convention where it had been planned that Carpenter would move up to the vice presidency. Now instead, though, with approval of the Farmers Union's legal counsel, the state board of directors appointed Carpenter "acting president until Mr. Christianson is able to perform all or a portion of the duties of the President", as provided in the organization's bylaws. As it turned out, he was to serve as "acting president" for the remaining one year of Christianson's term of the office to which the Gully leader had ascended nearly 20 years earlier.

Secretary Robert Rickert noted that the executive committee also made two other changes: Editor Milt Hakel was named to a newly-created position, "administrative director" and instructed "to work in close harmony with the vice president (Cy Carpenter)" during the period he would be acting state president. Hakel would be in charge in the absence of Carpenter, with full authority. The other change was the appointment of Ferne Simonson as acting interim state secretary.

And so exited Ed Christianson, unquestionably the greatest leader of all time in the Farmers Union in Minnesota, and into that post Christianson bore so well entered Cyril H. Carpenter, then 48 years old. It wasn't the best environment for succession – a young state leader trying to fill the shoes of a widely-recognized and acknowledged national veteran leader who had been robbed of more good years doing what he had done so well for a score of progressive years for the farm organization and for the cooperatives, its members and others who together had pioneered and built this remarkable marketing structure.

But in short order Carpenter would demonstrate in no uncertain terms that his priority was to be a real, outspoken champion for farmers, and that no one could be more dedicated to a family-type agricultural system based on the cooperative system, the power of legislation and education. Carpenter felt strongly that farm people stood a better chance of battling and surviving the odds of having to leave the farm and the land.

After all, Carpenter had grown up during an era when farm numbers were at their zenith. He had witnessed first-hand the migration from the land during his lifetime, because of drought and short crops, poor returns, low market prices, and such, despite "soil banks", "set-asides", conservation reserves or whatever farm program was available. His were the grassroots credentials of one baptized on a family farm, and who had communed through sweat, stress and hands-on labor and experience, and who had through the promise of organized efforts sought for years to solve problems and enhance opportunities for rural people.

Early in his new office as both state vice president and acting state president, Carpenter was to encounter change. At the regular meeting of the state executive committee Mar. 14, 1972, only a few months after taking over, Carpenter suggested hiring a public relations director to work with cooperatives. This related to disturbing news that officials of the Farmers Union Central Exchange had made the decision to change the agribusiness cooperative's original name, and were in the processing of enacting a new "trade name" eliminating "Farmers Union" wordage, and making its new name simply "CENEX".

This was the first of several changes that would come for cooperatives, and their longstanding, close working relationship and identity with the Farmers Union.

Without question, Cy Carpenter brought to the presidency the one thing no one could acquire any other way – an inimitable link to the land, and a bond with those close to and a part of it. It was a heritage he shared with his wife, Frances, a native of Polk Co. and born and raised on a farm.

Later in his career, he would call for "a national food policy" predicated on the amount of food needed to sustain the nation and meet its foreign food and fiber obligations. It happened in 1973, only a short time after Carpenter had become state president of Minnesota Farmers Union. The nation then was struggling with an energy crisis, a foreign oil embargo and related problems. While consumers in the cities waited patiently in long lines of cars to get precious fuel on odd or even numbered days, simultaneously, the farmers anxiously pondered how they could obtain sufficient fuel and supply items to plant and harvest crops.

A CALL FOR A "NATIONAL FOOD BUDGET"

What the nation sorely needed, Carpenter told delegates at the 1973 state convention, was "a national food budget" – outlining and prescribing the desired amount of food production Americans needed, and then rationing and budgeting the necessary gallons of gasoline, diesel fuel, distillate propane, kilowatts of electrical energy, fertilizers and chemicals needed to guarantee the estimated national food supply.

The Minnesotan further suggested budgeting "reserves" because agricultural needs could vary 20 to 30 percent, depending on weather, crop conditions, and other factors. In other words, he advised opting for plenty rather than possibly coming up short in supplying the nation with food.

Such planning for agriculture, Carpenter pointed out, was necessary to prevent the kind of crippling fuel shortage then gripping the nation. He blamed the energy shortages of the early 1970s on "the stubborn refusal of federal officials to plan ahead, and their failure to listen to those who offered realistic direction and to foresee the consequences of their actions".

He also suggested "rationing fuel now to avoid food rationing in the not-too-distant future". The "seeds of the food production problems of today (1973) were planted in the Agriculture Act of 1970", a farm bill he charged could be interpreted as a policy under which farmers didn't need price and income adjustments in line with rising costs and further, "that family farmers were expendable". The USDA policy-makers seem to believe that freedom and efficiency will pull us through, that farmers can do an expanding job without assurances of price and income, fuel, electrical energy, fertilizer, chemicals, and other inputs and transportation services".

Carpenter added, "If farmers operated their farms without planning ahead, like the USDA runs the nation's food policy, there would be starvation and disaster." At the same time he said he didn't advocate putting "farmers or the nation's economy under straight-jacket government controls, but when things become so muddled and precarious as they now are in the energy field, the temporary expedient of rationing available supplies is unavoidable".

His suggestion was "to move rapidly and vigorously through a rationing period". While this never came to pass, and the energy crunch waned, the challenge provided an opportunity for Cy Carpenter to demonstrate his leadership abilities – to not only clearly identify the problem, its causes and effect, but to also in the same breath communicate proposals that if adopted could not only rectify the situation but safeguard the nation from experiencing an equally-stressful encore episode as well.

"BETWEEN YOU AND HUNGER" – A FOOD POLICY CONFERENCE

"We are standing today at the crossroads of a world food crisis... the threat of widespread famine already is at hand... "
U.S. Senator Hubert H. Humphrey, June 7, 1974.

One of the first major programs engineered during Cy Carpenter's state presidency was a "Midwest Conference on Food Policy" held in June, 1974, a half-year before the first World Food Conference held in Rome, Italy, by the Food and Agriculture Organization of the United Nations.

Carpenter co-chaired the two-day conference with David K. Roe, the president of the AFL-CIO. The sponsors of the session also included the World Hunger Action Coalition and Bread For The World. Minnesota Farmers Union editor Milton D. Hakel was the coordinator of the Midwest Food Conference. In outlining the purpose of the session, Hakel observed:

"Ten years ago, five years ago, even two years ago, it would have been unthinkable to suggest food shortages and scarcity prices could occur in the United States. To be sure, there could be trouble elsewhere in the world, but we in the United States would use our bounty to feed ourselves and continue the 'war on hunger' as we had in the past 20 years... so most of us Americans dismissed... those who had begun to warn of famine as 'alarmists'...

"Then, all of a sudden, we found ourselves in 1974 not only with a continuing and intensifying food crisis in the world, but in a very real sense with 'a crisis in the bread basket of America'... the Russian grain deal (of 1972) and our energy crisis did not precipitate the food crisis, they only helped accelerate it... farmers had never farmed in a more precarious time...

"We had a national food policy in this country only in the sense that the policy was not to have a food policy... "

The idea for a World Food Conference was first suggested in the fall of 1973 by Senator Humphrey to U.S. Secretary of State Henry Kissinger. He took the idea to the United Nations, and the first conference was held in Rome in Nov., 1974.

At the Midwest Conference on Food Policy, co-chairman Cy Carpenter appealed for a national "food policy", telling delegates that "like an heir who has inherited great wealth but never worked to earn it, our nation has taken its food supply for granted". The consistent and sustained flow of food from farm to consumer had allowed the country to not develop an "overall food policy". This Carpenter said is "something we no longer can do... today both producer and consumer are alarmed at the wild gyrations in food supplies and prices... "

National Farmers Union President Tony T. Dechant told the conference that famine was rampant in Ethiopia, and in parts of Tanzania and Kenya in Africa, and starvation plagues were imminent in Bolivia, Syria, Yemen, and Nigeria. Dechant added that poor harvests could also create famine conditions in other populous nations like India, Bangladesh, Indonesia, as well as many other warm-climate countries. Food shortages existed in 1974 for 3.5 billion around the world, it was estimated.

The United Nations estimated 500 million to 600 million people were "undernourished", or about "two-sevenths of the human race", or about a billion people, considered "malnourished". "The world food reserve now on hand is only enough for 27 days," Dechant said. He said that the farmers of America can respond "to the pull of rising demand for food, particularly if they are protected and encouraged by reasonable measures to avoid collapsing prices when they occasionally produce short-run 'surpluses', as surely will happen... "

U.S. Rep. Bob Bergland suggested establishing "food reserves" which could be tapped when needed to combat hunger. But he cautioned that such reserves must be insulated from the market, and not allowed to cause "catastrophic price drops". Bergland urged "a policy that provides stability, that eliminates the booms and busts of agriculture, and that does not work to the disadvantage of the working man and women, their families and the producers".

Senator Humphrey also called for both national and international "food reserves", saying that such stockpiles would provide critically needed "stability in the world food economy". For 25 years, Humphrey said, "we have been waging a battle to improve the quality of life in the developing world... today that battle against global poverty and disease is in danger of being lost... "

To Cy Carpenter, the most significant figure in Farmers Union history was the legendary James G. Patton. Patton was born in 1902, the same year Farmers Union was founded, and then served as its national president from 1940 to 1966.

"Jim Patton had a very important combination of traits – an eloquence as a speaker, and a great ability to communicate, while at the same time having the foresight that let him anticipate conditions and problems so he could sound an alarm and create concern for them," Carpenter said in a 1995 interview.

Specifically, Patton was an outspoken critic and advocate for all sorts of matters pertaining to rural people in the U.S., as well as overseas. He championed programs like the famed Food For Peace measure which allowed the sharing of American farm productivity with needy, hungry humans abroad.

"He was an intelligent, effective leader strong in his convictions and ready to fight for farmers, and what he believed in," Carpenter added.

THE NATIONAL PRESIDENCY 1984-88

It most likely wasn't an ideal time to become the national president of Farmers Union. It was 1984, and the year that Cy Carpenter moved to the top post of National Farmers Union. His ascension to that post came at a time of real economic stress – the nation's farmers were experiencing the toughest financial crunch since the Great Depression of the 1930s.

There were a number of rural banks closing their doors, the nation's Farm Credit Banks were threatened by unexpected record losses, and stood in peril of bankruptcy at several locations, and the nation's fabled "lender of last resort", the Farmers Home Administration, was being buried under a bumper crop of unprecedented debt and bad loans.

There were headlines and television reports about farmers committing suicide. And in Iowa a farmer shot and killed his banker, a neighbor, his wife, and then as a state trooper approached his pickup truck, the distraught farmer used the same gun on himself.

The times cried out for leadership. And the man from Minnesota was one of those who responded, and stood out, front and center, in a leadership position that brought with it the expectation that in some way, some kind of solution or solace, might be produced there.

DARK DAYS FOR FARMERS

During his 12-year's leadership as the Farmers Union state president, Carpenter recalled that the over-riding problem confronting agriculture was what he described as "constant economic stress". That grim financial scenario was darkened even more by the farmers' constant declining political influence in the wake of steadily declining numbers of farmers and fewer votes represented by farm people. Further compounding this uncertain picture for farm people and rural communities was the fact that at the same time new technology was causing massive, far-reaching and disturbing changes in agriculture – not only at the farm level, but in the agricultural processing field as well, in the markets where fewer and fewer bids originated and while buying concerns grew more powerful.

"We went through periods when we lost thousands and thousands of good operating farms in Minnesota," recalled Carpenter. "These were farms that never really should have folded – farms that were victims of a national food system where all those participating diligently work to force the price of farm products down."

Escalating financial failures of farmers, especially the young farmers, were linked directly to the problems of lenders in a couple of ways:

Their own individual financial problems brought on by the combination of inflated land values, high interests rates, fluctuating prices for farm commodities.

Over-zealous agricultural lenders, following the advice of too many agricultural economists, professors and politicians wanting "cheap food", reflected reckless lending policies where young and beginning farmers wound up with excessive loans for what was actually needed to start farming, or to expand farming operations.

Carpenter felt that "too often some of the best, brightest, hardest-working, young farmers seemed to be the ones that failed", a development he saw as especially bad for the kind of "family farm" agriculture he and Farmers Union supported. Some years turned out to be more difficult than others, depending on world food production as well as international demand and trade, and on political developments including the use of grain embargoes. Then in some years, other factors such as the drought of 1987-88 compounded financial problems and made an already critical condition financially fatal to many while severely damaging others.

His tenure as Minnesota president included years highlighted by several victories for family farmers, and for young farmers, including military veterans returning home to farm. In 1972, shortly after becoming state president, Carpenter led the battle to make available farm management training classes for veterans under the G.I. Bill of Rights. Before he and Farmers Union became involved, there had been only two farm G.I. Classes in the state for returning veterans. Thanks to action led by Farmers Union, working with Service organizations, the State Legislature, and others, the number of returning GIs with farm backgrounds desiring to participate in a farm management class were identified, and 82 of these special classes were established in Minnesota.

A brochure distributed by the Farmers Union in Jan., 1973, alerted veterans to the GI Farm Management Training program established by the Veterans' Readjustment Benefits Act of 1966. "Now you can catch up with time and opportunities lost during military service," the brochure read. "GI Farm Management Training will pay you while you improve your farming skills and your future income potential." It pointed out that the new Veterans Bill provided "vocational training and education for service men and women whose careers have been interrupted or impeded by active duty after Jan. 31, 1955" and "aid to those in attaining the vocational or educational status which they might normally have aspired to and obtained had they not served their country".

The GI brochure also pointed out GI Farm Training could help farmers "learn and earn", with up to $3,504 in yearly payments for tuition and family living expenses. Class subjects and activities for the three-year program were listed: Farm Management, Animal Science, Farm Mechanics, Soils and Agronomy, Agricultural Economics, and other subjects including tax accounting and computerizing farm records for analysis.

A map of existing GI class locations was included, along with a request from whereby Farmers Union would provide enrollment information. The Minnesota map listed classes across the state, from Roseau in the northwest to Spring Grove in the southeast, and from Northome in north central Minnesota to Luverne in the southwest.

The brochure really encouraged military veteran-farmers to enroll by asserting graduates could expect nearly a 60 percent increase in family net worth during the three-year course of farm studies. In short, the campaign was well-promoted, and attracted and generated widespread interest.

Credit also must be given to Ralph Whiting, hired by the Farmers Union in 1972 to work with the Minnesota Department of Education and with instructors and the farmer-veterans. All told, there were 12,000 potential GI Training Class candidates identified in the state. About 10,000 enrolled, and received something like $5 million yearly in special government payments to the farmer-students. "This program helped keep hundreds of young farm couples on the farm who otherwise might not have been able to do so," Carpenter reported at the 1973 state Farmers Union convention.

Then, in 1977, the Veterans Administration initiated a new policy where it could deny farm management classes if net farm income failed to meet some undefined minimum. This requirement, not imposed on any other GI training participants was "crass discrimination", Carpenter and Farmers Union pointed out.

Veterans who wanted to farm, they rightfully held, should not be deprived of their benefits available under the G.I. Bill. After a two-year battle of letters, meetings, Congressional testimony, and four trips to Washington, D.C., by leaders of the Farmers Union and others, the VA finally agreed to change its eligibility requirements and then sought recommendations from its chief critic.

This was purely a campaign carried out by Minnesota Farmers Union, working with farmer-military veterans, Minnesota agricultural instructors, and the vocational agriculture education oversight staff at Ft. Snelling. The agricultural education officials were "honest enough and bold enough" to help Cy Carpenter and Farmers Union identify wrongful actions that needed to be corrected. But it was the Farmers Union which carried the fight, paid for travel expense for four trips to Washington, D.C., for key people to attend and take part in meetings and the VA hearing.

This was, Carpenter felt, an example of "Farmers Union being more than a farm organization". And while this project represented only Minnesota, it helped draw attention to the inadequacy that existed nationally.

WHITE HOUSE ACTIONS ARE "IMMORAL, ILLEGAL"

When the Nixon Administration cut short the disaster loan program that was designed for farmers who because of excessive and prolonged rain had not been able to plant crops, Cy Carpenter was livid. He issued a statement Jan. 8, 1973, calling "The White House actions of recent weeks not just mistaken policies, but policies that are immoral and illegal as well. The (Nixon) Administration has broken faith with farmers in terminating a disaster loan program which it had officially announced would be available until June 30, 1973, by ending the measure more than six months early in some counties and by mid-January in the remaining disaster-designated counties.

The early closure of the disaster program was part of a budget-trimming action by the U.S. Department of Agriculture. Carpenter reported that something like $2 billion had been cut out of the USDA budget, crippling or ending prematurely several programs ranging from the crop disaster program to ending direct 2 percent loans for Rural Electrification and Rural Telephone programs. "We don't know where the axe will fall next," Carpenter said in his statement issued after the board of directors of the Minnesota Farmers Union met and drafted a special statement as well.

"It is a question of priorities – whether to continue to invest in needful agricultural, conservation and human activities – or to allow the diversion of $2 billion into a few more weeks of bombing and (Vietnam) war or paying tribute to those who profit from a new round of inflation, tighter money and higher interest rates," the board statement said in part. "What the Administration is doing will not curb inflation, it will punish farm and rural citizens for an inflation which they are not causing."

There had been 26 Minnesota counties designated as "disaster areas" by either the USDA or the White House via presidential designation in a new federal program announced in August. The deadline for filing for "disaster emergency loans" (at one percent interest) was set for June 30, 1973. Farmers Home Administration officials even advised against filing early for disaster loans pending completion of harvest and a more accurate appraisal of crop losses caused by bad weather during the growing season.

So when the news came out in December, or shortly after the election, that the disaster application period had ended much prematurely, and unexpectedly, Carpenter and Farmers Union were the first to go public with their objections to such a breach of faith. Especially were they irritated knowing that two Minnesota political figures – Senators Humphrey and Mondale – had authored and sponsored a farm amendment to the new federal Disaster Relief Act.

Led by Carpenter, a lawsuit was filed charging Nixon's agriculture secretary, Earl L. Butz, with illegal and arbitrary action in ending the crop disaster loan application period. It didn't take long before Butz reversed his decision, and the disaster program was reinstated for the full application period as announced originally. Clearly, it marked an important victory not only for Farmers Union and its then new state president, but for the farmers in those 26 disaster-hit Minnesota counties. The total paid to these farmers was about $25 million.

Maybe more important was the fact the Minnesota action changed the entire attitude of those directing the USDA by making it very clear that farmers, and Farmers Union, would stand up and fight for their rights, Carpenter said later.

Then in Oct., 1974, Carpenter again initiated a lawsuit against Secretary Butz. This time the charge was that Butz was not complying with the directives regarding dairy price support policy as mandated by the 1974 Farm Act. Under that law the U.S. Department of Agriculture was ordered to maintain price supports on manufacturing grade milk at not less than 80 percent of parity.

Cy Carpenter's suit charged that by refusing to adjust milk support price levels, Secretary Butz was violating the intent of the 1974 law and "thereby injuring both producers and consumers".

While the successes for veterans and the successful lawsuits against the USDA represented major victories for Carpenter and his Farmers Union legions, what might be his finest hour as state president likely had its origin in the dust and heat of the drought of 1976. As the prolonged period of relentlessly hot, dry, rainless days persisted, nearly one-half of Minnesota was given a Presidential "drought disaster designation" as crops withered severely. This caused Carpenter and his staff to spearhead an effort to help farmers, and enlist financial support of regional cooperatives.

They successfully organized a "hay-lift", a program whereby they first located plentiful supplies of forage, and then established a means of moving thousands and thousands of tons of hay to the Minnesota farms where this feed was desperately needed and where countless herds of dairy cattle were much in jeopardy. Farmers Union was able to negotiate reduced railroad transportation rates, while in a 24-hour period arrange for a revolving fund of $100,000 to help purchase the hay in Montana, other states, and Canada, and then ship it to Minnesota farmers. While the money was raised in a day, it took another 10 days to two weeks to start moving emergency shipments of hay to forage-short and drought-stricken Minnesotans.

The first shipment of Montana hay under the Farmers Union "Haylift" program has arrived in Minnesota. Above, farmers unload the good-quality hay from a truck van that rode "piggyback" on a railroad flatbed car. The first 40 tons were delivered to St. Cloud.

Figure 97 The Haylift Action - 1976

This start-up of hay-lift money was provided by the Minnesota Farmers Union, the Farmers Union Central Exchange, the Farmers Union Grain Terminal Association, the Farmers Union Marketing and Processing Association, and the Minnesota AFL-CIO.

With the help of Minnesota Gov. Wendell Anderson, state agricultural commissioner Jon Wefald, and U.S. Senator Hubert Humphrey, Farmers Union obtained federal assistance in the form of partial reimbursement of hay transportation costs. Wefald was extremely helpful and supportive, working with state and federal officials and in communicating to the public about the effort. He also played a role in obtaining special freight rates in transporting the hay.

The supply of hay that year was so limited that many times hay-lift coordinators had to divert a load from one party who had ordered it, to another farmer who was on the verge of running out and desperately needed that shipment instead.

While a few unscrupulous persons tried to take advantage of this unfortunate situation, most farmer-producers were very cooperative and respectful of their fellow-farmers and Farmers Union. While a scant few rascals sought to capitalize on the hay shortage situation, those "cheaters" caused extensive auditing later. However, none of the wrongdoers were part of the Minnesota organizers who made the hay-lift a success, and who kept good records of this effort.

In a report to the state board in Dec., 1976, hay-lift coordinator F. B. Daniel reported 10,000 tons of hay had been delivered to 500 farmers in the 40 counties designated as Presidential "disaster areas". Daniel said that orders were coming in from hay-short farmers at the rate of 250 to 300 tons per week. Hay was being shipped to Minnesota from four markets in Montana, three in Manitoba, Canada, as well as from single locations in Idaho and North Dakota.

Then in a final report after the program ended in 1977, it was noted that nearly 23,000 tons of hay had been delivered by 25 suppliers from eight states and two Canadian Provinces to 752 farmers. Further, the hay-lift took place with a $1.25 million savings in reduced rail rates and through freight reimbursements paid to drought-hit farmers.

This effort helped threatened dairy farmers maintain their foundation herds rather than be compelled to pay exorbitant prices for precious hay because of the short supply, while also thwarting potential distressed, drought-induced sale of animals. And the bottom line, in addition to the heartfelt gratitude of the hay-short farmers, was the fact all of the emergency hay-lift loans were repaid in full, including the cash advances by the three regional cooperatives.

At the 1976 state convention, Carpenter paid tribute to the successful hay-lift. "With no established program, no office in every county and the capability of the USDA, our people in Farmers Union put together the resources provided by our cooperatives, some old fashioned 'farmer-knowhow' and a willingness to work, and helped hundreds of farmers... "

Carpenter publicly credited "the Homesteader attitude of F. B. Daniel, Margaret Schreiner, Elaine Radniecki, Carlton Anderson, Oskar Reimnitz, Bob Montbriand, Jim Blight and all of the others" with the success of the hay-lift program. Indeed, despite some factors Carpenter labeled as "some mistakes, and having to contend with some critics and spoilers", it was seen by many as a bright moment for the organization and testament to "making things work".

DEDICATE NEW FARMERS UNION HEADQUARTERS

On Nov. 21, 1976, Minnesota Farmers Union dedicated its new state office building located at 1717 University Avenue in St. Paul, occupying the Skelly Building that was remodeled to suit its new owners.

At the dedication ceremony, Farmers Union Editor Milton D. Hakel was a featured speaker. He pointed out that Minnesota was only one of the Farmers Union states improving its office facilities. South Dakota was making state office improvements as was Nebraska. North Dakota Farmers Union moved into a sparkling new headquarters as well.

Hakel said such developments indicated "the foundation and the solidarity these new buildings represent" for the organization. "It isn't lumber and bricks and stone and concrete which matter," Hakel said. "It's what the building means to people. People are the central thing, but the building gives them a solid base... in the 61 years since Farmers Union first came to Minnesota, nearly a thousand people have served as county officers, and perhaps 8,000 to 10,000 have served as local leaders."

Then Hakel used the occasion to note the history of Farmers Union, and the fact that in 1977, the National Farmers Union would be observing its golden anniversary. His research showed that Farmers Union activity began on April 20, 1915, when W. B. Evans of Bismarck, ND, was designated the organizer for the state of Minnesota. Records do not show exactly what happened, and only indicated that the initial movement took place in "western borders and southwestern Minnesota".

Hakel reviewed some of that early history, and then pointed out how Minnesota gained its first Farmers Union charter in 1929, and then a second charter in 1942. He recalled how Edwin Christianson was state president when the state office was moved to St. Paul from Willmar, and when in 1962, National Farmers Union President James G. Patton took part in the dedication of Minnesota Farmers Union's first own state office building.

Then Hakel quoted from the plaque that was placed in the lobby of the building at 1717 University Avenue in St. Paul: "Dedicated to the Continued Efforts of Farm Families to Work for Progress Through Legislation, Cooperation and Education." Having its own building for the first time in its history was indeed one of the brightest moments in the 12 years Cy Carpenter occupied the Minnesota Farmers Union presidency.

MEETINGS, MEETINGS AND MORE MEETINGS

If there is one thing farm organizations are known for, it is a generous offering of meetings. A sample of this kind of activity comes from an operations analysis compiled during 1976, five years after Cy Carpenter became state Farmers Union president.

There were 5,400 Minnesota Farmers Union local meetings held during the year, according to reporting form data compiled by the state staff . Then, there were 140 county meetings, with most attended by state staff. There were 83 county annual conventions, again with one or more state staff present. Two women's leadership conferences were conducted, and six regional meetings focusing on estate planning were co-sponsored with Congressman Richard Nolan. There were 40 "winter conferences", and another 50 meetings with consumer, church, legislative, and other groups. Finally, there were the inter-organizational meetings with other state Farmers Union organizations, with local and regional cooperatives.

The 1976 report also noted a legislative highlight – the passage of the Minnesota Family Farm Security Act. It was the first in the nation to provide assistance to help young and beginning farmers get started in their own farming operations.

At Farmers Union camp sessions, there were a series of classes which pertained to topics such as cooperative education, Farmers Union philosophy, nutrition, ecology, community leadership and creation of a mock cooperative. And 30 women attended the annual summer "Ladies' Camp".

In short, it was a busy year for junior members, regular members, their families, and the state staff of 36 officials and full-time and part-time workers.

A SOMEWHAT DIFFERENT FARM SCENARIO

Delegates meeting in New Orleans in late winter, 1984, elected Cy Carpenter of Minnesota national president of the Farmers Union. In his acceptance speech, Carpenter spoke of "an attitude of indifference and abuse that is applied to agriculture which is our real (farm) problem". He explained that "We in agriculture do not have 'a farm problem' – we have a problem... . imposed on farmers resulting from a lack of understanding at best, and pure exploitation in many cases... "

Carpenter voiced concern that the Reagan Administration had "a stated desire to destroy and remove farm programs, had bypassed both Congress and farmers in developing and implementing actions as costly and as radical as (the 1983) PIK (crop-share, payment-in-kind and , whole farm retirement program)... "As I have said repeatedly, as the world watches and compares Democracy and Communism, we cannot allow Democracy to be viewed as a form of government that allows growing hunger among its people while it complains about surplus goods... that is not Democracy; rather, it is the result of greed or abuse of power by a few within our Democracy ... that must be changed... "

Carpenter also called for renewing "our commitment to understanding and appreciation for others", noting that "when we are told that those unemployed stand in soup lines just to get a free meal, or when we are told those sleeping on sidewalk grates to keep warm do so because it its 'their choice', we must remember the same thing is being said about farmers... (that) when your son or daughter or neighbor gave up and quit farming, or was sold out, that 'that was their choice'... that's a lie... they didn't have a choice... you know it and I know it, and regardless of how you sugar-coat it with pious words, that's a lie ... "

He also attacked huge expenditures by the federal government for space exploration. "Before we spend billions exploring and studying the vapors around Venus and the particles surrounding Saturn, let's take proper care of our water vapors and our soil particles here on Earth... . we are told we cannot afford to provide a stable farm income, but we are told we now will build a manned station in outer space... before we spend additional time and money developing Outer

Space, we need to do a much better job of developing 'Inner Space', that undeveloped space between our ears... we cannot wait for better values and direction for agriculture... we need to go to the media, our friends among the farm broadcasters, rural and metropolitan newspapers and television and seek their cooperation in building a better understanding and appreciation (of farmers and agriculture)... "

Carpenter in his first presidential address suggested going to the people, "the work force, the consumers, the elderly and rural business friends" and promote a better understanding of "how we suffer or prosper together... that a bankrupt farmer cannot buy a tractor or a car or take a vacation, just as an unemployed worker cannot buy our beef and eggs and milk... "

However, when he became National Farmers Union president, Carpenter said he didn't notice as much economic stress in general for farmers. "The same economic pressures continued, but seemed less pronounced, I guess, because by then an awful lot of farmers had been weeded out. But the disparity and lack of reasonable returns continued, and in fact, continues today (1995).

"But it wasn't as visible because after much of the 'adjustment' had been made, and something like 2.5 million farms were eliminated, at least temporarily, some things seemed less critical (economically) for those who remained – simply because there were fewer failures and foreclosures and gross farm income for fewer farmers was higher." Still, he knew things could change, just as he had seen constant change over the years for farmers and agriculture. And new, meaningful relationships could be established – even with competing farm organizations.

When talk centered on ending federal farm programs, as it did during discussion of the 1985 farm bill, Carpenter's back stiffened. He strongly opposed a "free market" for farmers, as was being proposed then. "Improve – don't eliminate" farm programs, Carpenter wrote in a commentary article published in the Minneapolis *Star* and *Tribune* Feb. 26, 1985.

"The answer to imperfect farm programs is not to abolish them," he editorialized. "To do so is to eliminate a farm economy that has fed the United States and much of the world abundantly and reasonably for decades."

Carpenter suggested that farm programs be tailored – cutting out "those provisions that have given farmers who want to grow big for bigness' sake unfair advantages that force other, efficient farmers out of business".

A FARM PARTNERSHIP – NFU & NFO

It was during Carpenter's national presidency (1984-88) that a new, more meaningful relationship between the Farmers Union and the Iowa-based National Farmers Organization (NFO) had its debut. The two organizations launched a joint-marketing operation which began in Minnesota, and eventually was established in about a half-dozen other Midwest states.

"They (NFO) had their collective bargaining approach, and believed in withholding products, which made it hard to get a strong working relationship, especially at the top level. But that changed after I came in, and because they had a new NFO president (DeVon Woodland),." Carpenter related.

Until DeVon Woodland of Blackfoot, ID, became NFO president, Carpenter said the "differences" between the two organizations nationally were too substantial for any kind of merging or working together. The NFO and Farmers Union had basic political and philosophical differences which had precluded working together, he recalled, even when efforts were made to develop a "National Farm Coalition".

Consequently, up until Carpenter assumed national leadership and Woodland headed the NFO, there was little likelihood of a "working relationship" between the two organizations.

At the Minnesota level, though, a relationship blossomed earlier. When Cy Carpenter was state president, the NFO and the MFU held a "joint legislative day" at the Minnesota capitol in the spring of 1983. The effort was an attempt to show legislators there were issues existing which concerned all farmers. Representing NFO was Bill Christensen, the state president from Battle Lake, while Carpenter represented Farmers Union. Together, they pointed out, they represented "the largest farm constituency in the state".

"I was very pleased with the joint legislative day, and look forward to working with other groups whenever possible to make sure farmers are well represented in our state capitol," Carpenter commented. "While we will continue to use our own policy statement in working toward a better agriculture for Minnesota, I was extremely pleased that we were able to show those areas where we have a common interest."

A fellow Minnesotan, Gene Paul of Delavan, was both a long-time member and leader in NFO while also a member of Farmers Union. He played an active role in the development of the initial NFU-NFO joint marketing agreement in Minnesota adopted later, when Willis Eken of Twin Valley was Minnesota Farmers Union president. Then, in 1995, Paul was elected national president of the NFU.

The new "partnership" between the two organizations emphasized two of the recognized strengths of the two farm groups: the NFO's expertise in marketing farm commodities, and the Farmers Union long-standing effectiveness in legislative programs and representing farm people politically.

In addition, Carpenter said NFO's Woodland shared an interest in reaching beyond just farm groups to meet, and to organize a coalition-type relationship with non-farm groups, like consumer organizations, church interests, and labor. "This was something I had long been interested in," he added. Out of their meetings came an effort to foster a better understanding between various groups and their interests. NFO and NFU worked closely toward this goal, meeting with non-farm groups and others including politicians and government officials in promoting a better relationship between diverse interests. One example was the Rural-Urban Action Campaign (RUAC), which represented an organization of church, consumer, labor, women's groups and farm interests.

Then there were times when new groups emerged, such as the American Agriculture Movement in the late 1970s, and the Farm Unity Coalition and Minnesota Groundswell organizations during the "farm crisis" years of the 1980s.

Often, these new voices and new faces, because of their "newness" and emotional outpourings for farmers, commanded more and immediate media attention than the older, more entrenched and experienced farm organizations. Support of the established farm organizations usually proved essential, however, because such established farm groups were the ones who endured and persevered after the crisis of the era had ended or alleviated.

THERE WERE LEGISLATIVE SUCCESSES

There were successes in the legislative arena during Cy Carpenter's reign as state president in Minnesota. Basically, the Farmers Union provided leadership in recognizing and calling attention to the need for a family farm system and structure in agriculture.

"We have one of the most rigid laws of any state prohibiting the corporate ownership of farmland," Carpenter explained. One of the key persons in gaining such a law was a legislator named Willis Eken, a lawmaker and member of Farmers Union, and a leader who several years later would succeed Carpenter as president of the Minnesota Farmers Union.

Carpenter continued, "Some critics charge that you should not restrict ownership of the land. But we think the importance of this is that we're saying farmland is not just another commodity, and that the state and the people have a right and a need to give direction, to protect it. Much in terms of understanding grew out of this as well as the restriction (on land ownership) itself."

Farmers Union was much less successful in gaining "a favorable tax consideration" for families to use in transferring land ownership from one generation to the next, Carpenter recounted. In particular, the organization hoped to lessen the financial burden for borrowers who took on huge debt loads in acquiring the family farm assets through a proposed interest rate concession proposal.

"We wanted a system favorable to encouraging and allowing good, young farmers to take over. We recognized the thing that killed many farmers financially was the drain of interest money. Land became more valuable, machinery became much more costly and the operating capital literally went through the roof. The Extension Service and our Land-Grant College experts were advising us to 'get bigger and leverage your assets'. Consequently, many farmers, especially young farmers, obligated themselves in a manner that the farm wouldn't bear or sustain.

"We thought that everybody would benefit if there was a relatively short-term period where they could get a favorable or reduced rate of interest or a tax consideration. Quite frankly, this is not too different from the incentives often granted to businesses and industry to induce them to locate in a community. Communities often say that they will provide land, water, roads, tax breaks, grants or loans to attract new businesses or manufacturing and processing plants.

"Our thought is that we ought to be doing these same things for the best and biggest business we have in this state – and that's production agriculture."

Regretfully, Carpenter feels, "the truth of the matter is, rural America lost many of the best and brightest young people because they wouldn't settle for the restricted economic or social life that they could have farming, considering the long odds they faced". By that, Carpenter added he did not want to imply "that those who stayed in agriculture were less competent or less committed, but the truth of the matter is that those who left in many cases to become lawyers or doctors or whatever could have served well or served themselves well had they had a reasonable opportunity in their community."

He continued, "Much of the value of the rural community is the social and community environment. And I don't think anyone can argue very seriously but what the social environment, the values and all which still prevail in the rural community, are some of the most desirable in America today."

Farming also is a "profession, a very necessary and demanding profession, but an overlooked and unappreciated profession, simply because it goes back to the beginning of time and is so commonplace," Carpenter believes.

SOME MESSAGES FROM CY CARPENTER

An insight into Cy Carpenter and the times and challenges during his state presidency can be gained by reviewing some of his presidential addresses presented at the state convention. Following are some examples, selected at random:

"There's a 'spirit' of Farmers Union that is something very real... you cannot touch it, and you cannot see it. But you can feel it, and you can sense it, and it can move mountains... " Cy Carpenter, Dec., 1972.

CARPENTER PUTS PRIORITY ON HELPING FARMERS

In his first address as state president, Cy Carpenter noted the convention theme – "Family Farmers Speak Through Farmers Union", and then explained, "It means if we will put it to work, farmers can speak through Farmers Union, and apply some 'farmer thinking' to every problem we face, and every opportunity we have... I was asked last week what is the number one priority in Farmers Union? My reply was that our number one priority is putting Farmers Union more effectively to work... we need to speak for more farmers by increasing our membership... "

Carpenter also noted significant developments during that first year succeeding the stellar Edwin Christianson as state president, prefacing his report with these words, "Perhaps some of our actions did not have 'the master's touch' that only Ed Christianson had... ".

Then he told of negotiating a contract with the Minnesota state officials (the Departments of Natural Resources and Transportation) which increased the activity of Green View, a non-profit organization designed to employ several hundred low-income, older persons in the maintenance of highway rest areas, state parks and recreation areas.

Carpenter also reported Farmers Union had helped 200 rural disadvantaged children attend a training and development camp sponsored by KIDS, Inc. KIDS is a non-profit organization then housed in the Farmers Union building which raised money to provide scholarships for children with learning and emotional disabilities to attend a special session at Camp Buckskin. The camp featured education, stressing academic achievement and personal development. Cy Carpenter was chairman of the board for KIDS, Inc. A second organization, called KIDS Korps was formed to provide speakers who could tell the story of KIDS, Inc., and to help raise funds for scholarships.

Other special projects started during the Carpenter years focused on challenges facing young farm families through another new organization, the Young Farmers Opportunity Council, working to get more young farmers enrolled in the GI Farm

Training program with more locations for trainee schools, and, finally, proposing the establishment of a "Farm Family Communications Center" somewhere in the state.

In the summer of 1972 a "Young Farmers Opportunity Newsletter" announced the formation of the "Young Farmers Opportunity Council". The State Young Farmer Opportunity Council Chairman was a Meeker Co. farmer, Jerome Carlson of Grove City. In this Newsletter, Carlson wrote a column introducing the Young Farmer group, and giving credit for its formation to the new Minnesota Farmers Union President, Cy Carpenter.

"Our main purpose will be to get in contact with all young farmers in the state because we want to know the concerns and the problems of young farmers under 35 years of age... we are trying to put all of the problems together and come up with reasonable solutions, whether the problem is income, taxes, housing, or whatever... "

Vice Chairman of the group was Vere Vollmers of Wheaton, while Margaret Daniel of Mendota Heights was secretary. The Council consisted of persons from a dozen counties: Carlton, Chippewa, Dakota, Houston, Jackson, Kandiyohi, Meeker, Otter Tail, Polk, Renville, Steele, and Traverse.

Cy Carpenter explained in the 1972 Newsletter why he came up with the idea of establishing the Council. "Young people face some tremendous obstacles in hoping to establish themselves in a lifetime career in agriculture," Carpenter wrote. "Farmers Union has been notable among rural organizations for its concern about the problems of young farmers... we have worked for a Young Farmer Credit Program, supported vocational agriculture programs, worked for restoration of an effective G.I. Farm Training Program... supported the reduction in voting age... it is the hope of this project to involve young farm couples and individuals prominently in the Farmers Union organizational structure and in rural community leadership... "

In the same issue it was announced that a "Young Farm Couples Camp" would be held at Farmers Union's Education Center near Bailey in the Rocky Mountains west of Denver. Eligible would be any couples in their twenties (with baby-sitting provided). However, only five families could attend from Minnesota.

A year later, addressing the 1973 state convention, Carpenter told how Farmers Union had hired two special people – Ralph Whiting as a consultant to help young farmers enroll in G.I. Farm Training Classes, and Marcia Nygaard to conduct numerous discussion meetings for farm wives during the summer months. He also noted the leadership of Farmers Union in organizing a five-state "Fuel Crisis Conference" as supplies of strategic fuels became scarce during the energy crisis, and how the organization successfully sued Agriculture Secretary Earl Butz to reinstate the crop disaster program which brought $25 million to Minnesota farmers that otherwise would not have been paid.

"As we look at farming tomorrow, we must not make the mistake of staring so hard and long at the problem that we overlook or fail to see, the strengths and opportunities we have," Carpenter said. "One of these is our cooperatives... .without our GTA, would farmers have had wheat to sell when the market hit $4? Or $2 for last year's corn? And without our supply cooperatives, what would have happened to fertilizer? Or the supply and price of our fuel?"

At the 1975 state convention, Carpenter reviewed a year in which hogs sold for record prices (10 cents more a pound than cattle at one point) and in which farmland prices had soared (a 160-acre Minnesota unit sold for a third of a million dollars). But Carpenter noted the bicentennial year of the American Revolution with these words:

"The spirit of '76 is not new to farmers. When this country declared its independence 200 years ago, farmers were in the front line. And they still are. They are in the front line for change, for progress, and a better way of life... Many are asking who will control agriculture? I say we are no longer asking, we are declaring: we will control agriculture – our farm families, our cooperatives and our rural communities... here tonight we are offering 'A Declaration of Entitlement' on behalf of farm families as we move into our bicentennial year...

"We do not make this 'Declaration of Entitlement' in the spirit of belligerence", but rather in the interest of people, in the service of our nation, as we look forward to a better nation and a better world... "

At the 1976 convention, Carpenter focused on philosophy and people, identifying some of the latter by name. "In the span of one lifetime," he told the delegates, "Farmer judgement and Homesteader determination have moved farming from horse and man-powered operations to the mechanized marvel of food production we have today...

"When I think of the 'Homesteaders' we have in Farmers Union, I think of the Ray Watsons, Otto Olsons, Ellsworth Smogaards, Herb Botzs, and hundreds of others standing shoulder-to-shoulder with the likes of the Tony Dechants, Jim Pattons, Hubert Humphreys, Jerry Tvedts, Barney Maluskys, M.W. Thatchers, the Frank and Ethel Krukemeyers and the Fred Groningers... all we really need to do is display the same judgement, leadership, dedication, and compassion that has been the heart and muscle of Farmers Union and our cooperatives... and, we all can be truly standing proud of our organization, of ourselves, and a better world we will help build."

IT'S MORE THAN A FARM ORGANIZATION... MUCH MORE!

"Farmers Union in total is almost more than an organization as such. In many ways it's almost an extension of the people involved, of their concerns, of their values, of their commitment," Carpenter said in an interview in 1996.

"Two things make this true – it's comprised of people who are neighborly, friendly, oriented from the lives they have lived, and people who feel a need to work together to help make things better. It's the attitude of the things behind their joining – not as much with the idea of getting visible, tangible, personal benefits, but of joining as they would in an old threshing bee or something where they know there's a need to work together and they are willing to be involved and willing to contribute something of their time and effort... to do what they would do as a farm neighbor, or farm family...

"This doesn't mean that they don't recognize the importance of specific programs, proposals, legislative action and the rest of the things we do in Farmers Union, but in their minds, they are involved as good neighbors doing what they should do to help each other.

214

"This, I think, is what made the farm cooperatives that Farmers Union people developed so successfully. It was people who believed in cooperation, because they could see it simply as coming together, working together, and finding solutions to whatever challenge might come along.

"Over a period of time they patronized and were loyal to that cooperative far more than just what should have resulted from a favorable price or a favorable situation to them personally. They favored their co-op because it represented the community, the common interest, the things they believed in, the things they stood for, people working together.

"I have observed and have been involved in a number of organizations, and I don't know of any that causes, in me, the genuine respect for the purpose and objective, that Farmers Union has."

CARPENTER'S ALLY & "EQUAL PARTNER" – HIS WIFE, FRAN

In the 1996 interview, Cy Carpenter spoke warmly of the many years of untiring support he had received from his wife and "partner" of so many years, Fran Carpenter. She, too, shared the Minnesota farm legacy, born and reared on a farm near Crookston in Polk Co. with her three brothers. They met at a Farmers Union meeting.

Consequently, Fran Carpenter, too, had the same brand of feeling, compassion, and respect both felt and expressed for the people of agriculture, sharing a close kinship to the land, and the people who worked on it, and made it their future and home.

"Fran has been an equal partner in my career as well as in my life," Carpenter said. "She's fully knowledgeable and committed to the interests of farm families as myself and others. She is the wind beneath my wings."

Fran Carpenter's chores in addition "to sharing" the outstanding career of her husband, included being a homemaker, a master gardener, working with the school lunch programs in Thief River Falls and Bloomington, helping deliver meals on wheels in Bloomington, and serving her Lutheran Church wherever Cy Carpenter's career took them over their many years together.

BATTLING AGAINST CORPORATE INVASION OF AGRICULTURE

"Fortunately, in Minnesota, there is still time to act to retain a family farm system of agricultural production... a law should be adopted... to maintain a viable, individually-owned family farm system... "
Farmers Union Statement, Mar., 1971, St. Paul Hearing,
Corporate Farm Subcommittee, House Agriculture Committee.

A call for adopting a public policy in Minnesota concerning the kind of agricultural structure desired for the state was first sounded by the Farmers Union. In testimony heard on Mar. 16, 1971, the message came clearly and loudly in favor of an "individually-owned family farm system of agriculture". The speaker was Minnesota president Ed Christianson, who then had been head of the farm organization for nearly 20 years then.

Because agriculture and farming is so important in the economic base of Minnesota," It would be appropriate for the (state) Legislature to concur in stating a family farm policy", his testimony read. It suggested that state government could do more than the federal government in fostering a family-type agriculture, including setting policies for not only land ownership but also regarding land-use, zoning, and other activities affecting the quality of the state's land, water and air resources.

Farmers Union called for "a creative partnership of people and their government" at local, state and federal levels in developing and instituting an effective policy for agriculture and rural communities. The statement presented to the corporate farm subcommittee of the Minnesota House of Representatives Committee on Agriculture noted how there was ample precedent in the actions of the U.S. Congress for taking a policy stand on behalf of family farmers, through the periodic federal farm programs, as well as through special legislation like the soil conservation act of 1936, the Agricultural Adjustment Act of 1938, the Commodity Credit Corp. Act of 1948 and so on.

The timing was such, the Farmers Union pointed out, that "the corporate take-over of agriculture in Minnesota is in its beginning stages and can still be halted if the Legislature acts at this (1971) session". A survey taken by the U.S. Department of Agriculture indicated 7 percent of the land in the United States was owned by corporations (this was land both owned and operated by foreign entities, and land where the corporate identity was disclosed by the name of the firm or firms involved).

The USDA also reported the average size of these "foreign farms" was 4,000 to 5,000 acres, far bigger than the 300 acres that then was the average size of a Minnesota farm. If Minnesota's farms were to average the size of a corporate farm, it would require only 8,000 to farm the state's 32 million acres of crops. And there would be an average of only six farms per township statewide compared to an average of up to 70 to 90 farms per township in southern Minnesota, according to U.S. Farm Census figures of 1969.

At the same time there were substantial amounts of farmland owned by foreign interests in some states like Florida (31 percent), Utah (28 percent), and Nevada (22 percent).

The Farmers Union testimony concluded with comments about the impact corporate farm take-overs could have on rural communities. The "corporate giants" would by-pass those local businesses and agribusiness depending on farmers, dealing directly with manufacturers and supplies and gaining discount prices. "Mark it well, there is no place in a corporate agriculture for most of those who now service our family farmers," concluded the statement. "The decision needs to be made now on whether we will keep our family farm system, or whether the people of the state and its Legislature will stand on the sidelines as a new type of agriculture takes over."

Then, in Jan., 1972, Cy Carpenter reactivated the Minnesota Farmers Union Citizen's Task Force on Corporate Farming. "We expect the state Task Force to spearhead the effort for comprehensive legislation restricting corporation farming when the 1973 Legislature convenes," Carpenter said in a press release.

Carpenter reviewed the 1971 Minnesota law which required farming corporations to file annual reports with the Secretary of State. But Carpenter and others felt this law fell far short of their goal of enacting restrictive legislation that would thwart agricultural ownership and activity by non-farm entities.

State Senator Winston Borden was appointed chairman of the Corporation Farming Task Force, while Milt Hakel, director of research and publications for the Minnesota Farmers Union was named executive director. Others picked by Carpenter were Arnold Onstad, conservationist and Houston Co. farmer who had served as chair of an earlier Task Force on Corporation Farming; The Rev. Giles Ekola, an Alexandria clergyman and writer on rural affairs; Meredith Haslerud, manager of the Roseau Rural Electric Cooperative; Melvin Hamann, a Rock Co. livestock feeder; David Haugo, a Mahnomen Co. grain farmer; and, Robert Tongen, local cooperative manager from Red Lake Falls.

SOMETHING NEW – "SIP-AND-SCHEME" COFFEE SESSIONS

Like many farm organizations, the Minnesota Farmers Union tended to be pretty much a male-activist organization even thought it was officially open to, and did encourage and invite female participation. However, as true in most other agricultural organizations, there was a reluctance by rank and file women in seeking larger roles in those early years. That male-activist domination continued until a novel effort blossomed during Cy Carpenter's presidency.

"Someone – probably one of our lady-members – came up with an idea of holding an informal, women-only coffee-klatch," Carpenter recalled. "It turned out to be one of the best things we could have done to get more people involved in Farmers Union."

The sessions were called, "Sip-and-Scheme", and were known to be mini-conferences for ladies but involving major issues and problems. A whole page feature story in Minnesota Agriculture dated Oct. 31, 1974, told about four of these conferences held that month at Alexandria, Granite Falls, Mahnomen, and Rochester.

"As a family farm organization, Farmers Union has always offered an opportunity for farm women to be active... ," the article noted. "A new spurt in such activity has taken place since the Minnesota Farmers Union launched an annual series of 'Ladies' Fly-Ins' to Washington, D.C., in 1965. The past two summers, MFU has held a series of "Sip-and-Scheme" discussion meetings around the state... small group meetings designed to help individuals confront the major problems besetting farm families... "

The article explained that each of the min-conferences was structured like a mock Farmers Union meeting where the members were given a hypothetical situation and asked to come up with a solution or a method to improve the situation. Topics ranged from farm income, high prices at the supermarket for food while farmer-producers struggled with low returns, spiraling farmland costs, difficulty for young farm families to obtain capital, and public relations with consumers.

Sometimes local politicians. agribusiness and community leaders and officials, met with the women at the coffee sessions. These mini-conferences attracted media attention as well, and provided another opportunity to expound on the philosophy of Farmers Union and its deep-seated dedication to family-type agriculture.

Another innovation during the early Carpenter era were "Young Farmers Sound-Off" sessions, where participants were invited to engage in discussions about farm matters, problems and the like, and, more importantly, what might be done or recommended to deal with such issues.

During an era when high beef prices attracted national publicity, including "beef boycotts" by consumer groups, and campaigns for "meatless Tuesdays" and things like that, Cy Carpenter's response at a U.S. House of Representatives hearing was simply "I'm in no mood to apologize for higher beef cattle prices". Then he went on to explain that beef producers were still receiving prices below 100 percent of parity, and that he felt more attention ought to be focused on packers and retailers and the fact when cattle prices increased by "a cent more for the farmer, they charged (consumers) a nickel more".

CARPENTER SUPPORTED "BIG THREE" CO-OP MERGER

"Farmers should carefully consider all possibilities for strengthening their cooperatives in an economy dominated more and more by large, multinational corporations... " Cy Carpenter, July, 1980, on proposed co-op consolidation.

"If this merger takes place," wrote Brian Jones in the Minneapolis *Tribune*, "It would create a huge agricultural conglomerate." The super-cooperative in effect would:

Sell the farmer the fuel for his tractor (plus tires, batteries, etc.).

Supply the fertilizer, seed and chemicals for his fields.

Then, after harvest, buy, store, and market his grain.

This was the agricultural scenario when in July, 1980, the blueprint was drawn up that would create a gigantic farm cooperative out of the merger or consolidation of three big Upper Midwest regional cooperatives all based in the Twin Cities – the Farmers Union Central Exchange (then commonly identified by its new advertizing name "CENEX"), the Farmers Union Grain Terminal Association, and Midland Cooperatives.

The newspaper article quoted Minnesota Farmers Union President Cy Carpenter as strongly in favor of the three-way merger. Carpenter urged this step, explaining that he viewed it as a means of giving farm cooperatives more economic clout in the grain marketing and agribusiness world dominated, he said, "more and more by large, multinational corporations".

Combined sales of the three in 1979 totaled nearly $3.6 billion in the states of Minnesota, Wisconsin, North and South Dakota, and Iowa. GTA, with its 150,000 farm family membership and 700 local grain elevators, had sales of $1.66 billion in its fiscal year ending May 31, 1979. Cenex had 1979 sales totaling $1.4 billion, while Midland marketed $518 million worth of supplies to 300,000 eligible members that same year.

But later in 1980, the board of directors of GTA voted unanimously against the unification plan, explaining that it was felt that "our direction has been marketing, and that is where we should be going", according to a statement made by GTA board chairman Gordon Matheson of Conrad, MT, and printed in *THE LAND*.

IN RETROSPECT: THE CAREER OF CY CARPENTER

It wasn't the best of times to become president of Minnesota Farmers Union when former farmer Cy Carpenter took office in 1972. He inherited an era when farming and agricultural production were in the grips of likely the greatest change in history.

The wartime-spurred farm markets of World War II and the Korean Conflict, coupled with the post-war rebuilding of Europe and Japan, was gone. Simultaneously, farming itself was in the throes of a technological revolution – new, bigger machinery, improved seeds and fertilizers, new families of chemicals that could kill weeds or add pounds to growing livestock invaded the farm sector. And production exploded.

"This new ability to over-produce, in the face of declining world demand, and our gross inadequacy to deal with world hunger in developing nations, drove down farm prices – often below the cost of production," Carpenter wrote years later.

"With razor-thin margins of profit – if any – farmers had to scramble in every conceivable way to meet expenses. This meant increased volume, adding acreage – which drove up the price of land to buy or rent to unrealistic artificial and unsustainable levels. We saw many farmers fail, or quit. Some young farmers listened to the 'professional' farm advisors who said, 'Get big or get out', and over-extended in buying land or machinery or both."

Then followed a difficult time when farm foreclosures were commonplace, and the number of farmers declined dramatically. The government made what Carpenter described as "limited efforts" to control production and price support programs, while the "political voice" of farmers eroded and the influence grew for agribusinesses and international conglomerates bloated by profits from farm over-production and low prices.

FIRST DROP IN FARMLAND VALUES IN 22 YEARS

What might have been the first real indication that severe economic problems for farmers was occurring came in a Feb., 1982, U.S. Department of Agriculture report on farmland values. Minnesota's average value of an acre of farmland declined by 3 percent in 1981 – the first decline in land values in 22 years.

Further, it was only the second time that Minnesota farmland values had gone down since the Great Depression of the 1930s, according to noted University of Minnesota economist Philip Raup. Farmland had declined from an average of $1,231 an acre to $1,197 from Feb., 1981, to the same month a year later, or 2.8 percent. Previous statewide declines had taken place in the 1950s (from $107 to $105 between 1952 and 1953, and from $157 to $155 from 1959 to 1960).

The Minnesota decline of nearly 3 percent was matched by the drop in neighboring Wisconsin, and was small compared to other Upper Midwest states. Farmland values had declined 15 percent in Ohio, 13 percent in Indiana, 7 percent in Iowa and Missouri, and 5 percent in Nebraska. There was no change reported in South Dakota, while North Dakota values represented the only increase in the midlands (up 3 percent).

During that same year, in Mar., 1982, Minnesota President Cy Carpenter came up with a three-step proposal to help "pull agriculture out of its present crisis condition" in a statement given to Congressional agriculture committees:

An immediate hike of 50 cents a bushel in the Commodity Credit Corp. Loan program and the reserve program as an incentive to farmers to participate in USDA cropland set-aside programs.

Provide an additional $600 million in loans to be made by the Farmers Home Administration (FmHA) that had been authorized by Congress but which was being held up for budgetary reasons by the Reagan Administration.

Permit the deferral of one-half of each payment due on FmHA loans for up to five years, based on the discretion of local Farmers Home officials. Carpenter told Congress that his proposal was based on a recent series of public meetings held across Minnesota with farm people, bankers, business people, farm managers and credit experts. "It's been made clear that improvements must be made in the farm program and in credit available to farmers to provide some minimum income stability for farmers and thereby the rural businesses they patronize and depend on," Carpenter said. "It's clear also that these things should be done immediately to allow farmers time to arrange for their capital needs before spring fieldwork starts."

CARPENTER CRITICIZES AG TEACHER CUTS

When Farmers Union learned of planned or actual cutbacks in vocational and adult agricultural education programs in Minnesota in May, 1982, Cy Carpenter and others became immediately concerned.

In a letter to Board of Education Members, and reproduced in Minnesota Agriculture, Carpenter wrote: "We must not miss the fact that the economic problems facing our state and nation today were preceded by a troubled farm economy, just as in times past. As agriculture goes, so goes the nation and – perhaps even more rapidly – so goes Minnesota.

"The most significant difference between the present growing farm recession and those of the past, is that this one is almost certain to take out a much higher percentage of the remaining family farms than any in the past... one business on main street is boarded up for every six farmers who go out of business in that community... school and community are diminished... Our first question is: Can we afford (this)... We are alarmed that adult agriculture and vocational agriculture have been among the first casualties in some rural school districts faced with budget problems... agriculture requires a high level of skills and training... those programs are the source... every consumer has a stake in a healthy American family agriculture... in these very trying times, community understanding, appreciation and support could prove vital to the future of family farming... "

CARPENTER CALLS FOR REAGAN FARM CHIEF TO RESIGN

Saying that "sooner or later almost every (U.S.) Secretary of Agriculture is called on to resign" that post, Cy Carpenter called for the resignation of President Reagan's farm chief, John R. Block, in Aug., 1982.

Carpenter cited several reasons why Block, who took the job in Jan., 1980, should step down, including a failure to negotiate a new trade package with the Soviet Union and the fact that local grain elevators were being bypassed in the marketing of government-owned, farmer-stored grain.

220

"It is a necessary act to call attention to the problems have," Carpenter explained about his statement urging Block to resign. He added that his intent was not to be "just another voice" critical of Block. "But the facts make it all too clear that something dramatic must be done to gain recognition of the magnitude of the problem and the vital need for solutions right now," Carpenter added.

A resignation would "draw attention to the fact that present conditions can no longer be tolerated by the Secretary of Agriculture or by American farmers", added Carpenter. In the end, Block was to resign the post several years later, and after the farm crisis of the 1980s became the worst economic period for farmers since the Great Depression of the 1930s, and after the Block-Reagan PIK (Payment-in-Kind) Program of 1983, and after the Farm Bill of 1985 was enacted.

When the PIK Program was announced (at the American Farm Bureau convention in Dallas), Carpenter immediately called for public hearings to be held to better acquaint growers with the details of the cropland-idling program never used before – or, as it turns out, not since 1983. PIK provided options to corn growers of idling one-half of their base acres and in some cases bid to not raise crops on any eligible acreage. In response to Carpenter's concern, U.S. Senator Boschwitz held three PIK hearings in southern Minnesota.

GAPP WITH GATT?

In 1988, shortly before he left the presidency of National Farmers Union, Cy Carpenter and the organization suggested and promoted a new proposal – "A General Agreement on Agricultural Production and Prices", or GAPP. This was a plan to be used in conjunction with the General Agreement on Tariffs and Trade, or GATT.

GATT for 40 years has been the mechanism for periodic international meetings where trade policies and problems have been addressed. The seventh round of GATT was the Uruguay Round begun during the Reagan Administration. Farmers Union considered GATT "weak" and ineffective "in coping with the chronic, fundamental problems of excess production and low prices in worldwide agriculture and the lack of sufficient food-buying-power in the hands of one-half of the world's population".

Farmers Union saw in GATT "structural weaknesses, loopholes, waivers, exceptions and a lack of enforcement powers". Further, GATT "has no track record in raising the level of world farm market prices and stabilizing them within a fair range". Consequently, the organization under Carpenter's leadership came up with its GAPP proposal which it felt could be used in conjunction with GATT.

"Let GATT continue to harmonize trading policies, and let GAPP address itself to measures to lift world farm prices to levels fair to producers in both industrial and developing countries and to consumers everywhere, while augmenting the food-buying capacity of the hungry nations... by addressing and solving the production and price problems of world agriculture, GAPP will eliminate some of the contentious problems causing great difficulty in the GATT talks and negotiations... "

In 1988, when he became 65 years old, and had reached the age where National Farmers Union bylaws precluded being elected, Cy Carpenter did not seek re-election to the presidency he had held for four years.

Instead, he successfully proposed eliminating this section of the bylaws. However, then Carpenter chose not to be the first to apply under the newly-enacted change.

Cy Carpenter was not ready for retirement at 65, for years the conventional time of life when many are ready to take things easy. In his post-presidential years he has continued to be a Farmers Union activist, serving as chair of the Minnesota Farmers Union Foundation, and a consultant always available to help.

One of his projects in recent years was helping direct the history project for Minnesota Farmers Union, putting together an accounting and record chronicling the events, the people, and the movement that gave birth to a period of unparalleled growth for the farm organization and the cooperatives it fostered. Cy Carpenter himself was a historical resource, having lived a career of Farmers Union from Stearns Co. to the office of National Farmers Union in Denver, CO, and having personally known so many of the leaders and pioneers involved in the movement.

9 ...A TIME OF CHANGE, OF LITIGATION...

"Bushels and dollars will 'show trends and variations' from year to year, but we remind you, members do not... "
Committee for Member Input, 1981.

"They're challenging the (GTA) decision because they say their rights as voting members were violated... "
Farmers Union Official F. B. Daniel, Mankato *Free Press*, June, 1981

"In this Northwest territory... cooperatives for the most part carry the Farmers Union name... and have grown and flourished where the Farmers Union Education Organization has carried on educational work... "
E. A. Syftestad, General Manager, Farmers Union Central Exchange, North Dakota Farmers Union Convention speech, 1944.

"There's a symbiotic relationship (between Farmers Union and Farmers Union Cooperatives)... in which each ... has grown better and stronger... than it would have done on its own... "
Tony T. Dechant, president, National Farmers Union, 1966-1980.

It is unlikely that years earlier few if any farmer-members of the biggest, most successful grain marketing cooperative ever envisioned the day when some in their ranks would bring a lawsuit against that organization. Nor that these same unhappy members would charge the cooperative with misrepresentation, and with violating its own policies and bylaws.

Few, if any, could have foreseen the day when many cooperative patrons might feel the bitter sting of what they considered was a breach, perhaps even betrayal, of their trust.

Yet, that is precisely what happened in 1981 within Farmers Union Grain Terminal Association, in the highly successful farm cooperative's 43rd year of operation. It was a profound but disturbing development to many dedicated, long-time cooperative faithful.

It became an era indelibly – and painfully – etched in the memories of some Farmers Union partisans, members and officials who were involved at the time, and to those who knew the history of the two organizations, and who felt the final outcome was not the best resolution of the matter. Years later these cooperative partisans find it difficult to talk about this troubling time without noticeable traces of mixed hurt and even anger in their voices.

Out of the controversy from that frustrating, litigating episode came a major change in the productive, long-standing relationship of the state Farmers Union organizations and this prospering regional cooperative.

That's because from the very beginning, the bylaws of these two Farmers Union cooperative businesses provided for an allocation of a partnership-like percentage share of annual earnings to respective state Farmers Union units in the states in which they did business. The bylaws prescribed for "a sum determined by the Board of Directors (of GTA) equal to, but not to exceed, five per centum (5%) of the gross receipts from member patronage... "

Bylaw Section 1(b)(8) concluded with this provision: "(this) sum (5 %) shall be used for the purpose of promoting and encouraging cooperative organization".

Legal language notwithstanding, that meant the Farmers Union Grain Terminal Association clearly required 5 percent of the cooperative's earnings to be allocated to state Farmers Union units for co-op education, promotion and such activity which would encourage cooperative organization as they saw fit to undertake. This procedure began in 1938 when GTA was formed, and continued without question or interruption until late spring, 1981.

NO EDUCATION FUND POLICY, NO ORGANIZATION AT ALL?

Records show, however, that the first major cooperative to allocate 5 percent of its earnings to Farmers Union state organizations in which it did business was not the grain co-op, but Farmers Union Central Exchange. When the Exchange was incorporated in 1931, the incorporators specified in the bylaws that 5 percent of net earning, prorated, would be allocated for payment to each state Farmers Union as designated by its board of directors. Then, when Farmers Union Grain Terminal Association was incorporated seven years later, incorporators made such a 5 percent allocation mandatory.

No one in those early days, of course, could have foreseen the time when these budding cooperatives of the 1930s might be billion-dollar-a-year businesses, and that the 5 percent share would represent huge annual checks payable to a single Farmers Union state organization. Like, 35 years later when the Grain Terminal Association would mail three education checks totaling over $1.7 million to the Minnesota affiliate like happened in 1974-76. That included one check for $820,188 in one year – 1974 – from that one regional cooperative.

Nor, possibly, did any of these early patrons-members ever think as well that some day the sanctity of the 5 percent bylaws would ever be in jeopardy?

After all, this principle of sharing was developed early in the development of the cooperative movement and that of the Farmers Union. A. W. Ricker, the first editor/founder of the Farmers Union *Herald* and a member of the Northwest Committee which blazed the path organizing Farmers Union affiliates, had written about this special, intertwining relationship in the late 1920s in *The Herald* Feb. 2, 1930 (the year before Central Exchange was incorporated): "Had not the Farmers Union here in the Northwest adopted a policy of using the earnings of its business activities for organization, we would today have no organization."

In a column about those early years published in 1988, F. B. Daniel wrote: "Labor Union men will pay liberal dues, but farmers have never done so. Organization and educational funds must, therefore, come from the only other place where funds may be obtained, namely, out of the earnings of the sales organizations, insurance, commodity purchasing associations, and local cooperatives, etc... "

Daniel added, "This policy, which we adopted in the Northwest Division of the Union, of setting up our business activities so that a part, or if necessary, all of the earnings may be used for organization, is what has made it possible for us to organize... we must continue this policy until such time as our members will pay dues into their organization in sufficient volume to finance education and more organization... "

MANY USES FOR "CO-OP EDUCATION FUND" DOLLARS

Such earnings historically apportioned to the local units commonly were called "cooperative education funds", and were used in general as the bylaws required to promote cooperatives, cooperative participation and broader understanding and appreciation for the role these organizations played in marketing farm commodities and making available to them farm supply and input items needed by farmers.

The expenditure of this money was handled completely by the respective state Farmers Union organizations for such things as conferences and meetings, summer camps and winter seminars, educational and legislative trips by bus and plane to the state capitol and to Washington, D.C., as well as a variety of activities including the publication of its own newspaper.

Initially, these funds represented a strategic and important source of finances but provided sort of inconsequential amounts of money. However, as the cooperative movement mushroomed – with that expansion and growth in large measure fueled by the wise and studied use of "education funds" – critics of this bylaw provision emerged. And at high cooperative places. These new cooperative leaders apparently felt that there was a need for significant, and controversial change, and questioned the future funneling of hundreds and hundreds of thousands of dollars into the coffers of one general farm organization.

So, with an emphasis on secrecy, a plan for change was put into operation. Even many of the most faithful cooperative patrons had no idea such a proposal was in the works. No, not in late spring when GTA matters were without fanfare and public pronouncement and usually placed conspicuously on the front burner. But 1981 was a different time. Truly much different. And a year destined for litigation, lawyers, and the courtroom.

THEY SOUGHT MORE "BUSINESS-LIKE USE" OF FUNDS

These critics of the time-honored system saw a need to end the "mandatory 5 percent of net savings", according to an "explanation of proposed amendments to the (GTA) bylaws" which accompanied a mail ballot. In a letter sent on May 18, 1981, to GTA members by Donald F. Giffey, secretary, he noted that this source of money had provided nearly $10 million distributed to state Farmers Union organizations the past 10 years.

This translates into $2 million annually to be divided among five or perhaps seven states (North Dakota, South Dakota, Minnesota, Montana, Wisconsin, Iowa and Kansas), or less than $400,000 per state. That left $950 million, or the remaining 95 percent, for use by GTA for expansion, improvements or whatever, during that same decade.

This total of $10 million noted by Griffey included $1.6 million in the most recent year, 1980-81, the explanation read. Then the printed information continued:

"These (educational funds) payments come at a time when members and patrons recognize the following changing needs for the most business-like use of the cash funds of the (Farmers Union Grain Terminal) Association:

A need for consolidation, replacement and/or expansion of grain marketing facilities and processing plants – this at a time of high interest rates.

A need for development of new and existing markets in the U.S. and abroad.

A need for improving equity redemption program of the (GTA) Association.

A need for a more business-like approach in the distribution of promotional funds, and for discretion of the board to make those determinations."

A printed explanation accompanying the June 10 mail ballot said in part that the vote "will still allow the Board of Directors to contribute to Farmers Union organizations for cooperative promotion". However, it continued, "Large compulsory distributions, not targeted to the areas of greatest need, will no longer be required... instead the Board will be making distributions consistent with the growth of the GTA and the best interests of its members and patrons... "

Later in the prepared explanation, it was stated, the "present mandatory deduction and disposition of cooperative promotional funds is objectionable to many members, patrons and potential patrons... these amendments providing for discretionary amounts and distributions will attract new supporters and be beneficial to all members and patrons of GTA... "

It wasn't spelled out who were some of "the many members, patrons and potential patrons" who found the relationship in effect for 40-plus years so "objectionable". Nor did it identify the "new supporters" who would be attracted.

Then, the final paragraph announced that GTA's Board of Directors "unanimously supports the proposed bylaws amendment and urges that you use your Mail Ballot (enclosed) to vote in favor of the resolution."

The vote would come at a "special meeting" of the members of the Farmers Union Grain Terminal Association June 10,1981, at GTA's headquarters in St. Paul, according to the letter mailed to delegates by Secretary Giffey dated May 18. It was made clear that the proposed bylaw amendment was "the only issue to be voted on" and that "for economy and convenience, you are urged to use the Mail Ballot".

Giffey's form letter sent, it appeared, to all voting delegates, stated bluntly the same message that had been included with the ballot: that there had been an "unanimous" recommendation by GTA's board that the proposal be approved.

Giffey's letter also contained "Exhibit A", where the resolution for amendments to the bylaws contain changes marked by deletions and additions. Significant to Farmers Union partisans was the deletion of all of Section 10 which spelled out how the special funds were to be disbursed.

Eliminated entirely was this paragraph from the GTA bylaws:

"The sums provided... shall be paid to state unions chartered by the Farmers Educational an Cooperative Union of America which are actively engaged in promoting and encouraging cooperative organization in the interest of this (GTA)

Association in the proportion which patronage of this Association originating in each state bears to the total patronage of this Association."

Also deleted was the wordage "equal to" which required the GTA board to determine "a sum... equal to but not to exceed five per centum (5%) of the gross receipts from member patronage... "

SECRECY PREVAILS AS VOTING MATERIALS ARE PREPARED

What was not known generally at that time was the fact that highly unusual and irregular activity had occurred in GTA's office while these "special meeting" items, including the mail ballot, were being printed and prepared. Reports later indicated employees were instructed "not to read" any of the printed material being prepared there, and that a top-secret kind of atmosphere had been created regarding the upcoming "special meeting" then just a month or so distant.

The secretive, hush-hush activity was definitely not in keeping with past open-door, public kind of atmosphere that normally prevailed at 1667 North Snelling Avenue in Falcon Heights, MN, the home office of the regional cooperative.

A big question, too, was why such a "special meeting" would be called during early June – traditionally one of the busiest times of the year on Midwest farms. Also, why the traditional post-harvest business meeting schedule that had been used previously was being completely ignored.

However, secrecy could prevail only so long on such a historic proposal. And once word leaked out what was at stake, there was hasty reaction. However, even months earlier, some had anticipated problems. A meeting at Moorhead was arranged by those irritated by the subterfuge surrounding GTA even before the recommendation for amending such a major bylaw. Out of that meeting came the formation of a new group, "Committee for Member Input".

"This (CMI) group is furious over the way the GTA's board of directors conducted a special vote to eliminate mandatory annual donation to the Farmers Union educational fund," wrote reporter Michael Flaherty in the June 11, 1981, editions of the Mankato *Free Press*.

"GTA was established in 1938 with 'seed money' donated by the Farmers Union... bylaws passed that year required that GTA contribute 5 percent of its net profits to the Farmers Union for its educational programs such as summer camps, its weekly newspaper, cooperative education programs, and legislative lobbying and research efforts... "

What the Mankato newspaper account didn't report was the fact that back in 1938 when the Farmers Union Grain Terminal Association was created, the track record for grain cooperatives hadn't been all that good. Time after time in the 1920s and early 1930s, and even earlier in the century, cooperative after cooperative had gone bankrupt, leaving behind indebtedness, distrust and natural skepticism about any new effort to organize a new marketing organization.

And on the heels of the Great Depression of the 1930s there was likely little reason for much confidence that any new grain cooperative would be any different, nor that someday it might become the great success it did.

So when Farmers Union provided the "seed money" to launch yet another new grain cooperative where others had failed time after time, it obviously was not without some question and uncertainty. And, surely there was no inkling shared by

227

the founding fathers that some day it would be so successful that the 5 percent share of its earnings mandated for "cooperative education" would surpass $1.5 million in a single year.

So in 1981, it is likely few on the GTA board of directors could recall how things really were when GTA was more of a dream and hope than anything else. Nor how dedicated GTA's greatest manager, M.W. Thatcher, was to the concept of this new cooperative working hand-in-hand with the Farmers Union. "In this economic war," Thatcher had told GTA delegates in 1949, "our greatest associate and ally is the National Farmers Union and its state units... "

Later, in 1953, he would portray how "people (are) bound together just as GTA is, by a great farm organization, the Farmers Union" and how "their common objectives are the same."

Such a deep-seated feeling for Farmers Union and its educational efforts to organize more cooperatives was at the roots of Section 10 – the original bylaw section which spelled out that the 5 percent share of patronage funds annually go to state-chartered Farmers Union units in the proportion of business originating in each state.

But a generation or so later, in the early 1980s, this same feeling about the working partnership between the grain-marketing cooperative and the Farmers Union which first breathed economic life into it, for some reason didn't exist at the top echelon of management in St. Paul. Instead, a plan to rupture this long relationship was devised under a cloak of secrecy and surprise, and scheduled to come to a vote at the time of year when the main business on the minds of farmers was farming – not amending the bylaws of one of the nation's super-cooperatives then marketing more than 260 million bushels of grain in a year for farmer-members in eight states..

One of the GTA members upset by the proposal and the way it was handled was Leo Fogarty of Belle Plaine, who became co-chair of the newly-formed Committee for Member Input. He shared the feeling of many GTA members that the election was not held properly and that such an important matter should be voted on by elected delegates in the organization's annual November meeting, rather than in a hastily and clandestinely arranged mail ballot.

Where some critics of the bylaws amendment (passed by what was described in the Mankato newspaper as "an overwhelming majority") saw the move as one to disassociate GTA from Farmers Union, Fogarty saw it differently: "Five percent of the profits today is a lot more than it was in 1938." However, he added, "But then we've got many more people, and a dollar today is not what it was back then."

Fogarty's group, the CMI, took the matter to court. The lawsuit was filed in Ramsey Co. about 10 days after the controversial "special meeting". In part, the complaint claimed the actions by GTA "were an unfair, unauthorized, but carefully orchestrated scheme (by GTA management) to wrest control of GTA from its members". CMI also asked the court to declare as "null and void" the bylaws proposal vote, and "to enjoin" GTA from further such special meeting-mail ballot activity.

The June 18 edition of *Minnesota Agriculture*, the state Farmers Union weekly newspaper, reported how "a growing group of farmer-patrons of GTA has vowed to seek a full court hearing on their charges that member rights have been violated by GTA management" in the way the bylaws matter was handled.

This publication told how Judge E. Thomas Brennan had ruled on June 8 – two days before the controversial special GTA meeting that the time interval was too short to give proper attention to all aspects of the case. And he declined to halt the June 10 GTA meeting.

It was at the June 8 court session that GTA attorneys responded to some of the charges in the lawsuit. Many of the complaints by CMI, it told Judge Brennan in a memorandum, were "blatant hearsay". GTA said it had met legal requirements of notice by publication in the St. Paul Legal Ledger and that it made additional notice in the *Co-op Country News*. "Every effort to promote fairness and give notice to a wide range of persons was made", GTA told the court.

It was also claimed that "individual members of (GTA) affiliated associations are not members of GTA" and that there had been no discrimination between voting abilities of members of GTA patrons' associations and members of affiliated associations" as charged in the lawsuit.

Further, it defended the time of the special meeting noting it corresponded with GTA's fiscal year. Regarding the charge that employees lobbied "yes" votes on the bylaws proposition, the memorandum stated that GTA denied instructing employees to influence votes although it could not "state with certainty that one or more of its employees has never acted outside the scope of their authority".

DANIEL ON SPECIAL GTA ACTIONS: "DAMNED MISLEADING"

One of those still upset by the GTA actions many years after they occurred in 1981 is career Farmers Union employee F. B. Daniel. He also was a GTA patron at the time of the bylaws lawsuit. When interviewed in 1996, Daniel still terms the controversial change as akin to a "betrayal of trust".

About the farmer-members suing their own cooperative, Daniel had this to say: "They were challenging the (bylaws) decision because they felt their rights as voting members were violated." Daniel said the legal battle was not the decision the board made to change the way funding of Farmers Union is determined.

In a lengthy deposition taken Oct. 5, 1981, F. B. Daniel said, "It doesn't take a great mind to figure out that this time (May-June) is a very inconvenient time for farmer-members to take time out to study and to vote on a bylaws change... "

Concerning the argument that the GTA funds could be "better used" elsewhere, Daniel stated, "They were damned misleading, to suggest that those funds could be better used for facilities and so on. I think that was... bologna... those are net savings... you can't use your net or expend it before you have it... now, you could carry it over to next year, and expend it that way, but the suggestion that this amount of money is going to be significant in new facilities, and so on, is pretty misleading... "

About the use of GTA fieldmen dispatched to influence the bylaws vote, Daniel asserted, "There are two areas of activity pertaining to cooperatives in which employees have no damn business – writing or amending policy and bylaws, and the election of directors. These are member concerns, pure and simple! When they (GTA fieldmen) get into member matters, that's serious."

Also, Daniel felt strongly that "employees calling on individual delegates and suggesting how they vote – and even persuading them how to vote is something clearly and completely outside the realm of the responsibility of an employee". He added in his deposition, "That is not what they were hired for. Their job is to help run the cooperative... "

The long-time Farmers Union official also criticized how the "legal notice" announcing the special meeting and the proposed bylaw was published in the St. Paul *Legal Register*, and not in a publication normally received by farmers. "To the best of my knowledge not one single (GTA) member receives this publication, and (besides) that publication is at great variance from what has been traditional, namely double-publishing such information in the (cooperative) newspaper all members receive... " Previously such official notices have appeared twice in the *Co-op Country News* or its predecessor newspaper, Farmers Union *Herald*, Daniel testified.

Further, Daniel noted that publication of this special meeting notice came in spring, a time when farm people are extremely busy and do not have much time for reading.

Daniel, of course, hated to see the downsizing of the education funds. "They are needed for a continuing, maximum effort, in a way of member education in cooperative matters, in cooperative promotion. Limiting these funds directly affects the capability to do that job."

He saw the bylaws proposal and the way it was handled as "a very clear-cut violation of member rights... (something) of concern to us all... it was pretty evident when we found out what was going on, that there wasn't even a remote possibility that we could alert enough people to realize the significance of this particular action... it was evident that we couldn't do that because the voting was already in progress, and some were marking the ballots long before the word ever got out to members what was at stake here... so the possibility of reaching them (in time) and accomplishing anything was pretty remote... "

F. B. Daniel recalled that it was about May 19 when he first learned what was going on from concerned members who telephoned from opposite ends of the state. Both callers wondered what was going on, Daniel said. Both had received mail ballots that day, and both had been contacted personally by a GTA employee "soliciting their positive vote, or a 'yes' vote, on the ballot and were urged to get it (the marked ballot) in the mail promptly".

Then word of what was described as "a top-secret operation" in GTA's offices was leaked to Farmers Union leaders, Daniel indicated in his court deposition. GTA office workers were puzzled by unusual orders quite unlike anything done previously, such as "not to read any of the materials being printed and mailed out of the office". The instructions not to read, included even waste paper, and scraps of paper, used printing plates, etc., at the GTA printing office – a highly irregular, unprecedented development, the staff members confided. Of course, all of the emphasis on secrecy and the special instructions naturally piqued the curiosity of these office workers who had never been a party to anything like what was going on in the GTA office in the spring of 1981.

Daniel also made it clear that he felt "it is not bushels of grain, but the members who vote in cooperative matters... commodities don't vote; members do."

An employee of the Farmers Union Grain Terminal Association, Frank Bergenheier of Dickinson, ND, admitted in a deposition that he had promoted the controversial bylaws change which ended the automatic, mandatory 5 percent educational fund allocation.

Bergenheier began working for GTA six years before the vote in June, 1981. He was an assistant regional manager or fieldman who had responsibilities for working with four "line" elevators and a dozen GTA-affiliated grain elevators and "helping them in any way that I could".

He told how he had attended a meeting at Fargo Holiday Inn held in mid-May, 1981. One of those who addressed this GTA meeting was the president and general manager, B. J. Malusky.

He told the group of GTA fieldmen and others that the board of directors "was behind it (the proposed bylaw change) a hundred percent", Bergenheier recalled.

The fieldman said he had told the boards of directors of GTA-affiliated organizations that the bylaws change was recommended by the GTA board of directors and that he (Bergenheier) recommended it as well. Then Bergenheier added, "I might have even said that I felt the GTA board was intelligent enough to make a decision of what they should pay in educational funds, and to who it should go."

Bergenheier later said, "The (GTA) bylaws read that it was mandatory 5 percent and that it had to go to (Farmers Union) state associations. I believe the board should be able to decide this." When asked where he made this statement, the fieldman responded that it had been at Elgin, ND He recalled this meeting because, he related, there had been a tie vote. The deadlock was broken after his comment about the GTA board supporting the change one hundred percent and that it "was intelligent enough" to know how to use its earnings.

After he spoke, one Elgin director changed his vote and the bylaws change was approved by three to one opposed. But the official vote was recorded that all Elgin directors had voted "yes", Bergenheier said.

He told how the Elgin cooperative elevator had financial problems because of sustained losses in its operation, and that GTA had kept it from bankruptcy through a loan of about $300,000. The plight of this North Dakota elevator was so bad that the St. Paul Bank for Cooperatives would no longer finance them, Bergenheier added. Then GTA came to the rescue, and in essence, took over control of the Elgin elevator.

Bergenheier recalled that the GTA loan was mentioned at the Elgin board of directors meeting where he lobbied for approval of the educational funds bylaws change. "I said that part of the reason for the change in bylaws was if GTA wanted to pay a debt, they could use the money to pay the debt instead of putting a million-six-hundred-thousand dollars into educational funds in one year... " He added later, "When you have earnings, maybe at times it would be better to pay on the debt than it would be to go to work and pay it out in educational funds or whatever... "

When asked by CMI attorney Gerald E. Sikorski point-blank if he had not only recommended the ending of the 5 percent mandatory educational funding, but had used as an example allocating educational funds to pay back their the debt the Elgin elevator owed to GTA, Bergenheier responded that he had, and added, "Or to whoever they owed money to."

CONCERNED GTA MEMBERS ORGANIZE

There had been some cooperative and Farmers Union members who suspected something was in the works several months earlier. That attitude was reflected by the meeting held in Moorhead in November, 1980 – a good half-year before the "special meeting" where the historic bylaws change was put to a vote – where "The Committee for Member Input", was created.

It was at this meeting that F. B. Daniel and several others learned that the new committee was organized because of what was perceived as a serious "communication problem" between grassroots membership and the cooperative's hierarchy relative to a reapportionment of the representation on GTA's board of directors. It was feared by old-time and veteran members that what was happening meant a loss of traditional support from their regional cooperatives coupled with a growing unresponsiveness of elected leaders on the boards of directors and by a new generation of co-op managers who sometimes appeared both aloof and distant from member-patrons.

It was an uncomfortable scenario that was in sharp contrast to the warm, close relationship that had existed in the formative years of the cooperatives when their futures were cloudy and unsure, and the era when those same fledgling cooperatives had to battle the grain trade and other special interests opposed to cooperatives, and their efforts to gain better returns for farmers by cooperating with their neighbors and fellow producers.

In short, in 1981 there wasn't a M.W. Thatcher on the other side of the fence-row working for a better deal for both cooperatives and farmers as there had been back in the 1920s, the late 1930s, the 1940s, the 1950s and the 1960s. Gone in the spring of 1981 was a leader who saw benefits for both through a bonding, and a special relationship between cooperatives and the Farmers Union. A partnership mutually profitable to both, one that had spanned the years, survived criticism, courtroom attacks, the challenges of the marketplace, competition from international grain firms and even from defectors and disloyalists within.

Instead, it was a courtroom battle with angry farmer-members suing the cooperative they helped create – a day in court many years earlier neither party would ever have envisioned. Sue your enemies, like the grain traders who had denied cooperatives a seat on the Minneapolis Grain Exchange. Or take to court those opposed to a union of farmers marketing and buying collectively and cooperatively.

But sue your own cooperative? That was unthinkable. It was anyway, until 1981, that is.

A staff written story published in the Mankato *Free Press* on June 11, 1981, summed up the litigation between cooperative members and the Farmers Union Grain Terminal Association this way:

"The group calling themselves the Committee for Member Input charges in its lawsuit that the (June 10 bylaws amendment) election was improper in many ways:

"The special vote was set up and passed... during the time farmers are the busiest...

"That didn't give members time to properly consider the matter.

"GTA employees were used to attempt to influence delegates in the mail voting process.

"Farmers did not receive advance notice of the election as required by the co-op's bylaws.

"The board of directors (of GTA) did not act in good faith in presenting the issue... and the ballots had been printed secretly, the mailed out with orders to quickly return them.

The Mankato newspaper quoted an unidentified GTA spokesman as "noting that one reason for the change in (GTA's) bylaws is that fewer than half of the (GTA) members are Farmers Union members", a statement Farmers Union loyalists strongly contend is untrue, and should have been challenged.

Also, the newspaper reported that in 1980 GTA donated $1.6 million to the educational funds of Farmers Union in the states where it did business. It noted the Minnesota Farmers Union received $650,000 as its one-year share, making it the biggest state-chartered Farmers Union originator of grain marketed by GTA.

The Mankato newspaper also told how the Committee for Member Input had sought to stop the counting of ballots on June 10 at GTA's headquarters. However the District Court Judge refused CMI's request, ruling that any election result could be undone in later action either by the cooperative or the court.

In the lawsuit the list of plaintiffs included Leo and Marion Fogarty of Le Sueur Co., both members of the Farmers Union GTA of Le Center, MN, and both officers in the CMI organization. Leo was co-chairman while Marion was secretary. Other plaintiffs were Alvin Anderson of Epping, ND, a GTA member for 40 years; Roger Johnson, a member of the Durand Cooperative for more than 25 years and a farmer in Pepin Co., Wisconsin; Bruce Anderson, secretary of the Farmers Union Grain Co. Of Epping, ND; Werner Hehn, a GTA member for more than 30 years and president of the Elgin Farmers Union Elevator Co. in Elgin, ND; and, Jerome Delvo a member of the Langdon GTA, in North Dakota for 32 years.

The lawsuit pointed out that the defendant GTA was operating in a dozen states, with a network of 16 regional marketing offices, 21 feed plant service centers, 644 local affiliated cooperatives and GTA "line elevators", and 114 Great Plains wholly-owned subsidiary yards. GTA marketed 360 million bushels of grain nationally and internationally for thousands of American farmers "whose farms stretch from Wisconsin on the east to the Pacific Ocean on the west".

CMI charged that GTA provided no explanation for the special meeting called June 10, 1981, and did not indicate any "emergency" condition which warranted such "unprecedented action". The lawsuit charged that "the entire procedure surrounding the preparation of the one printed notice was conducted in a highly secret and clandestine manner by administrative staff members of GTA... management was so concerned that members would learn of the special meeting and the voting procedure proposed that employees charged with printing notices were forbidden to read them... the notices, scraps of printed paper, the plates of stencils... were wrapped in plastic and removed by members of the administrative staff of GTA... GTA directors themselves were not allowed by management to remove minutes from the board of directors room... "

It was also alleged that "certain GTA managers, fieldmen and employees were instructed to and did influence the voting process by actively garnering votes, making statements perceived as threats and physically marking and mailing ballots for certain delegates". And, "contrary to GTA policy, GTA agents have misled patrons and others... " Further, it was charged that GTA employees had called "upon sympathetic delegates" for votes at the same time formal notice and ballots arrived, and before the general membership of GTA was informed.

CMI felt that Minnesota law had been violated since members were denied "effective notice" of the meeting and the voting procedure, that GTA bylaws had been circumvented, and that GTA did not follow "historic practice" and had failed to see to it that "unofficial notice" of the special meeting was posted in a conspicuous place for the benefit of patrons and in facilities frequented by patrons.

In short, CMI charged that the subterfuge and secrecy was part of "the calculated scheme of the GTA management", and that GTA had committed unlawful acts to achieve its goal.

The 11-page complaint asked the Ramsey Court to rule as null and void the election results on the controversial bylaw change, and at the same time enjoin GTA from any similar meeting-voting procedure in the future.

The court never had the opportunity to rule, however. That's because on Oct. 23, 1981, an out-of-court agreement was reached. While it did not undo the elimination of the 5 percent education fund, the agreement did end the lawsuit and spelled out in detail how GTA would call meetings and amend bylaws.

The *North Dakota Union Farmer* on Nov. 26, 1981, carried a report concerning the bylaws change and the settlement reached by GTA and CMI approved by delegates to GTA's annual meeting in Minneapolis. "The bylaws change left the payment of educational funds completely up to the discretion of the GTA board of directors," read the newspaper account. "Previously, the bylaws had mandated a 5 percent educational fund payable to the Farmers Union organizations in which GTA operates."

About the agreement which ended the lawsuit, the newspaper reported, "As part of the settlement, the GTA board has agreed to offer an amendment to the cooperative's bylaws:

"Future articles and bylaws changes can only be made by delegates assembled at the annual GTA meeting, unless compelling business reasons require action at special meeting... "

"Two official notices of proposed article and bylaws changes must be published prior to the vote in GTA's official publication, or as an alternative, by a personal notice to each GTA member".

"GTA employees are prohibited from selecting delegates to future GTA annual meetings."

"Votes cast by affiliated local co-op boards of directors on behalf of the local co-op's members will be compiled in proportion to the affirmative and negative votes cast by each local board member... "

But the adoption of these bylaws changes may not have been the biggest news coming out of GTA's annual meeting in Nov., 1981. That's because delegates also learned of the plans of GTA president and general manager B. J. Malusky to retire the following year. Also retiring at the same time would be Malusky's aide, GTA assistant manager Lowell Hargens.

This news was greeted with the passage of a special resolution "expressing our deepest appreciation to them (Malusky and Hargens) for the skill and dedication with which they have fulfilled the responsibility of their positions since June 1, 1968, and for their many years of faithful service that they have given to cooperatives".

The resolution likely was sincerely meant. But there seemed little question that it quite likely was far from being "unanimously" enacted.

B. J. "Barney" Malusky carried with him excellent credentials of grassroots leadership. He was a native of Stutsman Co., North Dakota, and got his start in the grain business as an assistant to his father, Alfonse J. Malusky who managed an elevator in Wimbledon, ND.

Malusky began his career with the Farmers Union Grain Terminal Association in 1947 when he was hired as an auditor at its St. Paul headquarters. Two years later he returned to North Dakota as a GTA fieldman for cooperative elevator affiliates in the Williston area. Then, in 1956, Malusky moved back to St. Paul with financial responsibilities. He became head of GTA's field services in 1958, and then proceeded to move higher up the cooperative ladder – a vice president in 1963, assistant general manager two years later, and then general manager after M.W. Thatcher retired in 1968. Malusky and his wife LaVon lived in the White Bear Lake community of St. Paul with a daughter, Colleen.

GTA'S FIRST LOSS IN 31 YEARS

On Mar. 4, 1970, B. J. Malusky told of a real disappointment – the first loss for Farmers Union Grain Terminal Association since 1938 – the year when the cooperative was re-organized as GTA. Blamed by Malusky were several things – low margins, higher expenses, reduced grain storage earnings, losses in soybean processing because of a strike, and, again, a flood which hampered river marketing.

Figure 98 B. J. "Barney" Malusky

Malusky said GTA had three goals for the 1970s: A revitalization of cooperatives by getting more young farmers and young skilled employees in the organization, increase GTA's market power through research and planning and more advanced processing, and, reshaping the cooperative structures to better fit modern farms and the rural communities of tomorrow.

He predicted fewer and bigger elevators serving fewer but larger grain farms, with the threat that railroads would be abandoning branch lines. Such changes will "centralize the grain business at a few key points on main lines by the end of the 1970s". In the supply co-op field, Malusky said there would be "one-stop marketing and supply service" emphasized by the most progressive cooperatives.

There would be more mergers and joint offices and joint operations ahead, Malusky added. "Instead of some 20 regionals, we need only a few – one for each main grain area," he explained. "Today we have nine cooperatives competing in the Corn Belt, three in the Northwest spring wheat area, and three in the Southwest winter wheat area." In addition to increased mergers of grain regionals and joint sales offices, Malusky said cooperative officials needed to consider more integrating of marketing and supplies at the regional level.

EDUCATION FUND HISTORY

In his first year, B. J. Malusky reviewed the Farmers Union Education Fund history when he spoke at the annual state convention. GTA volume was down 7 percent for the year, Malusky reported, with Minnesota producers delivering 29 percent (48 million bushels) of the 164 million bushels delivered, and North Dakota the source of 38 percent. Savings amounted to a little over $1 million, or about a fourth of what they were the previous year.

As a result, the 5 percent Education Fund total would be "only $30,000" shared by four states, including Minnesota. In the past 30 years, Malusky said, payments from GTA to state Farmers Union organizations has totaled over $3.5 million. "I am sure that this has been a big help in financing Farmers Union organizations which also have received part of the savings from many local elevators who pay 5 percent educational funds," he added. "This is a fine record for 30 years of working together."

Malusky closed his speech at the Farmers Union convention in 1968 with this: "I believe the family farm, in order to survive, must have the help of strong, modern cooperatives... we need to give special encouragement to younger farmers to join, and to help make our co-ops into what they most need. We need more young farmers on our boards of directors, and in other positions of responsibility."

Shortly after Malusky became assistant general manager of GTA in 1965, he also addressed the state Farmers Union convention. At that time he praised the relationship between the two organizations. "Looking back over the history of this farm organization, and Farmers Union Grain Terminal Association, it truly reads like a success yarn in a story book," Malusky said. "It is a story of how dreams have been realized by people working together.

"I can imagine how the cynics and the by-standing critics must have scoffed and scorned when this great union of farmers began. Yet, you made it a success... I believe we could all benefit by an even closer relationship and an expanded interchange of ideas, and I would like nothing more than to see this happen in the years ahead... "

However, the "closer relationship" Malusky alluded to in 1965 was far from what was happening 16 years later.

"LOYALTY" – THE KEY TO PROGRESS, GROWTH

"Loyalty to the cooperative movement and the institutions built by Farmers Union members must transcend and supersede the individuals who serve as temporary custodians of these institutions... "
Stanley M. Moore, President, North Dakota Farmers Union.

"Time is essential to the workings of a democracy," pointed out long-time Farmers Union leader Stanley M. Moore of North Dakota in an editorial following the decision made to change the bylaws of the Farmers Union Grain Terminal Association. "Such questions cannot be decided quickly... "

Moore spoke then as president of the North Dakota Farmers Union. Like other prominent leaders, Moore wrote both carefully and constructively in his editorial about what happened in 1981. "It is deeply regrettable that there was very little time to discuss and to ponder the many facets of the issue created by the board of directors of GTA in initiating their recently approved bylaws change," Moore observed.

"The affiliation between cooperatives and the Farmers Union has deep historical roots and has been beneficial to the growth and effectiveness of both of these farmer institutions."

Moore stated that members of Farmers Union and cooperatives in the northern plains states "have provided for a common identity and purpose for their farm organization and their cooperatives".

The bylaws change, he added, "raises some serious questions as to the future relationship between these farmer institutions. "To Farmers Union members that organized, nurtured, patronized, and sought a coordinated and unified system of seeking their economic and social goals of equity and justice for agriculture, the GTA bylaws change in effect questions their loyalty, their wisdom and their foresight."

Every cooperative needs to maintain a core group of loyal supporters if that cooperative is to progress, attract new patronage, and build upon its base, Moore wrote. "That loyalty and the Farmers Union heritage it represents cannot be underestimated," he concluded. "Loyalty to the cooperative movement is what patriotism is to a country."

"HOLY COW" IN POLISH?

On Oct. 14, 1974, Minnesota Farmers Union president Cy Carpenter wrote a letter to B. J. Malusky, president of the Farmers Union GTA, beginning like this: "Dear Barney: How do you say 'Holy Cow!' in Polish?"

This was his response to receiving an "Educational Funds" check in the amount of $820,188 from the regional cooperative. Carpenter said he was "somewhat surprised at the size of this contribution" but added that the check came "at a most opportune time when we are attempting to make some substantial changes in building for greater outstate promotion and activity in behalf of cooperatives and all farm families, and at a time when we are sorely pressed in terms of building needs here in the state office... "

In 1975, Carpenter wrote to Malusky that an agreement had been reached to purchase "the Skelly Building" at 1717 University Avenue in St. Paul, made possible largely by the "very substantial support" of GTA.

In Oct., 1975, Carpenter acknowledged receiving a check for $610,000 for the Educational Fund from GTA. He told Malusky that the Farmers Union was looking forward to moving into its new headquarters on University Avenue in St. Paul. In Nov., 1976, Carpenter wrote to Malusky a letter of appreciation for GTA's check for $321,911 for the Farmers Union's Educational Fund.

Then in Oct., 1977, Malusky wrote to Carpenter informing him that the Internal Revenue Service "has become more critical of cooperative associations... part of its criticism involves close scrutiny of items deducted by the regionals (cooperatives) as expense... ". Enclosed with this letter was a check for $150,697 for Educational Funds, and a statement that "our (GTA) bylaws authorize us to pay a portion of our net savings to State Farmers Unions 'which are actively engaged in promoting and encouraging cooperative organization in the interest of this (GTA) association'... ". Malusky's letter seemed to warn that a re-thinking of the traditional 5 percent Educational Fund share normally allocated by regional cooperatives was taking place.

Then, on Oct. 27, 1978, Carpenter acknowledged receipt of a GTA check for $144,315, adding, "I note this is a slight decline from last year, which causes me a real concern because I know it reflects the continuing problem of very narrow margins and an extremely competitive position in grain handling and marketing, as well as processing agricultural products". Malusky again expressed concern about how IRS viewed such "expense" items, and added, "We have an obligation to see that these funds are expended by the state Farmers Union in 'promoting and encouraging cooperative organization in the interest of' GTA... " Malusky added that GTA's "counsel recommends, and has recommended for years, that we ask each state organization to keep records showing the detail of the expenditures made by it in carrying out this mission."

In 1979, GTA sent Minnesota Farmers Union an Educational Funds check for $476,042, and again expressed concern about how IRS viewed such "service expense" items.

Records also show a flow of dollars from Farmers Union Central Exchange to Minnesota Farmers Union. In 1981, Cenex president Darrell Moseson sent a check for $90,876.60 to Cy Carpenter in a letter reading in part: "This represents your proportionate share of the 5 percent service expense... your share is based on the volume of business coming from the cooperatives in Minnesota in proportion to the total patronage of Cenex... "

Another 1981 check for $109,325.74 came from Farmers Union Marketing and Processing Association of Redwood Falls, MN. General manager Ed C. Wieland said in his letter accompanying the check, "We know that these funds will be put to good use in promoting Farmers Union cooperatives."

A third 1981 check came from GTA. It was for $502,210, and represented the fiscal year ending May 31, 1981 – or about the time the cooperative's bylaws were to be changed at a special meeting.

In summary, a total of more than $3 million had been allocated in Educational Funds by GTA to Minnesota Farmers Union for the period of 1974-81.

The annual financial statement of Minnesota Farmers Union for 1978 showed a grand total of $657,431 in Educational Funds received that year from five regionals ($645,940) and from 24 local cooperatives ($11,490).

Here are the local Farmers Union co-ops by community and type of business (e = grain elevator and o = oil): Badger (e) $298. Balaton (o) $48. Brooks (o) $500. Climax (o) $143.

Crookston – two co-ops, (o) $400 and (o) $100. Drayton – three co-ops, (o) $1,187; Drayton (o) $1,067; and, Drayton (e) $400.

Fairmont (o) $2,682. Flom (o) $805. Gary (o) $449. Grygla (o) $250. Gully (o) $250.

Luverne (o) $68. Mahnomen – two coops, (e) $408 and (e) $50. Montevideo (o) $200. Moorhead – two co-ops, (o) $200; (o) $200. Roseau (o) $250. Sherack (o) $119. South St. Paul (o) $858. Thief River Falls (o) $553.

Here are the totals for the regional cooperatives: Farmers Union Central Exchange – $339,139. Farmers Union Central Exchange (deferred) – $633,795. Farmers Union Grain Terminal Association – $150,697. Farmers Union Marketing and Processing Association – $92,190. Farmers Union Accounting Services – $118.

In summary, it was an unmatched flow of dollars from regional cooperatives and affiliated organizations which never again would be duplicated. It was a source of funding that would shrink dramatically from the years when a 5 percent allocation from net savings for state Farmers Union organizations was mostly normal and routine, and the major question for the state units was how big would their check be this year.

Now there would be another question – not only a question of how much, but, indeed, if there would be anything at all for the farm organization which had been sort of a godfather to the cooperative offspring which now seemed not to want to be a providing part of the Farmers Union family as in the past.

It would be a time when hard feelings began to boil and fester. Some who were a part of that era would never overcome the pain of the experience. Years later there would be noticeable traces of bitterness as they mentioned the 1981 meeting, and what to them represented a betrayal in trust born of a time when Farmers

Union and the cooperative movement were much younger, and struggling not to prosper, but, indeed, just to survive.

SYFTESTAD: "CO-OP EDUCATION BOTH DESIRABLE & NECESSARY"

Strong support for the Farmers Union Education Fund was voiced by E. A. Syftestad, general manager of Farmers Union Central Exchange in a North Dakota speech at the state convention in 1944. "In this Northwest territory cooperatives for the most part carry the Farmers Union name and have grown and flourished where the Farmers Union Educational Organization has carried on education work," Syftestad stated, according to a published text reproduced in the Nov., 1944, Farmers Union *Herald*.

"Cooperative education and practical cooperative operation are both desirable and necessary if the best results for both education and cooperation are to be attained. As the people gain a better understanding of the aims, purposes and accomplishments of the cooperative movement, the more support the cooperatives receive... and the more the cooperatives demonstrate through practical operations the economic value of the cooperative way of doing business, the easier it is to carry cooperative education to more people... "

Syftestad's comments were made at a time when he saw cooperatives under attack from a new group, the National Tax Equality Association. He reminded his audience that from the origin of cooperatives, pioneer cooperators had to battle for the right to exist and to compete in both the grain-marketing and the farm supply business arenas.

Co-ops, Syftestad said, were bigger in number, members and dollars at a time when co-op enemies also were bigger in number, more widespread and better-financed in their campaign against the cooperative movement.

A TELEVISION MEDIA ATTACK

However, there would be yet another problem regarding the Farmers Union Educational Funds in Minnesota – a television station's attack on the organization and targeting state president Cy Carpenter.

This incident apparently was set off by a disgruntled former cooperative employee who fueled the campaign by feeding "inside" information and half-truths to headline-hungry, investigating reporters at a St. Paul television station about "the millions in the Farmers Union Educational Fund". (This same individual complained to IRS about alleged "improper use of funds". However, after a hands-on audit of MFU by IRS which lasted several months, it was determined there had been "no improper use". In addition, there was no criticism or question raised concerning the use of these funds from the regional cooperatives or their managers and management.)

On Jan. 4, 1980, KSTP-TV in the Twin Cities aired the last of three copyrighted reports which raised questions about "nearly a million dollars a year in tax-free income" in education funds provided to Farmers Union organizations in seven states.

The reports didn't accuse the Farmers Union of illegal activities, but hinted as much with phrases like "some members of the 'Farm Union' are wondering" about the use of "nearly a million dollars a year in tax-free income", and,

"*Eyewitness News* has talked with farmers on both sides of this growing controversy". And, "some outstate farmers say they've tried in vain to find out how nearly a million dollars a year is being spent in Minnesota alone."

The report mentioned how "one farmer ran for the Cenex board of directors and lost" on a platform of "stricter accountability for the education funds". Also, it was suggested that some people lost their jobs when they asked how the money was being spent. One former Cenex employee, the TV news reporters told viewers, had received $200,000 severance pay. But the report made it sound more like the employee, Ed Beiber, was bought off with "hush" money.

When Farmers Union President Cy Carpenter was interviewed by the television news people for broadcast the following night, he was asked if any of the education funds "helped buy hay for starving cows"? When Carpenter replied that some of the funds were used in an emergency Farmers Union Hay-Lift program, the news reporter asked, "What has that got to do with education?"

Carpenter's response was, "I didn't say it had to do with education. It has a great deal to do with the well-being of that farmer and that local cooperative." The next comment came from the newsman who told his audience "(Former Cenex worker Ed) Bieber says the funds are tax-exempt only if spent on education. He has written to federal tax officials... "

Later in Jan., 1980, when U.S. Vice President Mondale spoke at the annual meeting of Cenex, KSTP-TV again brought up the "multi-million-dollar education fund spent by the Minnesota Farmers Union", adding that "we broke this story some time ago". Cenex Board of Directors Chairman Joseph Larson of Climax, the report said, "allowed only two farmers to ask about the money just as the convention ended. And then he shut off debate". Larson said on-camera, "It would not be fair for me, as chairman of the board, to start answering individual questions in the presence of a biased TV news media."

Then the reporter quoted Larson as saying "Cenex will go on giving the educational money to the Farmers Union as always... ", and then added, "One farmer who has questioned the education funds for several years says co-op members are afraid to ask about the money". The news segment ended with the TV anchor saying, "The Cenex board also refused to answer several questions from *Eyewitness News* reporters about that fund"

Complaints were filed by Farmers Union, including one made to the Federal Communication Commission concern the "biased, one-sided and unfair" reporting done by the station, and relating how efforts to cooperate and contribute to their broadcasts were not successful. And, of course, this kind of publicity created concern among some Farmers Union members about the "image" such televised reports presented of the organization and its leaders.

"SOME FELT (5 PERCENT)... WAS TOO MUCH MONEY"

For 30 years, from 1960 to 1990, Joseph Larson of Polk Co. was a member of the board of directors of Cenex. He served as its chairman of the board for one-third of that time, including the era when questions were raised about the Farmers Union Education Fund, and when the end came to the automatic 5 percent of the profits allocation to state units.

Larson feels that the cooperative pioneers and founders never envisioned a time when the 5 percent share of earnings would run into hundreds of thousands of dollars per year, as was the case in the early to mid-1970s. "Some felt that 5 percent was just too much money," Larson said about the change made in 1981. "That policy began back when the profits of our cooperatives were small. But then, that changed, and 5 percent became a lot of money."

"Change," he added, "took place in everything, in Farmers Union, in the cooperatives – a lot of it because a younger generation was coming along, and that always makes for change. And, some get caught up in the 'bigness is always better' philosophy.

"Farmers Union was a great achievement of our generation. It didn't just represent agriculture either in what it stood for. It represented many other different things, including business and organizations. It seemed that agriculture was going to go the way it was no matter how hard we were trying to fight for the family farm. But Farmers Union has been in the forefront of that battle for many, many years. But change happens in all things, including agriculture, farm organizations, and cooperatives. I've seen it all my life."

Larson's association with Farmers Union began after he started farming in 1946. He was fresh out of 3½ years in the U.S. Army during World War II in Europe, and bought a farm near where he was born and raised. He recalls joining after a visit by a neighbor and active Farmers Union leader, Jens Erickson, enrolled Larson and his wife, Blanch, into the Neby Local.

"We had a very active local for many years, held our meetings in an old, two-room school house, put on a minstrel show for fund-raising with big crowds of people there – you can't get that many people together these days," recalled Larson about the post-war period in an interview made in 1996.

He liked Farmers Union as an organization because "what they were working for was more than agriculture, a lot of other programs that affected people who worked in agriculture, too – it was not just an agricultural organization... "

After becoming president of Neby Local, Joe Larson moved up. He was elected to the board of directors of Farmers Union Central Exchange in 1960, a post he would hold for 30 years and chairman of the board for one-third of that time. It was a historic period, with rapid growth for the Exchange. And an era of change.

One change was in the name of the Farmers Union Central Exchange, as it had been known from its beginning officially in 1931. And it became simply, "Cenex". Why? Because, "One of the ideas was that it (the name) was too long," Larson recalled. "But, also, I think too many people had reached the point where they thought that it was bad to be a part of Farmers Union and to be a business organization. I thought this was unfortunate because I thought the name had served us very well."

Larson voted against the name change, but added, "Like anything else, if you haven't got the votes, you lose." He said it was close to "a 50-50" decision.

Why did he serve 30 years on the Cenex board? Joe Larson has a ready answer: "When I started farming, I just decided co-ops was the way to go. So both my wife and I became much involved with Farmers Union and with cooperatives. She was a youth director. Our two younger children went to youth camp."

One son, Jerry Larson farms and is a patron and a director of one of the co-ops that hasn't dropped the Farmers Union name – the Farmers Union Oil Co. of Climax", Larson said. There were some benefits in being active leaders, such as the trip Joe and Blanch Larson made to England in 1982 and to Norway in 1990, with both made to participate in the conference of International Federation of Agricultural Producers.

When Joe Larson was on the Cenex board, there were 13 directors, and he was one of two from Minnesota. Now, however, Larson feels farmers "are losing control of cooperatives because of business activity".

He explained this feeling: "As business gets bigger, you seem to have less commitment of people, both locally and regionally. One thing that happens, and it is showing up very rapidly, is that regional cooperatives are doing more local business than what they used to. Local cooperatives have gone out of business and the regionals have picked up that business. So along with that, you're going to have less farmer-control of our regional co-ops."

Another result of such change, too, Larson added, is that many local cooperatives had discontinued providing educational funds to Farmers Union as in the past. That's also true of some regional cooperatives as well. "Some have cut back, and some don't pay anything," he reported about regional cooperative support to the Farmers Union educational fund.

THERE WERE "SOME DARNED GOOD MEN" BACK THEN

"We'd been debating for four months, so I said, 'Listen here, fellas, let's quit fooling around like this so we can start working again'... "
Olaf Haugo, member of the board, Farmers Union Grain Terminal Association, 1983.

Olaf Haugo remembers it all, almost word-for-word, like it just happened yesterday. It was, the Waubun, MN, farmer said, "one of the biggest days of my life".

There were, he recalled 14 years later, almost as many lawyers in the board of directors meeting room as there were members of Minnesota Farmers Union and Farmers Union Grain Terminal Association combined.

"We met around a long, round table, with GTA on one side and Farmers Union on the other," recalled Haugo, a veteran Farmers Union member and nearly an 18-year member of the board of directors of Farmers Union Grain Terminal Association. Born in northern Iowa Jan. 11, 1912, he had farmed near Waubun all of his life and still lived on the same farm when interviewed in 1997. Farmers Union and cooperatives, he said, had been a large part of his life – for over 50 years. And, his fellow-members had elected him to leadership position in most of those to which he and his family belonged.

But there was a controversy in 1983, and a near-crisis in the relationship of two of the most important things in his long career – Farmers Union and GTA, the cooperative it helped build.

"We'd been debating for four months," Haugo continued about the long-ago discussions which pertained to the "Educational Fund" – the 5 percent share of profits that bylaws earlier had proclaimed to be paid to state Farmers Union organizations by the regional cooperative. The critics felt 5 percent was "too much money", Haugo explained. They sought to reduce the allocation to 2 percent instead.

Haugo was in an uncomfortable situation. He had been an active Farmers Union member for years, Mahnomen Co. chairman and president and board member of the local cooperative elevator for 21 years. And he had been president of his local rural electric cooperative as well. Yet, he also had been one of the members of the GTA board of directors since 1965, representing farmers and their viewpoints in helping make policy decisions for that growing regional cooperative.

And here he was, not all happy with the issue that brought the group of officials – and all those lawyers – to the negotiating table that historic day, the proposal to sharply reducing the educational fund allocation to Farmers Union.

Haugo recognized what was at stake, and the stiff odds against maintaining the status quo. About the outcome of the GTA-Farmers Union debate, he was very realistic: "I knew we couldn't win fighting for what we had – the 5 percent (share)," he related. And Farmers Union (leaders) told me that if we could get 2½ percent, that would be good enough."

So that was Olaf Haugo's first proposal, after patiently waiting nearly four hours to speak. He could remember when he was elected president of the Mahnomen Co. Farmers Union and the first time he got up in front of an audience. "I felt like I was shivering," he recalled about that memorable initial time when he was shaking in his boots and the pressure of having to speak before a group. Unlike some, he had not gone to high school after graduating from the eighth grade, and had received no formal speech schooling in the FFA or such.

However, that day of the big meeting between GTA and Farmers Union, Haugo wasn't shy, or timid. When he suggested a 2½ percentage apportionment, he said, a board member from another state immediately stood up, looked him in the eye and yelled out, "Haugo, you're crazy! You're not going to get 2½ percent!"

This threat didn't phase Haugo, he said. Instead, he retaliated quickly, stood up and told his critic, "The hell we won't!". Then he directly addressed his fellow-board member, "You sit down and shut up. GTA is fed up, and so are Farmers Union. So let's settle this matter right now. That's what I told them."

Then Haugo, realizing he had full attention, said he appealed, "Listen here, fellas, we have some darned good men here. We have been debating for four months. Let's quit fooling around so we can start working again." And all of a sudden, he related, the meeting room got quiet, and then, people started clapping – and on both sides of the table. "I knew we had won... and then people started coming up and saying, 'Congratulations', and shaking my hand... even the lawyers," Haugo said "I think everybody was glad to get it all over with, and get on with business as usual... ." In the end, however, the Farmers Union state organization did not "win." They did not get 2 ½ %, and today they get no help from the regional they built.

Afterward, Haugo told others, "By God, it worked. Even the president of the Farmers Union (Cy Carpenter) came up and congratulated me." And there were those who urged him not to retire after 18 years on GTA's board of directors, as he had indicated were his plans. "But I was tired," Haugo, then 71, explained. "I only had another couple of months left on the board. That was enough."

During his career, Haugo provided much leadership elsewhere, including serving as chairman of the county Agricultural Stabilization and Conservation committee for many years. And he declined an invitation to move up higher in the USDA agency, made by a pair of USDA interviewers whom he said were "amazed to find out I hadn't gone beyond the eighth grade in school". Then he added, "They told me that didn't matter, that they were going to recommend me anyway".

In talking about the GTA-Farmers Union controversy, Olaf Haugo voiced concern in an interview. "I've been a member of Farmers Union for over 50 years. We had a lot of darned good men there on the (GTA) board," he said. "They would stand up and fight. It seems like it's not that way any more. You've got an entirely different group now. And maybe not the kind of guts it took back in the old days."

Maybe, he added, it's just that the times were different back then, back when he, Minnesota, Farmers Union and GTA all were younger.

MORE CHANGE – NEW NAME FOR FARMERS UNION *HERALD*

A new name – *Co-op Country News* –was given to the Farmers Union *Herald* in Oct., 1974. While divorcing the publication's name from Farmers Union, the editors did acknowledge the role of Farmers Union in the long history of the publication, and the cooperative movement in the Upper Midwest and the Northwest states.

For nearly a half-century, or since 1927, the *Herald* carried the "Farmers Union" name in its title. It had become the successor publication to the Cooperators' *Herald* (a house organ for Equity Co-op Exchange which began publishing in 1913) and *The Farm Market Guide* (the house organ paper of the Producers Alliance published from 1924-27).

"*The Herald* resulted from the merger of three groups (the Equity Co-op Exchange, Producers Alliance and Farmers Union) to bring Farmers Union into the northwest states," read the new name announcement in *Co-op Country News*. "That led to these present-day marketing and supply cooperatives and to the Farmers Union state organizations which together have brought great benefits to two generations of farmers in these states.

"However, as time went by, younger persons not familiar with this history have understandably mistaken the *Herald* as an official organ of the state Farmers Union, which it has not been since the state Farmers Unions set up their own papers many years ago. The new name, *Co-op Country News*, reflects what this newspaper is all about – a publication devoted to providing information to, and news about, cooperatives' patrons in some of the greatest co-op country in the United States.

"These are the Upper Midwest and Pacific Northwest states served by Cenex and GTA. Statistics show 33 percent of the nation's farmer cooperative memberships are found in these states (Idaho, Iowa, Minnesota, Montana, Nebraska, North Dakota, Oregon, South Dakota, Washington, Wisconsin, and Wyoming), along with the headquarters for 42 percent of the country's farm supply, marketing and related service cooperatives. Indeed, this is 'co-op country' throughout... "

The announcement also promised that the new masthead of the newspaper would include the logos "for the two regional co-ops which for many years have published the *Herald* for their cooperative owner-patrons – GTA (Farmers Union Grain Terminal Association) and Cenex (Farmers Union Central Exchange)".

But once more, it had been decided, the "Farmers Union" wordage would not appear as prominently in print in that same publication as it had for nearly a half-century.

10...MANY BATTLERS FIGHT FOR FARMERS...

"Cooperatives are the true form of private enterprise... there's a doctrine that competition brings out the best... and they proved it... "

Bob Bergland, Former U.S. Secretary of Agriculture, and, Minnesota Farmers Union member, and organizer.

BRIEF BIOGRAPHY OF BOB BERGLAND:

Figure 99 Bob Bergland

Born: July 22, 1928, Roseau Co., Minnesota

Education: Roseau public school; Univ. of Minn. (2-yr. agriculture course).

Employment: Farmer, Farmers Union (part-time), U.S. Department of Agriculture (1961-68), U.S. House of Representatives (1972-76), U.S. Secretary of Agriculture (1977-81), CEO, National Association of Rural Electric Cooperatives (1981-94).

Community Service: Congressional advisor, United Nations Food and Agriculture Organization, Rome, 1973; Congressional delegate, United Nations

Trade & Development Conference, Nairobi, Kenya, 1976; member, Congressional Rural Caucus Committee.

Married: Helen Grahn of Roseau, 6/24, 1950. Seven children, one deceased.

Honors: First farmer to serve as U.S. Secretary of Agriculture since Roosevelt Administration (Claude R. Wickard, 1940-45). Won re-election to Congress in all counties in 7th District of Minnesota in 1976 (but resigned to be appointed Carter Administration's Secretary of Agriculture).

"People wonder sometimes if we are short of everything but trouble," Bob Bergland told his farm audience. This unusual comment was made during Bergland's third year as a U.S. Congressman, at the 31st annual convention of the Minnesota Farmers Union. It was 1973, and there existed deep concern by Bergland and others about the future of the Presidency and Richard M. Nixon in particular.

"These are troubling times," Bergland pointed out. "There is the shortage of fuel and transportation and chemicals and fertilizer... " Then he shifted gears, and spoke about the 1974 farm program, or more accurately, the lack of one.

"The lids are off, no set-aside, no soil-bank, no diversion program, no nothing... every cropland acre in the U.S. is going to be called on to produce food and fiber, not only to meet our own domestic needs, but to sell so we can get money to buy foreign oil... ," Bergland added.

This was the unusual scenario credited mainly to several factors, including a tight world food supply as evidenced by the historic billion-dollar grain sale to the Soviet Union in July, 1972.

All told there was a bullish atmosphere that would impact in the price of grain, land, and farm inputs like fuel, farm chemicals, and fertilizer.

Bergland, never known to mince words, attacked the school of thought by those who seemed to feel that the upturn in the farm economy, and the farm prosperity it promised, was not a passing fancy, and that it would prevail as it had in the wake of severe world-wide food shortages: "I've heard some say, 'We don't need a farm program now, our troubles are gone, everything's fine.'," Bergland reported.

Bergland made it clear he felt such sentiments were sadly short-sighted, and that some people were blinded by the prospect of an extremely food-short world which in turn would lead to the first "World Food Conference" held by the United Nations Food and Agriculture Organization in Rome within a year.

Bob Bergland then suggested that over-production might soon plague farmers, just as it had in the United States since the days of Abraham Lincoln:

"I think as a national government we have an obligation to guarantee against disaster if you (farmers) respond as a patriot to this call to increase food and fiber production... we need to increase the (price support) loan rate, and increase the target price... in case the bottom falls out... "

THE NEW CO-OP'S SUCCESS GOT ATTENTION

Until the Great Depression, Selmer "Sam" Bergland had little interest in politics or public affairs, recalls Bob Bergland about his father. "He had voted for Herbert Hoover," Bergland remembered. "But then when the depression came and after everything went down the drain, Sam Bergland changed. He became a 'Populist' politically and took a real interest in public affairs."

One activity that caught his attention came when he and others helped organize local cooperatives, including the first Farmers Union Cooperative Oil Co. in his home community of Roseau in 1929. It was the first fuel cooperative in that far northern region of Minnesota, and competed against Standard Oil Co., the only fuel dealer in town – and a firm that then refused to make farm deliveries.

The new cooperative organized, and promptly bought a pair of delivery trucks. This caused immediate predictions from community skeptics that what was happening was "a sure way to the poor house" for the fledgling co-op, Bergland recalled. "But they were wrong," he added. "Not only did the co-op pay for the trucks, but it came up with a cash dividend to its patrons as well, and in a short time. It was a real, eye-opening success."

Next came a Farmers Union Grain Terminal Association elevator, organized in 1938, Bergland said. Up to that time farmers had been told they couldn't grow malting quality barley – let alone get paid a premium for it. Again, the farmer-growers proved wrong the local, and vocal, critics. And, for the first time, the member-farmers pocketed a premium that first year that was almost as big as the price of the barley itself.

A third classic example of the value of cooperatives came in wheat. Growers had been led to believe that their wheat wouldn't qualify for a protein premium. But that first year their co-op was on the market the local wheat price included a premium payment on every load hauled to their GTA elevator.

"Those sort of things made my Dad a true believer, and me, too, like those of us who grew up in that kind of environment, " Bergland remembered. "We believed that cooperatives are the only true form of private enterprise, that there's a doctrine that competition brings out the best, and my father and those other cooperative pioneers proved it."

Another memorable lesson learned from his father was spawned during the depression of the 1930s, when Selmer Bergland – like countless others – lived in poverty. "Politics is power," the elder Bergland told his progeny. "Get in it." It was advice son Bob Bergland never forgot and in later years, put to excellent use.

In the 1940s, young Bob Bergland became involved with Farmers Union, attending "youth camps" then held in the Detroit Lakes area. The highlight of his junior career came in 1948 when he was singled out for the coveted "torch bearer" recognition, and earned a trip to Estes Park, Colorado, with other cream-of-the-crop Farmers Union youth from Minnesota and other states.

Bergland said he remembers being exposed to "some pretty radical stuff" at the Minnesota youth camps, quickly adding, "I don't think it was communist stuff, but it was a brand of socialism which for the time in the eyes of some was unacceptable." Bergland believes this "radical" gospel came under the Farmers Union administration of Einar Kuivinen.

"There was kind of an ugly chapter in Farmers Union history back then," Bergland recalled. "The farmer-labor and the Democratic Party were merging (into the DFL), and the farmer-labor faction had been affiliated with the Non-Partisan League, the Farm Holiday Association, and other groups. They were pretty militant.

249

"Einar Kuivinen came out of that wing, and was denounced by some as 'a communist' which I don't think he ever was, but back then in those years, that was a pretty fashionable thing to do. My Dad and others decided that Kuivinen had to go."

Young Bergland was asked by a neighbor, Roy Wiseth, if he would help him campaign for the state presidency. "It was a mean convention, representing a big change in the basic structure of Farmers Union, and a more moderate approach to things," Bergland recalled. Wiseth won, and the Kuivinen faction understandably took the defeat of the long-time Finnish leader bitterly, Bergland recalled.

A PERIOD OF PROGRESSIVE GROWTH

After Wiseth's victory, the new state Farmers Union president asked Bergland to join his team, and to work building membership. His pay would be $10 a day (with another $10 for daily expenses) paid out of the $5 annual Farmers Union dues. He worked only during the winter months when his farm chores weren't as demanding of his time, and found as often can be the case in northern Minnesota, Old Man Winter didn't cooperate at times in a membership-building campaign.

Bob Bergland was one of the members who became part-time Farmers Union insurance agents from 1950 to 1953, according to Floyd C. Borghorst's History of Minnesota Farmers Union Insurance. The man destined to become a popular Congressman, then U.S. Secretary of Agriculture during the Carter Administration, and finally the executive director of the National Rural Electric Cooperatives Association, was one of several part-timers marketing the new Farmers Union insurance.

Another part-timer then was Robert Wiseth, the son of former Farmers Union state president Roy Wiseth, for whom Bergland had campaigned in the Farmers Union election earlier. It was a time when the organization marketed three main insurance types – auto, health, and life.

However, Roy Wiseth's presidential reign was short-lived – only a year. "He was a good person, but not the best organizer," Bergland said. So the next year, the state delegates elected Ed Christianson president. "He was the successful manager of a large cooperative and I suspect had a lot of support from the two big regionals (GTA and the Central Exchange)," Bergland added.

Again, Bergland recalled bitterness and controversy surrounding the Wiseth-Christianson presidential contest. However, after the election of the highly-successful Gully cooperative manager, Farmers Union "really took off".

Bergland thinks the phenomenal growth to a record membership happened for a couple of reasons: First, the regional cooperatives poured much-needed financial resources into the Farmers Union. For one thing, M.W. Thatcher and other key cooperative figures liked Ed Christianson personally, and felt he was an excellent choice for state president. Consequently, they were eager to support him politically, and just as importantly, financially. Secondly, there existed a friendly sort of "rivalry" between the two cooperative giants.

The two rapidly-growing organizations openly put their business muscles and know-how into supporting the Farmers Union, providing moral support as well as finances. It was a profitable relation, a combination of "education" money and automatically-deducted Farmers Union dues plus the highly-conspicuous presence and help of regional cooperative fieldmen-organizers combined with on-the-road membership workers like Bob Bergland of Roseau.

Another important factor, Bergland indicated, was that GTA was led by the legendary Bill Thatcher, a charismatic, gifted, dedicated farm leader who also was an avowed Democrat. He relished being an associate of key political figures like Hubert H. Humphrey and others who built into Minnesota politics a force that was to last a long time through the development of the Democrat-Farmer-Labor triumvirate, Bergland added.

All of these things translated into unparalleled growth for Christianson's Farmers Union, from a Kuivinen state membership of about 10,000 members to an all-time high of about 41,000 in the 1950s-60s. But Ed Christianson was not a political partisan in the manner of Thatcher, Bergland recalled.

Instead the "cooperative giant from Gully" was recognized as "a steadfast progressive" politically, who maintained the traditional Farmers Union-Democratic affiliation but seemed instead "rather circumspect about it – he (Christianson) didn't fly the Democratic flag like Bill Thatcher was known to do".

Ed Christianson also asked Bob Bergland to do membership work in the slack winter months, along with other duties like participating in state commodity meetings, county fairs and special conferences and the like.

Then from inside the state staff emerged an anti-Christianson movement led by Russel Schwandt of Sanborn. He was a skilled and fiery speaker who through his membership work activity had developed quite a following among the membership. The challenge by Schwandt was at best a professional gamble. After his unsuccessful bid to oust the popular state Farmers Union president, Christianson, Schwandt the challenger quite understandably lost his Farmers Union job. But then Schwandt was hired by the GTA. And Bergland assumed a portion of Schwandt's duties.

About that same time, though, state Farmers Union membership began to decline, just as was the number of farms in Minnesota and elsewhere. Bergland recalls that era very well: "In Roseau Co. we had 1,000 Farmers Union members. But we also had 1,800 farms in the county back then. Now, there's only 400 left."

It was in 1960 during the Kennedy-Nixon presidential race that Bergland attained special recognition. He had a special assignment – fighting potential anti-Catholic reaction to Kennedy's election bid. "One of my jobs was to fight anti-Catholicism," Bergland related. "I'd find a prominent Democrat in a Lutheran congregation and then have him help 'convince' the (Lutheran) preachers to back off. And it was a pretty effective tactic."

After Kennedy won, Bergland was appointed chairman of the Minnesota state Agricultural Stabilization and Conservation committee which oversaw the administration of the federal farm programs. Then, later, in 1963, he was appointed Midwest ASC director overseeing federal government farm program activity in 12 Midwest states.

In 1968, however, his career interests took aim at a higher target – a Congressional seat. But it was not to be – that year, and he finished second to the Republican incumbent. However, two years later, it was a different story. Bergland had learned a lot about politics during his initial Congressional race, and put some of those lessons to work in 1970 when he was elected to the U.S. House of Representatives.

Then, six years later, in 1976, Bob Bergland exhibited a rare peak in political popularity: he carried all of the 27 counties in his northwest Minnesota Congressional district.

However, despite this impressive re-election victory, Bergland did what some people (including his father) considered the unthinkable: he resigned from Congress to become a member of President Jimmy Carter's Cabinet – accepting the job as Carter's Secretary of Agriculture.

When his puzzled father asked why anyone would leave a comfortable seat in Congress to take what easily is one of the toughest and most thankless jobs in almost every presidential administration, Bob Bergland had an answer: "I told my Dad I did it because I was asked by the President," Bergland told farm audiences later. "I told him it's pretty hard to say 'no' to the President."

While his father felt it was "foolish" to accept President Carter's offer to be his farm chief, so did Bob Rupp, editor of *The Farmer* magazine. Bergland "would make a good Secretary", Rupp wrote in the popular Minnesota farm magazine, but his experience was needed much more in formulating farm policy for the nation's agriculture.

Bergland, the Roseau farmer, farm leader, and Congressman was not the first Farmers Union member to hold the office of U.S. Secretary of Agriculture. That distinction is owned by the late Charles F. Brannan, a National Farmers Union staff member for many years. Brannan was selected by Harry S. Truman after his upset presidential victory in 1948. Bergland, though, was the first farmer since the FDR years to be named Agriculture Secretary.

Unfortunately, Bergland's timing for the job as Carter's farm chief wasn't good. It coincided with the emergence of the American Agriculture Movement, the highly-publicized "tractorcade" protests in the nation's capital. It was a period time when farm commodity prices began to sag sharply in the wake of record prices and bullish markets that had followed the Russian billion-dollar grain purchase announced in July, 1972.

The mid-1970s featured sharp upturns in the price of farmland, and record prices for the big four of Midwest farming – corn, soybeans, hogs and cattle – as well. Evidence of a new wave of overzealous optimism emerged as President Nixon ordered a soybean embargo in June, 1973, after earlier imposing unprecedented peacetime price controls at the wholesale level for both beef and pork in March that same year.

In the aftermath of such economic turbulent times, Bob Bergland, the former youth star-Torchbearer in his Farmers Union camp days, and the former state membership organizer working winters to keep his Roseau Co. farm, himself became a target – a target for vicious verbal attacks from fellow farmers, the Tractorcade protestors, and other critics.

Part of the ire directed at Bob Bergland had its origin in a sincerely-made comment he had given in response to inquiries made by Washington, D.C., agricultural reporters about the tractorcade invasion of the nation's capital by unhappy, protesting farmers in the winter of 1978-79. Bergland had in part tried to explain that he felt many members of the American Agriculture Movement taking part in the protest "really represented local (farm economy) problems, that others had made bad business judgements, sought publicity or were simply driven by old-fashioned greed".

While there were undoubtedly solid elements of truth in what Bergland had told reporters in his appraisal of the Tractorcade protest, of course, his remarks didn't set too well with some of the protestors who then proceeded to focus their wrath and discontent on Bergland.

Things got so bad that one winter day in Washington, D.C., Bergland related years later in an interview with the author that Bergland had to be escorted by the Secret Service out a basement window in his U.S. Department of Agriculture office building for personal safety reasons after his immediate USDA office area in that government building suddenly was filled with angry farmers who at the moment looked disconcertingly "mob-like". It obviously was a time for complete and instant discretionary procedures.

So, Bergland and his personal security entourage escaped clandestinely and safely by way of the USDA basement window. It was a tense moment Bob Bergland will never forget.

Then later, on a trip to make a speech in Oklahoma, Bob Bergland became the target for more than critical words and profane shouts and jeers. This time he fell victim to hand-tossed rotten eggs and snowballs. Surely, those misguided individuals who lobbed such missiles lacked a full understanding of who Bob Bergland was, and where he came from. They must not have known how hard and long he had worked for so many years on his Minnesota farm to hold things together for his family, fighting for their future.

And, indeed, how hard he had labored in long days and late at night for so many years for Farmers Union selling memberships and insurance policies, and doing all he could in general to try and convince fellow-farmers that by joining together, and by working together through a farm organization, they all had a better chance at a better future.

Most regrettably, those who tossed the aged eggs and icy snowballs at Bob Bergland in Oklahoma surely did so in ignorance of such things. After all, Bob Bergland had solid ties to a family farm. He was not like two of his recent predecessors from agricultural campuses as had been the case a few years earlier when Richard Nixon picked Clifford Hardin of the University of Nebraska and then Earl L. Butz of Purdue University. "Oh, Hell, not another professor!," commented Iowa farmer and Republican Congressman Bill Scherle when he learned that Nixon had selected Purdue economist and agricultural dean Earl Butz to succeed former Nebraska chancellor Hardin in Nov., 1971.

Nor was he a former Governor like fellow-Minnesotan Orville Lothrop Freeman, a former Minneapolis "city kid" appointed by a President (John F. Kennedy in 1960). No, this U.S. Agriculture Secretary was Bob Bergland, a former family farmer and a former young farm boy who carried the Farmers Union's leadership torch at a national youth camp, and a former Congressman who had championed legislation supporting the family farm agriculture system. He also was the Agriculture Secretary who "fathered" the "farmer-held grain reserve" program where producers – not agribusiness-types – pocketed government grain storage payments.

Such a person indeed represented a most improbable target for rotten eggs and snowballs, or even voices of sharp criticism. Instead Bob Bergland had demonstrated time and time again a dedication and determination to all he possibly could for the family farm and for family farmers. Those lobbing rotten eggs and snowballs or critical comments in his direction surely did not know this about the soft-spoken, man from far northern Minnesota where creating a future in farming took a little more innovative effort and work than in that area of the nation where soil resources are richer and where Mother Nature provides a bigger window of agricultural opportunity.

BERGLAND SOUNDS WARNING – SURPLUSES ARE LIKELY

In 1974, Bob Bergland addressed the National Farmers Union convention, noting at the start that it was the 25th consecutive convention he had attended, and that he as a member of Congress was wrestling with helping draft a new farm bill. It was also Bob Bergland's third year as a U.S. Congressman.

He then told his Farmers Union audience of his long association with that organization, noting, "With each convention I grow more certain that Farmers Union is the most vital and dynamic force on the American agricultural scene... " One of the reasons, he added, was the fact the organization arranged for farmer-delegations to visit Washington, D.C., to make known their views on legislation. In Congress, Bergland said, there were only 35 members out of the total then of 434 who didn't come from cities and who know something about agriculture. These urban Congressmen are confronted daily with well-paid, high-power, slick lawyer-lobbyists.

That's why visits by "real farm people" are important, Bergland pointed out. "I don't intend to demean the legal profession, but when farmers, farm wives and others come to Washington, it has a special meaning. They speak in terms that are clearly understood, and it does make a difference," Bergland said.

BERGLAND ON POLITICS: "ONLY NATURAL"

There has been a close relationship between Farmers Union and political leadership over the years, exemplified by such figures as Charles F. Brannan and Robert S. Bergland. "The Farmers Union has always been politically active and always encouraged its members to become (politically) involved," Bob Bergland noted in a letter written in May, 1997. "It is only natural that so many over the years have succeeded in the political world as well."

While he had no access to records, Bergland recalled personally that Minnesota Congresswoman Coya Knutson (who served 1954-58) and Minnesota Congressman and former Lieutenant Governor Alec Olson of Spicer who served in Congress 1964-68, were Farmers Union members. "While I don't know for sure, I believe former Minnesota governors Floyd Olson (1930-37) and Elmer Benson (1937-39) were members of the Minnesota Farmers Union."

So was former Minnesota Secretary of State Art Hanson (1954-48), Bergland added.

"There have been dozens of Minnesota Farmers Union members who have served in the Minnesota State Legislature over the years, too," he pointed out. The list includes Dave Frederickson of Murdock, the state president since 1991, and Willis Eken of Twin Valley, who preceded him in that office 1984 to 1991, and who was a majority leader in the Legislature.

"There were a lot of 'Friends of Farmers Union' in the Minnesota Legislature," commented Eken who served a total of seven terms. "Some of these were active

Farmers Union members, but we had other 'friends' there who had no background with our organization, but wanted to help."

He recalled several, including the late George Mann of Windom who during his many years in the Minnesota House of Representatives was chairman of the House Agriculture Committee. "George Mann was a very active Farmers Union member, and was a member of the board of directors of GTA (Farmers Union Grain Terminal Association) for a number of years as well," Eken added.

Another legislator who shared a Farmers Union-cooperative heritage was Vic Schultz of Goodhue. A House member, Schultz was active at the local and state levels in the rural electric cooperative movement.

Some others whom Eken remembered came from Farmers Union ranks to serve in the State Legislature included Harry Peterson of Madison, a five-term representative in the Minnesota House (a post now held by his son, Douglas); Clarence Purfeerst of Faribault who served in the Senate; and, Glen Anderson of Bellingham, the chairman of the House appropriations committee and as member of a family active in Farmers Union.

Another Farmers Union member/legislator is Ellsworth Smogaard of Madison. He was elected in 1974 and served two full terms in the Minnesota House. Smogaard was a member of the first class of Minnesota Farmers Union Torchbearers following its reorganization and rechartering. Smogaard was honored at the second state convention held in Alexandria in the fall of 1943.

A current House official with Farmers Union roots is Edgar Olson of Fosston. This Polk farmer is in his seventh term of office and is executive director of the new Minnesota Agricultural Research Institute, an organization established to find new uses for farm products. Olson has chaired the House property tax and tax increment committee.

One of the most powerful House positions, chairman of the Ways and Means Committee, is held by another Farmers Union member, and former Torchbearer, Loren A. Solberg of Bovey. Solberg, from Dakota Co., is in his eighth term.

Other former Torchbearers among the more than 1,400 young Minnesotans who achieved that recognition include county presidents Sheldon Haaland of Renville Co., a member of the Harvest States board of directors; Ron Reitmeier of Polk Co., chairman of the Minnesota Farmers Union executive committee, Korydon Chervestad of Pennington Co., president of the MFU state board of directors; Merlyn Hubin of Cottonwood Co., former president of the MFU state board of directors and chairman of the state executive committee.

Torchbearer-winners who distinguished themselves include Carol Knudsen and Kathy Laurer, both of the state Farmers Union staff and directors of the youth and family services division; and, Julie Bleyhl, the MFU director of legislative services and recently appointed to the Minnesota board of regents.

Minnesota Farmers Union's certified public accountant and manager of its accounting service, Lyle Reimnitz, earned the Torchbearer honor, as did credit department executive Tom Kopacek of Cenex.

11... FARMERS UNION TODAY...

THE WILLIS EKEN ERA, 1984-91

"The strength of our country since it was founded is the concept that the renewable resources of our land are placed in the hands of those who produce food... I don't think this country can stand the loss of the Family Farm structure and continue as we have known it... "
Willis Eken, Presidential address, MFU Convention, 1984.

"Anybody who voted for that bill in Congress should be sent back home."
Willis Eken, commenting on 1985 Farm Bill.

"Farmers don't benefit from rising food prices... The consumer has to eat a lot of Wheaties before farmers make a dollar... "
Willis Eken, commenting on article critical of farm programs, 1985.

Figure 100 Willis & Betty Eken

A BRIEF BIOGRAPHY OF WILLIS EKEN:

Born: April 12, 1931, Twin Valley, MN
Education: Graduate, Twin Valley High School.
Employment: Farmer, Twin Valley, MN; Congressional ag advisor .
Community Service: Served seven terms in Minnesota House of Representatives, 1971-1984; Majority Leader, 1981-1984; President, Fossum

257

Local, Minnesota Farmers Union; State Farmers Union President, 1984-1991; State Chairman, Green View, Inc.; member, Board of Directors, National Farmers Union; member, Governor's Commission on the Economic Future of Minnesota; member, NFO-MFU State Livestock advisory board; National Farmers Union vice president for International Affairs; participant in trade talks and conferences held in Australia, France, Belgium, Zimbabwe, and Spain.

Married: Betty Skaurud, June 20, 1954. Three sons, Loren, Twin Valley, and, Lee, Ada, both farmers; and Kent, a teacher at St. John's University prep school, St. Cloud.

Honors: Torchbearer's Award, 1949; Special Minnesota Farmers Union award for dedication to the preservation of Family Farming,, 1978; Honorary American Farmer Degree, National FFA, 1986.

When Cy Carpenter moved up to the presidency of the National Farmers Union in March, 1984, three long-time members indicated an interest in succeeding Carpenter as Minnesota Farmers Union president. They were veteran state staff member Dennis Sjodin of Cambridge who for some time had been a special assistant to the Minnesota Farmers Union president, Ted Suss of Aldrich, the Todd Co. Farmers Union president, and Minnesota House of Representatives Majority Leader Willis Eken, a grain and former dairy farmer from Twin Valley in Norman Co.

However, before the actual vote, Suss and Sjodin withdrew their candidacy bids and indicated their support for Eken. Subsequently, Eken moved in to succeed Carpenter.

"When Cy Carpenter was elected president of the National Farmers Union, everyone knew his shoes would be hard to fill," Sjodin said in April, 1984, following the national election of Carpenter who had been Minnesota president for about a dozen years

Also, Sjodin noted that the name of Willis Eken as a possible successor had come up four years earlier when Cy Carpenter had shown interest in seeking the national presidency. At that time Sjodin expressed "full support for Willis Eken".

Suss told the state executive committee that as a former state legislator he "had seen Eken in action on the Minnesota House of Representatives floor working for agriculture, and knew the Minnesota Farmers Union could benefit greatly from his leadership". Eken had a 14-year career as a legislator, serving on committees on agriculture, education, taxes, rules and legislative administration.

Further, Eken had been singled out in 1978 for a special award presented by Minnesota Farmers Union "for dedication to the preservation of family farming". This honor came after the Legislature enacted a model anti-corporation farming law, called the Family Farm Security Act and since copied by several states.

In April, 1984, by what was reported to be "an overwhelming voice vote", Sjodin was named "acting president", to fill in until Eken could take over the presidency Aug. 1 after his legislative chores as majority leader of the Minnesota House of Representatives had ended.

Sjodin had joined the state FU staff in 1966, and had served as field representative, director of program development, and assistant to the president.

The Minneapolis *Star* and *Tribune* expressed surprise in an editorial April 14, 1984, that Willis Eken "would give up a powerful legislative position (leader of the House DFL majority) to step into big shoes (worn so well for years by Cy Carpenter)".

Then the editorial noted "Eken's firm, quiet style will be effective in negotiating farm legislation". It stated further that Eken now as Farmers Union president would represent something like 22,000 farm families, and one of many agricultural groups in Minnesota. Perhaps, the newspaper observed, "Eken's experience riding herd on the fractious House DFL caucus may be useful in increasing communication and cooperation between the diverse groups".

In his first major address as Minnesota's president, Eken spoke at the 20th annual South Dakota Farmers Union Picnic at Huron, SD He told his audience there that "a top priority in the 1985 farm legislation must be a major revision so that the benefits of the farm bill are directed toward protecting the family farm structure of agriculture". He lashed out at the escalating loss of family farms which he blamed on a variety of things, from Land-Grant institutions, farm publications, and agricultural lenders to the U.S. Department of Agriculture. These factions, Eken claimed, "have pounded and pounded us with the theme that bigger is better and somehow or other, biggest is best". It was a theme he, with three young sons on the farm back home, abhorred.

Instead of subscribing to the "bigger is best" philosophy, Eken urged tailoring farm programs to help "family farmers", saying, "We have to get away from the concept of total support for total production. If large, corporate-type farmers are as efficient as they claim to be, let them take their chances on the open market."

He also urged action to tackle the problem of a growing federal deficit. "The deficit must be brought under control or any farm program will have a difficult time being effective," Eken said.

In retrospect, there is simply no way the Twin Valley farmer and legislative leader could have picked a worse time to be state president. It was late 1984, and the worst economic crisis since the Great Depression of the 1930s was gripping the countryside. Rural banks were closing; scores of farm foreclosures were occurring, and the chant of the auctioneer at countless farm auctions reverberated across the wintery countryside. Farm families across the midlands of the nation succumbed to the economic woes which struck across all age groups at random, in proportion to their indebtedness – and especially debt linked to farmland acquisition.

Then, later that same year, at the 43rd annual state Farmers Union convention in St. Paul, Eken focused on the reported dire predictions being made about what was happening in rural America as presented in a pair of new studies of what now was being called "the farm crisis":

More than a third of Minnesota's farmers faced liquidation within two to five years. This startling forecast that one out of every three farm families in the state would have to quit farming came in a study made for the Minnesota Department of Agriculture.

In a second study, performed by the state Department of Economic Security's Job Training Program, researchers determined 375 to 400 farm families would leave their farms in a five-county southwestern Minnesota area (Rock, Jackson, Murray, Nobles and Cottonwood). This prediction meant an average loss of 75 to 80 farm families per county was likely. Equally disturbing was the prediction that such farm loss would have a huge impact on the number of farm-related urban jobs in those same five rural counties.

"Unless we stand up and fight for rural America," Eken told his convention delegates in Dec., 1984, "We'll see 10,000 state farm families and thousands of agriculture-related jobs lost by bankrupt businesses in the coming year (1985)."

He recommended "farmers and rural citizens get involved, get together and share their concerns and difficulties so that they together can build a united base of support for meaningful farm programs" in the 1985 federal farm bill.

"We have paid for the nation's cheap food policies through high interest rates, through tax laws that provide incentives for outside investors, through extreme, one-shot farm programs like the (1983) Payment-in-Kind that created a disparity in the system, and an increase in the cost of our inputs," Eken said. "We have suffered through continually skyrocketing costs and sinking prices. Now, we hear some people saying we should subsidize the rest of society even more by reducing our commodity prices to attract foreign buyers. I don't buy that. Farmers and rural people are hurting, and it's time we stood up and fought for rural America."

Delegates at that 1984 Minnesota convention heard National Farmers Union President Cy Carpenter tell of "a growing awareness by mayors, church leaders, governors of rural states, and others about the injustice and exploitation of agriculture". The goal, Carpenter added, "is that we fully intend to take a lead role in developing a cooperative effort between all of these groups and between food producers and consumers to stop this exploitation and bring back some semblance of equality to our country's farm families".

EKEN: FARM BILL OFFERS "LITTLE HELP"

Hopes for the 1985 Farm Bill fell far short of what Willis Eken thought should be done. "By pushing through this disastrous piece of legislation, the (Reagan) Administration and Congress have once again turned their backs on the American farmer," Eken said in a statement after the measure was signed into law on Dec. 23, 1985.

"Family farmers are crying out for enough income to meet their obligations and basic human needs, and this bill does just the opposite – it pushes farm prices even lower."

Then the veteran lawmaker and farm leader added, "Wealthy farmers are the winners in this bill. Top USDA officials, largely responsible for writing the bill, are leaving their offices to go back to their huge farms and collect fat paychecks from this program and avoid conflict of interest charges... the losers will be the struggling farmers and the taxpayer."

Eken's comments were published on the front page of the Jan. 16, 1986, Minnesota Agriculture, a column away from a news item telling of the resignation of John R. Block, Reagan's Secretary of Agriculture, who left his large hog farm in west central Illinois to accept that appointment in Jan., 1980.

The USDA reported Minnesota lost 3,000 more farms from June 1, 1985, to that same date a year later. This put the state's total at 93,000, or 8,000 fewer than two years earlier. The 11.6 percent decline in Minnesota farm numbers was nearly double the percentage loss in South Dakota (6.5 percent) during the 1985-86 period, but was smaller than the decline in North Dakota (17.5 percent),

Wisconsin (11.9 percent), Nebraska (12.3 percent) and Illinois (18.7 percent). Iowa lost 8.4 percent of its farms, while the smallest decline came in Missouri where farm numbers dropped by only 4.1 percent, according to USDA.

A FARM RALLY DRAWS 10,000

Tough times usually bring out new faces, new spokes-persons, and new groups. It was like that in the mid-1980s when the "farm crisis" prevailed.

Anyone around that sunny, but cold January day in 1985 will remember it well – a crowd of 10,000 took part in a rally at the State Capitol in St. Paul to dramatize the fear that Minnesota's farm and rural economy was collapsing. Their mission of this throng was to protest what was happening, and the hope that state government and the legislature might be able to help.

One of the organizers of the farm rally was Bobbi Polzine of Brewster, one of those who started the new Minnesota group called "Groundswell". She told reporters for the Minneapolis *Star* and *Tribune* that "We come united by one purpose, to reclaim rural America for the family farmer, independent businessmen and laboring people. We have broke Republicans and broke Democrats. We have broke members of every farm organization. We have one thing in common – we're broke."

BAUMANN: "WE NEVER LEARN ANYTHING"

One of the faces in the crowd that day belonged to a person to whom farm protests weren't new. He was Archie Baumann, a long-time Farmers Union activist. He told a reporter he had attended a farm rally in the State Capitol 60 years earlier, as a boy accompanying his father and others who in 1933 demonstrated in the same place in much the same way – with a crowd of unhappy, serious-minded farm people looking for help.

Baumann recalled how he and his father had slept at night in the Capitol, on seats and benches there, waiting for dawn and the farm rally to begin. And, a decade later, he traveled with his father to protests organized by the Farm Holiday Association.

"We never learn anything," Baumann told the reporter on Jan. 21, 1985. "This is the same situation. It may be worse now because years ago farmers didn't depend so much on borrowed capital."

Baumann has been a member of Minnesota Farmers Union field staff, a contributing cartoonist to its newspaper, *Minnesota Agriculture*.

GROUNDSWELL'S DEMANDS

The Minnesota Groundswell group came to St. Paul with an agenda of recommendations for the lawmakers. Some of the things they sought:

An emergency moratorium on foreclosures for farm, home and small business mortgages which would last 120 days, and be renewed until commodity prices improved enough to provide a 15 percent return on investment.

A state-guaranteed farm operating loan program for seed, fertilizer, fuel and other operating expenses of spring planting.

An emergency conference of farm state governors, Congress representatives and others to institute a "guaranteed minimum price" for farm commodities, with that meeting to occur within a month.

Some kind of state action be taken to stop farm creditors from further reductions in the paper value of farmland, livestock and machinery, with farm real estate values frozen no lower than 1981 values.

In speaking for farmers, Willis Eken had a flair for providing headline-grabbing quotations. Like his response to a 15-page, two-part report printed in the Minneapolis *Star* and *Tribune* Aug. 11-12, 1985. "Farmers don't benefit from rising food prices," Eken countered. "That distinction once again belongs to the various corporate middlemen. The consumer has to eat a lot of Wheaties before farmers make a dollar."

Then, in July, 1986, Eken blasted the Reagan Administration's farm policies which he charged "were burying the hopes of America's family farmers under a mountain of grain". Then he added, "I suggest President Reagan check with the Magician's Union for his next Secretary of Agriculture because the only way these surpluses are going to disappear is by using a magic wand."

About the 1985 Farm Bill, Eken observed, "Anybody who voted for that bill in Congress should be sent back home."

And when Minnesota voters turned in the highest "yes" vote in the nation in the wheat referendum, Eken termed it "a big one, a battle we couldn't afford to lose, and we didn't... we won despite powerful and influential opposition... farmers saw through the smokescreen put up by the USDA, Farm Bureau, the Chamber of Commerce groups, and the grain trade... they voted in favor of supply management... I am especially proud... as farmers, we're used to fighting against the odds, and we came through with flying colors... "

THE FARMERS UNION-NFO MARKETING PACT

As a first step toward "a cooperative effort", Minnesota Farmers Union and the National Farmers Organization (NFO) began working on a new, joint marketing plan. This led to the first intra-state joint venture by the two general farm organizations. The historic contract was negotiated in Minnesota by President Willis Eken and by DeVon Woodland, NFO president from Corning, the Iowa town where the NFO then had its national headquarters.

The new marketing agreement came about even though in the past these farm groups understandably had their disagreements. The two organizations differed openly on matters such as farm policy, where Farmers Union historically championed effective, legislated farm programs and commodity price supports, the NFO concentrated on marketing and seeking its primary agenda of collective bargaining for agriculture, and bargaining for the prices it wanted. There were other disagreements, including a dislike by some Farmers Union leaders of the NFO's periodic and controversial use of market "holding action" tactics, or strikes, where members refused to market commodities at less than "target" prices and where they sought price and production contracts with buyers and processors. They could recall well how twice in years past this was attempted but failed.

That first in the nation Farmers Union-NFO livestock marketing agreement went into effect in Minnesota in Feb., 1986. "We hope this action signals a new era of cooperation between major farm groups," Eken said after the pact was signed. Within a year, both parties termed the agreement successful. And the Farmers Union-NFO arrangement in Minnesota was copied by two other livestock states, Iowa and Nebraska.

In April, 1987, Eken was just as enthusiastic about the marketing arrangement:

"This agreement has allowed the two organizations to mesh our relative strong points to the benefit of all. Our support of NFO's livestock marketing efforts led to NFO lending its backing to our legislative initiatives. This has led to more success in getting additional farm groups to join together and speak as one on legislative issues."

One of those who worked closely with the MFU/NFO agreement was Dennis Sjodin, assistant to the Minnesota president. He estimated traveling 50,000 miles, attending more than a hundred meetings, and visiting with thousands in the 1986-87 period concerning the new arrangement. "The cooperation has been tremendous," Sjodin said. "Producers realize the value of working together to increase their leverage in the marketplace and the benefits that can bring."

When the joint marketing program began in Feb., 1986, there were 35 NFO marketing points in Minnesota. Then four were added – Thief River Falls, Rush City, Dovray and New Munich. "All of the farmers we contacted thought it was a good idea for farm organizations to be working together," Sjodin added in an interview in April, 1987. "It's been a good year, and we look forward to even more progress."

At national level, Carpenter stressed that "working together" was extremely important then because farmers were in the throes of an economic crisis and also, because it was a period when a new federal farm bill was being drafted. "Before we write the 1985 farm bill, we as a nation had better get down to the cause of our problem and decide which direction we want to pursue to preserve family agriculture," Carpenter said at the 43rd Minnesota state convention in St. Paul.

In response, Farmers Union voting delegates at this convention proposed "a five-point farm crisis plan":

1) Lower interest rates and restructure farm debts until farm returns provide a "dramatic increase" in net farm income.

2) Utilize immediately, all possible federal and private funds to aid financially-stressed farm families.

3) Provide a one-year moratorium on all regular and emergency Farmers Home Administration loans where delinquency was caused by a continuing poor farm economy or factors beyond the borrower's control.

4) Provide a four-month grace period on private loans to allow time for renegotiating long-term debt restructuring agreements with FmHA.

5) Reorganize FmHA to better serve the needs of farmers and rural communities, with a priority for family-sized operations.

As for himself, Willis Eken epitomized the very thing he and Farmers Union were fighting for – the family-type agriculture in which he had grown up, and which he and his wife Betty had tried to pass on to their three sons (all of whom were active in the FFA, or "Future Farmers of America", and the fact that two chose careers farming).

Further, Eken – thanks to his folks – "had grown up with Farmers Union", and recalled in a 1996 interview how very active his family and most of his neighbors were in this organization locally many years earlier. Young Eken had become a member of the local youth education program where local farmers and members were the "teachers".

"We had educational classes where we could earn the opportunity to go to (Farmers Union's summer) camp," Eken related. Unlike in later years, winning a camp trip was an awfully important goal, and represented a real achievement for farm kids of his generation.

EKEN A FARMERS UNION "TORCHBEARER"

The Farmers Union local to which Willis Eken belonged met in a local school house, with good-sized crowds, he recalled. Generally, kids played together while their folks gathered in the meeting hall. In the summer of 1949 when he was 18 years old, Eken earned the classic coveted youth recognition – he was named a National Farmers Union "Torchbearer". Such an achievement is a highlight and climaxes at least five years of classes and study, and of attending local camps, and meetings. This Torchbearer honor was presented at a special state convention session where the new crop of honorees represented the future of the organization honoring them.

For Eken there also was the once-in-a-lifetime trip to "Old Bailey", the Farmers Union's "All-States Camp" held at the facility southwest of Denver beyond the second range of Rocky Mountains where the farm organization owns several hundred acres of valuable Crow Valley land straddling the deep-blue, fast-flowing south fork of the trout-infested South Platte River. This splendid mountain retreat, so close to Denver and its burgeoning suburbs, represents a most valuable real estate property. However its value is enhanced even more by another resource – the water rights that are owned with the "Old Bailey" mountain valley camp.

Eken became a Torchbearer about the same time another Minnesota Farmers Union youth (and future U.S. Secretary of Agriculture), Bob Bergland of Roseau, attained that honor. In later years another promising Minnesota youth, Leland Swenson (a future president of the National Farmers Union) would gain the Torchbearer recognition.

Willis Eken was raised on "a typical" family farm of that era in northwestern Minnesota. His parents had purchased the farm with a Farmers Home Administration loan, and had a flock of chickens, milked a few cows, raised hogs, and, of course, grew crops as well. After marriage, he and his wife Betty established their own farm. In the 1970 elections, Eken became a State Representative in the Minnesota Legislature, and as a DFL member, he moved up to become Majority Leader in the Minnesota House of Representatives, one of the most powerful and influential state legislative posts.

One of his most satisfying accomplishments was the authorship and sponsorship of a corporate farm bill, which would curb or limit encroachment by absentee owners and investor-owned farming enterprises.

At the 45th annual convention in St. Paul, President Eken spoke of the consequences of the loss of family farms. "It isn't just you and me, and it isn't just our neighbors being impacted by what's happening," Eken told delegates. "It's our children, our grandchildren, our small towns and our cities. And it goes even beyond that. The strength of our country since it was founded is the concept that the renewable resources of our land are placed in the hands of those who produce the food. I don't think this country can stand the loss of the family farm structure and continue as we have known it."

Regarding the 1985 farm bill, Eken described it as the "disaster Minnesota Farmers Union and others predicted it would be". He expressed support for the Harkin-Gephart farm proposal which if enacted would have featured a "supply management" approach to farm policy.

Eken also pointed out that Minnesota Farmers Union had supported a state interest buy-down program, debt restructuring efforts and mandatory mediation activity which he now said served as models for other states. Then he urged that such programs be adopted nationally to help farmers during the farm crisis of the mid-1980s.

When Cy Carpenter made it known he would not be seek re-election to a third, two-year term as National Farmers Union president, and national vice president Stanley Moore of North Dakota also chose to retire, Willis Eken was one of three men who announced their candidacy.

Eken praised Carpenter's leadership in Minnesota and nationally, saying, "He has established our National Farmers Union in the leadership role for the future debate and implementation of agricultural policy, both in the United States and throughout the world."

However, after weeks of checking with delegates from several states, Eken said he became convinced "that it was in the best interest of our membership that if not me, then Lee Swenson (of South Dakota) offered the right attitude and personality to lead this organization through these difficult times... "

During Eken's years as state president, Minnesota Farmers Union moved its offices twice, including when it sold its own building in St. Paul. "It was a matter of finance," Eken said about selling the headquarters building. "We made the decision that we'd rather have our money in investments. We had this big building, well-located, plus a really good offer to sell. Our plan was to invest the proceeds (from the sale), invest it and draw interest on it, and rent other office space," Eken explained.

He acknowledged that some liked the idea of ownership better than paying rent. However, at the time, the state staff didn't utilize all of the space efficiently, and to the majority in charge at the time, "it made sense" to sell the headquarters (at 1717 University Avenue in St. Paul).

In announcing his resignation, Willis Eken said he looked forward to returning to his Twin Valley farm with his wife Betty after commuting for 21 years to the Twin Cities – 14 in the Legislature and seven as Farmers Union president. The trip totaled 274 miles – one way. Also, having five grandchildren nearby back home figured into his decision as well.

"The only credibility we have is our honesty and our rightness on the issues. That's what we bring to the State Legislature, year-in and year-out... "
David Frederickson, President, Minnesota Farmers Union, 1997

Figure 101 Dave Frederickson

Date (and place) of Birth: Mar. 2, 1944, Benson, MN.

Education: Itasca Community College; St. Cloud State University.

Employment: Farmer, teacher, cooperative education specialist.

Community Service: Member, Kerkoven-Murdock-Sunburg School Board, Lutheran Church; 4-H alumni; State Legislator, six years (1986-91); vice-chair, Minnesota Senate Agriculture and Rural Development Committee; vice-chair, Taxes and Tax Laws Committee; chairman, Senate Rural Development Committee; President, Minnesota Farmers Union, 1991 – ; president, Minnesota Farmers Union Service Corp., 1991 – ; Member, Minnesota Pork Producers Association .

Married: Kay Kennedy on Sept. 4, 1965. Children: Anna, Emelyn.

When David Frederickson of Murdock became president of the Minnesota Farmers Union, the Murdock farmer elected to delay his presidency in order to complete his second term in the Minnesota State Senate. Still, when the Minnesota Farm Bureau Federation rated the Senate voting record, Frederickson scored an "average 60 percent".

The Farmers Union leader said that he would have gotten a higher mark from the competing farm organization except that one of the votes used in rating the lawmakers called for a ban on flag-burning. Frederickson voted "nay", he joked, "because I thought flag-burning was a First Amendment issue – not a farm issue. But I know farmers get upset when people are free to burn a flag but they can't burn their weeds" off their fields.

266

This incident, reported by Lee Egerstrom in the St. Paul Pioneer Press, indicates the sense of humor possessed by David Frederickson, the eighth elected president of the Minnesota Farmers Union. The Murdock teacher-politician-farmer received a degree in elementary and special education at St. Cloud State University. However, he earlier attended Itasca Community College where he saved expenses by living with relatives, explaining that his folks ("who were dirt-farmers") had three children in college simultaneously. To further help his parents hold down his college costs, he worked part-time in two jobs –in a garage at night, and in a pizza parlor on weekends.

Frederickson, who has a distinctive, deep, broadcast style of voice, then taught for eight years in St. Cloud, and Mora in Minnesota, at Midwest in Wyoming, and lastly, was a cooperative education specialist working out of Willmar in 18 school districts. He turned to full-time farming in 1974. Always interested in politics, and raised in what he termed "a pretty hard-core DFL family", he became charged up politically when he was invited to go on a trip to Washington, D.C., with Governor Perpich to meet with officials and comment on what should be in the 1985 Farm Bill then being drafted.

FREDERICKSON SPEAKS HIS MIND

It was by a strange set of circumstances that young Frederickson was personally invited by the Minnesota chief executive to make the Washington trip. When Perpich had stopped in a local cafe in Swift Co. to promote his open-enrollment plan for schools, the session came on a perfect fall evening and during the soybean harvest. Reluctantly, Frederickson decided to shut down his combine and attended Perpich's meeting because he knew the Governor's education plan would be one of the items discussed. And, as a local school board member, Frederickson had strong feelings about this issue. Consequently, when the time came at the meeting, he sort of "unloaded" on the Governor, telling him that he was a local school board member, and then blasted Perpich's school plan as a scheme that was "bad for rural people".

It was this brash, young debating farmer that the Governor later thought probably would have something to tell the Minnesota Congressional delegation and others in Washington, D.C. Once there, explaining how farmers could use a good farm bill, Frederickson's comments brought what he described later as "a patronizing response from one of our (Minnesota) U.S. Senators, where he said something like, 'Well, young fellow, I can see you're doing some things wrong'."

The Senator then went on to lecture Frederickson, he related in an interview, about "learning how to pick a price where I could make a profit, share or lease (farm) equipment to lower costs, and, third, consider an off-farm job", and jotting these recommendations down on a note-pad. "When he got done, he folded up what he wrote and threw it across the table, hitting me in the forehead and it dropped into my lap," Frederickson recalled years later. "Hey, I thought, this isn't any way to treat a constituent. I got mad.

"What the Senator failed to realize apparently was that he was recommending 'moon-lighting' as a substitute for a price for my farm production and a return on my investment."

The memo-tossing incident at the Washington meeting with the U.S. Senator (Rudy Boschwitz) sparked Frederickson to do more than talk about politics. He became a candidate for the State Senate. He was to fill the DFL ticket where an old friend, and a long-time Farmers Union stalwart, Archie Baumann, had lost to the same opponent by 700 votes. A politically naive Frederickson thought at the time, "it should be easy to pick up 700 votes and win".

Little did he know how hard it is to earn votes in a region where folks pretty much vote the same way no matter what, election after election after election. But the Murdock farmer's timing was good. And like some people say, "Timing's everything." Looking back, Frederickson recalled, "I'd have to say that I was in the right spot at the right time."

That race took place in 1986 – the middle of the Farm Crisis of the 1980s. And, maybe being a farmer helped young Frederickson win by what he jokingly describes as a "small landslide" – 32 votes! He then represented Lac Qui Parle, Yellow Medicine, and Chippewa counties plus portions of Swift and Redwood counties as well.

Four years later when he was challenged by the same tough opponent, the Murdock man would keep his Senate seat by a bigger "plurality" – 260 votes. Frederickson laughs now as he looks back, and how he naively thought "it wouldn't be hard to pick up a few hundred votes".

EKEN RESIGNS; FREDERICKSON TAKES OVER

At the end of his second term in the Minnesota Senate, Frederickson chose not to seek re-election. It was then that Willis Eken, the president of the Minnesota Farmers Union, resigned in order to become a special assistant to a Minnesota Congressman (Rep. Collin Peterson, Detroit Lakes, MN 7th District).

To fill the vacancy, the Farmers Union board of directors (includes all county presidents) selected 47-year-old Dave Frederickson to fill out Eken's term as state president. However, he took a leave of absence as president in order to complete his State Senate term of office before assuming the Farmers Union presidency. This was not unexpected since Minnesota Farmers Union bylaws require any officer at any level to resign if elected to public office.

Then, in the regular Farmers Union convention election held in Nov., 1992, Frederickson was re-elected.

FREDERICKSON: MORE THAN PRICES INVOLVED

In his first President's Message column printed in Minnesota Agriculture in Sept., 1991, Frederickson told readers of his first tour on Farmers Union business. It was a three-day tour, to meet media, and to meet with farmers and members.

"The discussions... put a heavy emphasis on prices, as they usually do... prices for now at least are high enough to keep families on the land... somewhere down the road, fair prices for what we raise and sell... but talk went a lot deeper than that... it centered around making a community... made up of farmers... access to health care, educating our kids and jobs that will give them a chance to live somewhere near home, if they don't want to take over the farm, roads that don't wreck the pickup on the way to town... all those things that make up the quality of life we like to brag about in Minnesota...

"I guess that's what a general organization such as Farmers Union is all about... we'll be working with other farm groups... we have to expand our thoughts, be quick to make friends and slow to assume any group is our traditional enemy in our effort to maintain and improve life on the farm... "

In his column printed in Jan., 1992, Frederickson told readers his top priority was "membership", explaining how he wanted Minnesota Farmers Union to maintain its good reputation and influence as a "respected organization".

Then, in July, 1992, Frederickson applauded the proposed change in the Cenex cooperative bylaws to require any candidate for its board of directors to be "active farmers", resident of their respective regions they wish to represent, and that they cannot be full-time employees of Cenex or any of its affiliated cooperatives or any other corporation (except a family farm corporation). The proposal, Frederickson observed, "sends a message – you're important to us", and called for patronizing "our cooperatives because a cooperative system is a two-way street".

"MAKE PEOPLE THE MOST IMPORTANT PART"

A hard-hitting column was authored in March, 1993, when Frederickson focused on the annual report for agribusiness conglomerate ConAgra and its bottom line which company spokesmen said declined because of higher livestock prices. The truth was, Frederickson said, that cattle and hog prices the previous year were actually lower.

"The bottom line here is pretty simple. ConAgra is not in any sort of economic trouble, and producers are not to blame," he said. "Farmers simply produce the raw materials that the food giants turn into the myriad of products that fill grocery store shelves. Economics have forced us into that undesirable choice.

"That's not the way our system should work. Family farms and rural communities were not built for the benefit of corporate giants. They were built for people. And it's time we again made people the most important part of the system."

ABOUT "THE FARM LOBBY"

When the Minneapolis daily newspaper left the impression that "the farm lobby" still has a lot of influence in the State Legislature, Frederickson's response included the thought that the news article " left the impression farmers were getting something they shouldn't have".

From the Farmers Union perspective, he wrote, "Money and power don't talk in this organization – people do. And it is that grassroots support that makes Farmers Union the effective organization it is...

"Nothing speaks louder than a personal visit during a legislative bus trip, a talk at the local cafe, a phone call or a letter. That's how Farmers Union works. I know that as your president. But I also know from my six years as a State Senator. Farmers Union works because it is a people organization... "

ABOUT THE THREAT TO "FAMILY FARMERS"...

Dave Frederickson used his President's Message column in April, 1994, to address the efforts in the State Legislature to restrict corporate control of family farm agriculture, and to modify Minnesota's strict corporate farm law to accommodate new type of corporations being formed by groups of farmers such as feeder pig and market hog groups, or cooperatives owning livestock or poultry.

These developments concerned the Farmers Union president. "I see more and more encroachment of outside interests in something that has been a part of my family for four generations," Frederickson wrote. "But as farm prices stay low, I see more and more threats to farmer control of their own operations... finding ways to skirt laws designed to protect our very business is not a responsible way to adapt to the changing nature and methods of agriculture... "

Frederickson added that he could understand why some farmers have decided to pool their capital and resources in endeavors to keep farming operations financially healthy. "They see it as a viable alternative and a new way of competing," he added. "That argument has some merit. After all, Farmers Union started cooperatives for the same reasons.... but those cooperatives don't compete with my operation... they don't produce commodities and sell them on the market...

"Rather, these cooperatives supplement my farming operation with supply distribution and marketing outlets... it seems clear that a need for a new kind of cooperative endeavor exists today (1994)... Farmers say they want it, and some have taken steps to make it so.

"I don't have a problem with that... farmers must work together... but these cooperative endeavors must be structured so the interest of the farmer, not the corporation, remains foremost... and implemented in such a way that true farmers... control them."

ABOUT FARMERS UNION...

In Aug., 1993, Dave Frederickson commented on how Farmers Union answers critics who charge "farmers are living off the taxpayer" and other criticism surrounding farm programs.

Such charges, he wrote, must be answered. "Our organization's role is to advocate for family farmers and rural communities. And that is exactly what we do. We work on your behalf – both up front, and behind the scenes – on policies, legislation and programs that can help our rural communities grow and prosper.

"Our Legislative Services Department covers the bases at the State Capitol. Our Communications Department keeps the membership, the media and the public informed. Our Education Department works to ensure that an appreciation and an understanding for a rural way of life is passed on... the Membership Department strives to keep Farmers Union spirit and its goals alive and well in the country...

"We do those jobs for you... rural Minnesota, the family farms and the rural communities that comprise it are important to me. They are a part of me, just like they are a part of you. I believe in the fight. And I want to win. For me. For you. For our rural way of life."

When interviewed in March, 1997, Dave Frederickson had just celebrated his 53rd birthday. He laughed as he told how he "celebrates" his birthday every year at the National Farmers Union convention which is always held in early March.

"I have the dubious distinction of turning a year older at every national (Farmers Union) convention," Frederickson added. "I turned 48 in Des Moines (at the NFU national convention in 1992), 49 in Sioux Falls, 50 in Fargo, 51 in Milwaukee, 52 in Cincinnati, and 53 in Nashville. Looking ahead, God-willing, I'll be 54 in Albuquerque."

Looking ahead, Frederickson said he feels "the biggest challenge facing Farmers Union is membership", then adding, "We need to do a better job of boasting. Our track record is pretty darned good, those things we do for farmers and producers across Minnesota. Sometimes I think maybe we tend to be a little bit too shy. Maybe that's a Scandinavian thing that we deal with up here, sort of 'a we don't want to boast too much', you know; maybe it's just too 'Minnesotan', that we don't tell our story as we should."

Recognizing that there are fewer farmers and consequently fewer potential Farmers Union members in rural Minnesota, Frederickson said he's "tried to push the envelope a little bit, and to broaden the (Farmer's Union) agenda", emphasizing concerns like rural health care, public education, rural church issues, and such along with the primary farm policy and farm bill issues like corporate farming, industrialization of agriculture and other matters.

In its heyday, Frederickson said, the Farmers Union had 41,000 members. "Of course there were a lot more farmers then. Looking back in history, I saw some statistics compiled in 1958, Minnesota's state centennial year. It showed 193,000 farms, and here we are nearly 40 years later and we've had a drop in the number of farmers by nearly 100,000. We're at 85,000 to 90,000 farms today.

"So obviously, our struggle will be how to maintain our membership. We need to keep our traditional production agriculture as our 'centerpiece' but also we need to broaden our (membership) base to allow other folks who have an interest in rural issues to be part of the organization. So that's the challenge for us, the challenge for the organization.

"There's always some natural reluctance on the part of members who say, 'We're a farm organization'. I agree. We are, and we always want to maintain that, front and center. But as the (farm) numbers continue to drop, we're going to have to look at other approaches to issues so we can maintain a membership base and keep the organization vital and active."

"WORK WITH OTHERS"– A FREDERICKSON GOAL

One means of broadening the Farmers Union base is cultivating working relationships with other organizations like the National Farmers Organization. "We're constantly working with NFO to see if we can't develop some enterprise that would be of service to members of our organization and their organization," Frederickson said. "I think in a lot of cases there are dual memberships, with producers belonging to both organizations."

A good example of working together is the Farmers Union-NFO joint livestock marketing agreements first started in Minnesota in the Cy Carpenter – Willis Eken years, and since expanded into several other Midwest states.

Also, Frederickson said "new" cooperatives represent something else that might provide membership support and growth. "Look at what our parents and grandparents did," Frederickson said. "We in Farmers Union can look at a couple of fairly major cooperatives built by our grandparents and others. So maybe we today should stop and consider what we're good at doing, and that is building two huge cooperatives and they are Farmers Union Grain Terminal Association that today is Harvest States, and Farmers Union Central Exchange which now is Cenex."

Noting that the National Farmers Union's 1997 Meritorious Service Award was presented to Noel Estenson, president of Cenex, Frederickson said, "Noel understands the value of institutional memory, and he understands how Cenex was built and who built it. Our good, solid members built that co-op, we ought to be proud of that, and I know he is, too.

"And if we need to take a lesson from that, we ought to recognize that we were pretty good at that, so why don't we focus our energy on building some new cooperatives. I think that's always been part and parcel of this organization's mission."

In accepting the Farmers Union's highest award, Cenex President Noel Estenson noted how this organization had played a major role in starting and supporting local and regional cooperatives. "It's always great to be here," he told the NFU convention audience meeting in Nashville. "It's like coming home."

Cenex, Estenson said, represents 320,000 farmers and 1,400 local member cooperatives. He said that the cooperative had recorded a strong financial year. It had provided $650,000 in educational funds to various organizations in the past few years, plus $150,000 in scholarships in 1996 alone. Minnesota Farmer Union's share for its 1997 budget was just under $90,000, according to Frederickson.

In 1991 nearly $300,000 in Farmers Union educational funds came from four sources: Farmers Union Marketing & Processing Association ($169,617); Cenex ($121,840); Farmers Union Milk Marketing Cooperative ($5,322); and, local cooperatives ($2,895). In addition, nearly $5,400 originated from Farmers Union county, local units and cooperatives, averaging about $25 to $75 in contributions.

THE TOP 10 FARM COOPERATIVES

Three regional farm cooperatives based in the Twin Cities ranked in the top 10 nationally in terms of sales in 1993, according to a survey made by National Cooperative Business Association.

Farmers Union-originated cooperative Harvest States, was ranked No. 1, while a second cooperative fostered by the same farm organization, Cenex, was listed as No. 8. A third Minnesota-based cooperative, Land O'Lakes was fifth. Here is the listing, with sales totals for the most recent fiscal year:

1. Harvest States, Falcon Heights, MN $3.5 billion.
2. Farmland Industries, Kansas City, MO $3.4 billion.
3. Agway, DeWitt, NY $3.3 billion.

4. Associated Milk Producers, Inc. (AMPI), San Antonio, TX $2.8 billion.
5. Land O'Lakes, Arden Hills, MN $2.6 billion.
6. Countrymark Cooperative, Indianapolis, IN $1.9 billion.
7. Mid-America Dairymen, Inc., Springfield, MO $1.9 billion.
8. Cenex, Inver Grove Heights, MN $1.8 billion.
9. Gold Kist, Atlanta, GA $1.3 billion.
10. Ag Processing, Inc., Omaha, NE $1.1 billion.

FARMERS NEED "SAFETY NET"

The 1996 Farm Bill, Frederickson feels, presents a problem for farmers – the so-called "Freedom to Farm" program phases out traditional federal farm programs with their price support capabilities over a 7-year period. He is concerned about farmers not having an "economic safety net" in case of disasters, over-production, embargoes and such. At the same time he believes "it is incumbent upon Congress to help educate farmers" in the transition period.

"If we (farmers) are going to get cut loose from the Federal Government in terms of any kind of support system, any kind of 'safety net', then I think it's incumbent upon members of Congress to try to provide some marketing education," he added. "Everyone says we have to do a better job of marketing, so here's a chance to do something. I don't think a lot of those people who got caught up in 'hedge-to-arrive' contracts, for example, really understood totally what they were dealing with, and, on the other side, they didn't know corn was going to $4 and $5 (a bushel) like it did in the spring of 1996. They got caught – in the wrong spot at the wrong time.

"That kind of experience tends to make a lot of us a little cynical about forward contracting, hedging crops, and using the futures market."

Regarding Farmers Union, Frederickson said, "I wish it were possible for us to own our own building again. We need to own our own building. It was an unfortunate situation where they had to sell. We're a reputable farm organization in Minnesota, with deep roots and a long history, and we ought to have our own place. But right now, we're renting. Some day, I would hope, we'll be able to change that."

"Things change.", he added. "Just like for our church back home. Now, we're a 'two-point church', with two congregations – a country church (Bethesda near Murdock) and a church in town (Calvary in Murdock) sharing one pastor.

"At first it was farms, combining one, then two and then three farms (to make one farm). Then it was consolidation of schools districts, and now, in my lifetime, I have the opportunity to deal with the same issue – combination and cooperation within the (Lutheran) church community. We've seen it happen in health care as well. It's the alternative to closure. It's a frustrating thing to have to deal with that on all points, on all fronts in our local communities.

"We see it as well in our farm organizations. We see it in outreach to other farm organizations. What are some of the things that we can do together."

ONE FARM WHERE THERE WERE THREE

At home, Frederickson farms today what were three viable farms a generation ago, and what now have been merged into one farm unit of 800 acres devoted to feed grain and small grains production in Swift Co. near Murdock.

And he wonders what the future generation or generations will see in the way of change. "It's no wonder we've got problems with the lack of support at the church level, and a lack of kids for our public education level. Just take a drive out in the country, and look at what's happening. And, it's not happening just in Minnesota."

Two men stand out as individuals for whom Frederickson has tremendous respect and admiration – former Farmers Union presidents Cy Carpenter and his successor, Willis Eken.

"Cy was a mentor to me, as state president during my formative years. His name was a household word when I was growing up," Frederickson said. "I admire Willis for his leadership not just in Farmers Union but as majority leader in the (Minnesota) Legislature when he set such good direction for our state."

In the Legislature, Frederickson said, the Farmers Union has had excellent support over the years, with numerous members winning seats in both houses as well as in the U.S. Congress.

"Our message is well-received in the halls of the State Legislature," he added. "And the reason it is well-received is because it's 'right'. It historically has been 'right'. And that's why we have the respect we have over there. We speak out and advocate for the 'little people', and that's well-received. We've come out for things like legislation on corporate farming. Our influence is there. We don't buy it. Our little political action committee raises only a couple of thousand dollars a year. With 201 legislators, that figures out to be $10 apiece. So we don't buy any votes. The only credibility we have is our honesty and our rightness on the issues. And that's what we bring to the State Legislature, year-in and year-out."

The Farmers Union and the DFL party pretty much go hand-in-hand in Minnesota, and historically had this relationship, Frederickson added. "We in Farmers Union are probably the more partisan farm organization than some," he said. "South Dakota (Farmers Union) is more non-partisan. But definitely for the record, though, we are non-partisan. But that's a curse and a blessing at the same time because we are perceived by some as going the same way the Democrats go on issues. And that's not the case at all. We've worked with members of both political parties

"But yes, there's probably a lot of truth that in the end we probably drift more with the DFL than with the Republicans. But there's a price to pay for that obviously, but there also are some benefits as well."

When Frederickson became Farmers Union state president in 1991, Lee Egerstrom wrote a Sunday feature story in the St. Paul Pioneer Press about the fact Frederickson was a neighbor, friend and has similarities with another local farmer and farm leader, Al Christopherson, president of the major competing farm organization, the Minnesota Farm Bureau Federation. "Opposite Sides of the Fence" was the caption in the Sept. 22, 1991, article because it noted both similarities and sharp differences in the philosophy of the two men.

The similarities in the two young farm leaders of the state's biggest general farm organizations, as noted by Egerstrom:

Both farm organizations had about the same number of members (25,000).

Both men are fourth-generation farming sons of Norwegian immigrant homesteaders.

Both Frederickson and Christopherson operate similar farms, growing corn and soybeans.

Both head organizations that started cooperative businesses and insurance companies as member services but are best-known for their farm lobbying activity. This is where sharp differences come quickly into focus, with the two farm leaders easily identified as being on "opposite sides of the fence".

Farm Bureau long has stood out as "conservative" and more Republican-oriented, preferring "less government intervention in and regulation of agriculture". Farmers Union is identified as more "liberal" and Democratic-oriented, and historically has promoted legislated programs involving price supports, acreage controls, and market protection as such, as vitally needed by farmers.

Frederickson feels that farm people get their political and farm policy bent the same way that land is passed down from one generation to the next.

Dave is a fourth generation Minnesota farmer who was raised in the midst of active Farmers Union relatives, including his father, Rudolph, and two uncles, John and Donnell "Don" Frederickson. His grandfather, Theo Frederickson, was president when Swift Co. Farmers Union received its charter from National Farmers Union in 1941, or a year before Minnesota was re-chartered.

Later, his Uncle Don also became county president and also was a member of the Minnesota Wheat Commission for a number of years. Dave Frederickson's cousin, Jim, is the manager of the Farmers Union insurance operations in Minnesota. Dave Frederickson and his wife, Kay, have two daughters, Anna, 23, and Emelyn, 17.

As a state legislator for six years, Frederickson served as vice-chair of the Taxes and Tax Laws committee and vice-chair of the Agriculture and Rural Development Committee (and chaired its Rural Development Subcommittee). In addition, he was a member of the Senate Education Committee, its Education Funding Division, and the Governmental Operations Committee. He gained recognition for his efforts to pass legislation to promote new markets for agricultural products such as ethanol, as well as his work in the area of rural development.

As state president of the Farmers Union, Frederickson represents an organization of 25,000 Minnesota families and what is known as a general farm and rural advocacy group. He is a member of the National Farmers Union's executive committee which represents 250,000 members, and represents the national organization on the International Federation of Agricultural Producers' Environmental Committee. In addition he is a member of the Agriculture, Trade and Environment Advisory Panel for the U.S. Congress" Office of Technology Assessment.

Further, Dave Frederickson is president and chief operating officer of Minnesota Farmers Union Service Corp., a for-profit entity of the Farmers Union that operates a full-service travel agency and provides management services for

Green View, Inc., a company that performs maintenance of Minnesota's rest areas, state parks and other municipal facilities.

In his home community, Frederickson is a member of the Swift Co. Farmers Union and is active in the Swift Co. Co-op Oil Co., Murdock Restaurant Corp., Kerkhoven-Murdock-Sunburg School District, Murdock Bethesda Lutheran Church, and Swift Co. 4-H programs.

FIRST CONGRESSWOMAN HELPS IDEA BECOME REALITY

A dedicated Minnesota Farmers Union member and district supervisor provided Minnesota's first Congresswoman with the idea of the Federal Government providing low interest loans to college students, a concept which highlighted the career of U.S. Rep. Coya Knutson.

Knutson, who died in Oct., 1996, was a former North Dakota farm girl elected to Congress in 1954. It was a real breakthrough, being the first woman elected in Minnesota. At the time she won election to the U.S. House of Representatives, Coya Knutson was completing her second term in the Minnesota Legislature. Before that she helped operate a small hotel cafe in Oklee, MN, with her husband Andy Knutson. Reportedly inspired by Eleanor Roosevelt, she decided to run (successfully) for the Minnesota Legislature in 1950. Then she stunned even her own Democratic-Farmer-Labor party by winning election to Congress in 1954.

It was during her first term in Congress (she was to serve only two terms there) that she had a memorable encounter with Harold Johnson at the 50th anniversary celebration of Middle River, MN. When Johnson, a Greenbush farmer and then a Farmers Union district supervisor, told Knutson he had something he wanted to discuss with her, she invited him to do so while she rode in a convertible in the town's anniversary parade. It was then that Johnson told her of an idea he and his wife Carol had thought about for some time – if the U.S. Government could loan money for things like rural electrification and other interests at 2 percent interest, why couldn't it do the same thing for the nation's young people interested in receiving "a decent education".

Figure 102 Coya Knutson

Johnson's timing was good – the Soviets had successfully shot Sputnik into space, and a lot of people, including President Dwight Eisenhower, were excited

276

about education and keeping up with, or ahead of the Russians. In a 1956 letter he saved, Knutson had written to Johnson, thanking him for his suggestion of government student loans. "I am enclosing several copies of the student loan bill and press release which I sent out when I introduced the bill," Knutson wrote. "It was your idea in the first place and I thank you for giving me the inspirations... "

Then, in a second letter dated Sept. 5, 1958, Coya Knutson told the Johnsons "the President (Eisenhower) has just signed the National Defense Education Act of 1958 (which contained the student loan provision) into law so now eligible boys and girls will be able to get Federal assistance with their education... (it) goes into effect in June, 1959... "

Also, during a second-term campaign speech at a meeting of the West Marshall Co. Farmers Union, Coya Knutson pointed out Johnson in her audience, said he was responsible for suggesting the government student loan program, and asked him to stand and to be recognized by a round of applause.

However, the role played by Harold and Carol Johnson went unknown in an obituary in the St. Paul Pioneer Press Oct. 15, 1996, which credited Coya Knutson with being "successful in the passage of legislation allowing high school graduates to obtain college educations through loans from the Federal Government – and always considered it one of her most significant accomplishments... " The newspaper said also that Knutson had initiated "the first federal appropriations for cystic fibrosis research, income tax checkoffs to finance presidential elections".

In *Northern Light*, published in Thief River Falls, Harold Johnson who farmed for 20 years and worked for CENEX 20 years, explained the origin of his interest in loans to students. It dates back to when Johnson was an engineering student at North Dakota State University and "ran out of money". But his idea resulted in student loans to all five of his children, all of whom graduated with degrees.

"A HARVEST OF BENEFITS FOR FAMILY FARMERS... "

A membership in Minnesota Farmers Union provides a variety of things, according to a new brochure prepared by the organization, including:

1) The latest in farm news, brought by the twice-monthly tabloid, Minnesota Agriculture, with its Member Exchange advertising service section.

2) Group legal services, where local law firms are recommended by members and accepted into the group will provide legal services at negotiated, predetermined rates, plus a toll-free telephone number (1-800-542-4420) for members seeking legal assistance and locating a contracted attorney.

3) Family activities that include Farmers Union week-long camps which offer both recreational and educational opportunities for young people, from Grades 3 through Grade 12. Also, local youth classes provide the opportunity to learn about farm cooperatives, the philosophy and structure of Farmers Union and its history with cooperatives. Young people and young farmers/couples of all ages can earn trips to Farmers Union's "All-States Camp" in the Rockies near Denver, CO, and to state and national education conferences.

4) Because rural crime may be a possibility, Farmers Union provides a $1,000 reward program to help prevent theft, vandalism, and arson, if information supplied leads to an arrest and conviction.

5) Group insurance is provided through the Farmers Union Group Health and Life Plans which offers high quality protection and a choice of deductibles. Other standard insurance is provided by Farmers Union Insurances, such as farm, machinery, crops, auto, and home coverage.

6) Farmers Union Travel is a member-owned, full-service travel agency which caters to group or individual travel and which sponsors several member-group tours yearly. A toll-free telephone number (1-800-1966-TRIP or 8747) is available.

7) Farmers Union provides "a voice in government", talking in "the farmer's language" in the halls of government in the state capitol in St. Paul and in Washington, D.C., with a staff of legislative specialists. The top priority of this effort is "increasing farm income".

8) Cooperatives – Farmers Union historically has worked to build and strengthen cooperatives. Farmers Union does not "own" any cooperatives, but its members have organized and built more cooperatives than all other farm organizations combined. And, they do own hundreds and hundreds of cooperatives, including some of the biggest in the world.

Among those are:

CENEX – This is one of the largest farm supply cooperatives found anywhere. It is based in Inver Grove Heights in the Twin Cities, was founded in 1931 and originally was known as "Farmers Union Central Exchange".

Harvest States (Cooperatives) – Its history dates back to the early years of this century, and was known first as "Farmers Union Terminal Association" beginning in 1927, and then as "Farmers Union Grain Terminal Association" from 1938 to 1984. It is a St. Paul-based grain marketing cooperative which is the only co-op to boast three major waterway export outlets (the Mississippi River, the Great Lakes-St. Lawrence Seaway, and the Pacific Ocean.

Farmers Union Marketing & Processing – It is a rendering plant operation in Redwood Falls, MN, which features value-added processing and pet food manufacturing. Farmers Union Livestock Commission Co., which became Farmers Union Marketing and later merged with Farmers Union Processing, was actually the first regional Farmers Union cooperative, helping give birth to Farmers Union Central Exchange and Farmers Union Grain Terminal Association. In its present operation it is an outstanding example of recycling at its best.

Farmers Union Milk Marketing Cooperative – This co-op, based in Madison, WI, provides a market for milk for members and works to achieve the best prices possible.

There are other cooperatives which could be mentioned here. But these four are high-profile concerns which represent important agribusiness activities for farmer-patrons. "The Minnesota Farmers Union builds, strengthens, and works in tandem with the owner-patrons of these cooperatives to assure their continued service to farmers," states the final page of the recent brochure explaining what the organization represents.

The relationship of Farmers Union and the farm cooperatives is one that dates far back in time, to an era when they openly worked together hand-in-hand, and played significant roles in the development of both entities. It was a time when, the late, legendary "cooperative giant" M.W. Thatcher described the relationship like this: "We must never forget that our greatest asset, our greatest strength, lies with the people who conceived this great (Farmers Union Grain Terminal) Association, who built it, who use it, and who believe in it... we must never forget those people... bound together, just as GTA is, by a great farm organization, the Farmers Union... their objectives are the same... "

A MODERN VIEW OF FARMERS UNION

As Farmers Union nears its centennial, how do you define just what it is?

One answer comes in a few paragraphs inserted at the end of news releases issued by National Farmers Union from its headquarters in the Denver suburb of Aurora, CO. Here's a sample from a July 15, 1997, news release mailed to media (and also available on the Internet: http://www.nfu.org):

"Since its founding in 1902, National Farmers Union has provided cooperative know-how and taught cooperative principles to its members and the general public. Throughout its history, NFU members have founded a variety of cooperatives, including supply cooperatives, rural utilities and most recently, new-generation, value-added cooperatives.

"NFU is a general farm organization representing nearly 300,000 family farmers and ranchers nationwide. NFU serves its membership by assisting with education, by providing stimulus and know-how for farmer-owned cooperatives and by presenting the organization's policies to lawmakers at the local, state and national levels."

This news release announced a $39,000 grant from the Wisconsin state agriculture department to the Wisconsin Farmers Union for the purpose of creating "high-quality, regional cheeses from milk". This project came from Therese Tuttle, NFU's director of cooperative and economic development. She developed the concept of adding value to Wisconsin's milk supply through the development and marketing of specialty cheese items after visiting with Italian dairy farmers who offer European consumers a wide variety of localized, specialty cheese products they process locally.

Theoretically, Wisconsin farmers could gain value-added benefits as a portion of their milk supply is utilized in the production of new specialty cheeses that could be featured at Farmers' Market outlets, retail cheese/wine stores, local restaurants and cafes, and perhaps made available in mail order/catalogues.

ANOTHER DEFINITION OF "FARMERS UNION"

The Sept. 14, 1994 edition of *Minnesota Agriculture* printed a special article explaining what Minnesota Farmers Union stands for: "A team effort to preserve family farm agriculture".

"Minnesota Farmers Union – The Family Farm Organization" is a motto used by the organization on all of its promotional material. "It's more than just a catchy slogan," the article stated. "It is the foundation on which this organization is built and clearly directs our legislative involvement at both the state and national level.

"We are sometimes criticized as being only concerned for the small farmer... this is not an accurate portrayal of Farmers Union... Our concerns do not lie in the size of a particular farm , but rather in how that farm is structured and functions... we support the farm structure that allows the family on the farm to not only contribute labor, but also to make the decisions on its operation... "

This dedication to family farm agriculture is behind the Farmers Union leadership in enactment of the state's corporate farm laws, the article pointed out. "We continue to oppose vertical integration of agriculture where the farmer loses the flexibility to operate as an independent businessman and is only a provider of labor... usually in situations such as this, allegiance to community is gone, and the number one concern is to maximize profit for the corporation... society is the ultimate loser as farm families are displaced, along with the environment that produced responsible, dependable, hard-working, value-oriented children that have contributed so many positive things to it... This is why so many farm families have chosen to be members of Minnesota Farmers Union... ".

CY CARPENTER'S VIEW...

In Jan., 1986, National Farmers Union President Cy Carpenter outlined his views of the organization and its priorities in a letter mailed to members/leaders: "National Farmers Union has a proud history of working within the democratic system to right the inequities between rich and poor, urban and rural, majority and minority. We have a proud tradition of the individual activism of leaders like yourself.

"That personal involvement is even more important today, as we undertake a major counter-attack on the attempts to dismantle farm programs and eliminate our family farm system of agriculture... public awareness of the farm, rural and credit crises has developed in the recent past into one of clear opportunities... "

Carpenter noted things like "the generous response to the first Farm Aid Concert" held in Illinois in Sept., 1985, and "the growing economic and political awareness and awakening" which he said has caused new interest in protecting the family farm system of agriculture. However, he added, some national leaders continue to pursue policies that "would eliminate family farm programs, and drive hundreds of thousands off their land". These people, Carpenter charged, "have not awakened to the need for infusion of capital into our Farm Credit Services and other lenders, or to the impact this (farm) crisis will have upon our total U.S. economy".

Carpenter's letter ended with an appeal for support of the Farmers Union "Budget Fund", which is used to promote the organization's philosophy in the legislative arena. "We have the strength of the right cause, the right policy positions formed by our members," Carpenter concluded. "And we have strong allies... the time clearly is at hand for a counter-attack to protect these principles that have built a strong America and which can help maintain a strong America."

OFFICES AND LOCATIONS OF MINNESOTA FARMERS UNION:

1929 – Canby.
1942 – City Hall, New York Mills (1942-45).
Dec. 1, 1945 – MFU office moved to Branton Bldg., Willmar.
1951 – Twin Cities office, 2470 Univ. Ave., St. Paul
1963 – First Offices owned by Farmers Union, 1275 Univ. Ave., St. Paul.
1976 – 1717 University Ave., St. Paul.
1986 – 317 York Avenue, St. Paul.
Dec.1990 – Co. Road D West, St. Paul.

12... IT ALL BEGINS WITH THE LOCAL...

A 30-Year Look At Farmers Union

Isanti Co., Minnesota

Thirty years of records were provided to the author/researchers of this Farmers Union in Minnesota history by officials of Isanti Co. The purpose was to :

Trace activity at local level over many years via official minutes and records.

Summarize the broad range of activities ranging from picnics and dances to wild rice pancake suppers and banquet meetings with important public and private "name" speakers, to help publicize the concerns and problems of farm people.

Reflect in resolutions, letters and minutes of "action" and special meetings, and to compile an abbreviated look at the equivalent of a generation of Farmers Union happenings at the grassroots level.

What happened in Isanti Co., and to its three Farmers Union locals – Braham, Rum River and Isanti – during those 30 years likely was akin to what occurred in countless locals in the Upper Midwest states.

Perhaps, to some distant observers, some of those happenings and events recorded do not take on significance when viewed years later. Yet, the Isanti Co. record shows the Farmers Union units – small in number (membership peaked at about 500 families in the 1950-60s era) – attracted large crowds – bigger than its membership – at its "Town and Country" banquets/meetings/ dances/conventions.

Speakers annually included U.S. Senators, National Farmers Union officials, political candidates running for Congress, state offices and such, as well as top executives of the major cooperatives like Farmers Union Grain Terminal Association and Farmers Union Central Exchange, and other organizations

Resolutions adopted by Isanti Co. members over the years pertained to a wide variety of issues and topics, from demands for 90 percent and 100 percent of parity, to legislation protecting the family-farm agriculture system, and from concerns about "bigness" and corporate farming to opposing a state sales tax and urging improved farm-to-market roadways.

Like in 1957, Isanti Co. Farmers Union passed a resolution asking President Eisenhower to fire his Secretary of Agriculture, Ezra Taft Benson. Or, later, when Co. Chairman Franklin Lundquist in 1962 called for President Kennedy to consider a licensing program which would "protect family farmers". Under Lundquist's plan, only family-type operations could obtain a license to farm.

Such things brought attention to the Farmers Union locals, and kept policy-makers, would-be politicians and others alert to many of the concerns shared by farm people. Not all of the topics were strictly front-burner agricultural matters either. They covered a myriad of concerns from school issues and school consolidation to veterans' rights and the GI bill to local and state tax matters.

Resolutions adopted and then forwarded to the Minnesota state convention could be a mixed bag, from boosting dairy price supports and the "oleomargarine tax" to expanding the conservation reserve, and from ending daylight saving time to a "get out the vote" campaign for the upcoming election.

Minutes kept of county meetings tell of troubled times, like Dennis Sjodin's warning in 1985 that the economic troubles in rural Minnesota "are worse than many of us thought". Also, they relate how long-time cooperative booster Franklin Lundquist in 1979 expressed regret that "young people" lacked appreciation for cooperatives and what they have done, and are doing, for farmers, and "weren't active" in supporting co-ops.

Time and time again, the Farmers Union minutes indicate, county officers wondered what they might do to stimulate attendance at meetings, how to recruit more young members and how to reach membership goals.

Then, there were smaller, nagging "problems" to be solved – such as spending scarce and precious county organization dollars on a new mimeograph machine. Or, finding participants for the next Farmers Union bus trip or "fly-in" lobbying effort. Or matters like donating $50 for the State Farmers Union new building fund.

The minutes tell how meticulously plans were made for a Farmers Union booth at the Isanti Co. Fair (and interest in reserving a spot conveniently located next to the DFL booth there), how the annual picnic at Spectacle Lake turned into a five-county annual affair as well (with participants from Anoka, Chisago, Kanabec and Mille Lacs counties joining Isanti Co. members), and how annual conventions were carefully planned.

Countless hours and hours of donated time by volunteer, unpaid, committee leaders and others are represented in the Farmers Union county minutes, and carefully-kept records show how precious dollars were invested in picnics, dinners, lunch meetings, and such, for the good of the cause. And a variety of fund-raisers were staged to help raise needed operating funds.

Time and time again, county meetings included speakers, including many colored slide shows of trips taken by members or others overseas to visit foreign countries, on Fly-Ins to Washington, D.C., and such activities. While the bigger attendance meetings were held in local restaurants, motels, Legion Clubs, or Co-op Hall (if organizations were fortunate enough to have their own facility), other, smaller meetings took place in local country school houses, or farm homes of the leaders.

More often than not, there was entertainment – music, slides, a speaker, and such – and then the evening was topped off with refreshments or a serving, according to the minutes. And most offerings warranted a positive notation or adjective in the secretary's log, penned into place as carefully as were more important matters of the evening such as resolutions or committee appointments made earlier.

Over the years the list of speakers is impressive – Minnesota Gov. (later U.S. Secretary of Agriculture appointed by President Kennedy) Orville L. Freeman, the renown Minnesota Farmers Union President (and National Farmers Union vice president) Edwin Christianson of Gully (he spoke at Isanti Co. functions numerous times), Minnesota and then National Farmers Union President Cy

Carpenter (a frequent speaker), and others like Farmers Union editor Milt Hakel, fieldman Dennis Sjodin, and others on the state and national staff.

Then, in 1984 when Cy Carpenter becomes national president, Dennis Sjodin of Cambridge, becomes "acting" Minnesota president. Sjodin is a long-time state staff worker and special assistant to President Carpenter, and president for many years of the Farmers Union Rum River Local in Isanti Co. His appointment by the state board is temporary, until Willis Eken of Twin Valley can assume office Aug. 1 when his important state legislative chores have ended.

Sjodin's responsibilities grew at the 54th annual convention of Minnesota Farmers Union in Nov., 1995, when he defeated incumbent Norma Hanson of Goodridge in a contest for the state vice presidency. Sjodin's credentials are plentiful, going back to 1966 when he became a part-time field representative. He also headed the Green View program in Minnesota, was director of the joint MFU-NFO livestock marketing program, and has for years been in charge of the MFU Building at the Minnesota State Fair.

Studying the detailed Isanti Co. secretary's recorded minutes and notes, one can't help but be impressed by the selfless dedication those thousands and thousands of words represent, the thought and discussion behind what is written, and all of the work and duty it suggests took place over the years, from the time the county charter was granted in 1954 up to 1990, the period pertaining to the records made available to the writer of this book.

You can't help but be impressed by the concise, well-written and well-reported summaries of county meetings that commanded their presence, and attention, for four or five hours each session. Those carefully recorded paragraphs communicate a sense of sharing, building and growing, and a strong feeling that more should be done in behalf of family farmers – especially "young farmers", as reflected by a resolution adopted in 1973 calling for low-interest, long-term loans to those beginning young farm couples.

Or the sincerity behind Joanne Sjodin's appeal to campaign for a "no" vote in the soybean checkoff referendum in response to a 1984 letter sent to the county by Minnesota Farmers Union President Cy Carpenter urging opposition to the measure.

Or, the concern which developed for a new "Farmers Union" oil cooperative, and led to the opening in 1959 of the Rum River cooperative featuring products supplied by Farmers Union Central Exchange.

In addition, when one reads county convention minutes showing how a candidate won or lost a bid for an office by only a vote or two, there is the feeling of keen competition to represent the members. And a sense of admiration for the author as you study the neatly-recorded minutes carefully preserved in the pages of the Secretary Report Kit (which, according to the cover, cost 50 cents) by persons like Roman Jaloszynski, Violet Anderson, Myra Westphal and Marian Eklund.

Those pages provide a real sense of the kind of dedication, loyalty and support given to Farmers Union over the years. And, especially when one considers that such records represent only one eighty-fourth of what took place at the county level in Minnesota.

THE 30-YEAR CO. RECORD

Here's a self-styled summary of Isanti Co. Farmers Union activity over a 30-year period with reported events and happenings selected at random from information and records generously provided but not entirely complete:

1954 : July 19 – The first record of a meeting of the Isanti Co. Farmers Union was held in the Coop Hall in Cambridge, with Alden Anderson, president of the Rum River Local, the only member absent. Secretary Doris Blomgren did not state in her minutes how many Locals were there. The main order of business was a vote to ask the state office for a county charter. The leaders also made plans for an exhibit at the upcoming county fair.

Oct. 11 – Franklin Lundquist of Braham was elected President at the first Isanti Co. Farmers Union meeting. Jack Laase of Rum River Local was elected vice president, and Mrs. Reuben Johnson of Rum River Local became secretary-treasurer. The main action was to adopt a resolution calling for government price supports at 90 percent of parity. Other resolutions opposed ending the state "colored margarine (oleo) tax", opposed a state sales tax, and urged building better farm-to-market roads in rural Minnesota. Speakers included Farmers Union Editor Milt Hakel.

1955: April 11 – Representatives of the three Farmers Union Locals met in Co-op Hall, where the chief action pertained to a county picnic.

April 27 – What was described as "a very inspiring talk" by Minnesota Farmers Union President Edwin Christianson was the highlight of a meeting in Co-op Hall.

May 19 – Guests included officers of Farmers Unions in Chisago, Anoka, and Mille Lacs Locals. Plans were made for a three-county Farmers Union picnic June 19.

Oct. 3 – Alden Anderson of Rum River Local is elected county president. Harold Pearson of Braham Local becomes vice president, and Roman Jaloszynski of Isanti Local is named secretary-treasurer. Farmers Union organizer Archie Baumann was a speaker. A resolution which suggested 100 percent of parity supports on first $25,000 of farm income was adopted.

Oct. 14 – Invitations for picnic speaker are sent to Rep. Coya Knutson, Senator Humphrey and Senator Thye.

1956: Mar. 16 – Gov. Orville L. Freeman is speaker at Isanti Co. Farmers Union fund-raising dinner (Tickets cost $1.50) in Cambridge High School Auditorium. Attendance was 480. Isanti Co. has 407 paid-up members (Cambridge Rum River – 179; Isanti – 126; and, Braham – 102.).

Oct. 13 – State Farmers Union President Edwin Christianson was Isanti Co. annual meeting speaker. He reported Farmers Union membership nearing 40,000 in Minnesota, compared to 12,000 when he became president six years earlier. He said Minnesota Farmers Union now was the largest farm organization in the state.

1957: Oct. 26 – Delegates ask President Eisenhower to fire his controversial Agriculture Secretary, Ezra Taft Benson. Carlton Anderson of Isanti Local is elected county president, while Arnel Beckmann becomes vice president.

Nov. 25 – Secretary Roman Jaloszynski announces Farmers Union Grain Terminal Association bus tour for Isanti Co. members.

1958: Feb. 11 – Discussion centers on how to get more business for the Isanti Co. Co-op Oil Co. Suggestion is made for it to buy half of its petroleum supplies from Farmers Union Central Exchange.

Oct. 25 – John P. Anderson is elected county chairman, with Franklin Lundquist of Braham Local selected vice president. Roman Jaloszynski continues as secretary-treasurer. Annual meeting speaker was Clinton Hess, Minnesota Farmers Union secretary.

1959: Rum River Farmers' Co-op Oil Association is open for business, with advertisement promising "prompt delivery" of Farmers Union products. Former county president Alden Anderson is manager.

Feb. 28 – Fieldman Archie Baumann reports 355 Farmers Union members in Isanti Co., and announces 1959 goal of 460.

July 19 – Five counties (Isanti, Anoka, Chisago, Mille Lacs and Kanabec) take part in annual picnic sponsored by East Central Farmers Union Picnic Association at Barr's Resort on Spectacle Lake.

Oct. 3 – Membership goal is surpassed, with 464 enrolled in Isanti Co. (Rum River–179; Isanti – 198; and, Braham – 87). Resolution calling for a minimum wage of $2 per hour for farm labor is adopted.

1960: Franklin Lundquist assumes county presidency, while Forrest Westphal of Isanti Local becomes vice president. Violet Anderson is new secretary-treasurer.

Later, Forrest Westphal of Isanti is elected county president. Other county officers are John P. Anderson, vice president, and Violet Anderson, secretary-treasurer.

1961: Mar. 21 – State Farmers Union President Edwin Christianson reports membership is now 41,000 farm families, representing 30 percent of all active farmers in Minnesota. He was speaker at Isanti Co.'s annual Town and Country Meeting in Cambridge.

Oct. 21 – Farmers Union Editor Milt Hakel was county convention speaker and warned that a sales tax bill would be introduced in the Legislature "under a new name such as gross income tax". All officers re-elected, including Mrs. Deloren Sandstrom, secretary-treasurer.

1962: Feb. – Alden Anderson resigns as manager of Rum River Farmers Co-op Oil Co. to take USDA job; Gale West of Hutchinson Farmers Union Co-op takes over.

Feb. 23 – Co. Farmers Union membership is a record 495 families. State president Christianson is speaker. He comments on successful space flight of Astronaut John Glenn, the first American to orbit the earth, and praises Kennedy Farm Program which increased farm income in Isanti Co. by $570,000.

Mar. 26 – Co. Chairman Franklin Lundquist sends a letter to Agriculture Secretary Orville Freeman suggesting legislation which would provide a "license to farm" for only family farmers. This would help protect them, and "keeping big business out of farming".

Nov. 2 – Resolution adopted at annual county convention calls for publicizing names of 20 persons receiving the biggest government farm program payment to see if any are going to non-farmers.

1963: Mrs. Elmer Bodeu becomes county secretary-treasurer.

June 7 – Co. officers write Minnesota Governor raising questions about "commercial chicken" business interests and activity.

Oct. 19 – Decision is voted to donate $50 out of county organization's treasury for new Minnesota Farmers Union building fund.

1964: Aug. 28 – Co. officers vote to support livestock market withholding action of National Farmers Organization (NFO).

Oct. 19 – State President Edwin Christianson was featured speaker – again – at annual Isanti Co. convention held in the Legion Hall.

1965: Nov. 6 – Cy Carpenter spoke on "The Future of Farming" at the annual county convention. Mrs. Franklin Lundquist becomes secretary-treasurer.

1967: Oct. 21 – Carl Bergstrom is elected county chairman. Other officers: vice president – Ralph Blomgren; Secretary – Edna Lundquist. Cy Carpenter spoke on "The Sales Tax".

1968: Feb. 5 – Chairman Carl Bergstrom reported on Feb. 23 Farmers Union bus trip, Legislative Fly-In to Washington, D.C. Feb, 18-20, and delegates to National Farmers Union convention in St. Paul beginning Mar. 17.

1969: Oct. 18 – Marvin Wallin of Rum River Local is elected vice president.

1972: Oct. 21 – Discussion at county convention about having one meeting for all Farmers Union locals "because of poor attendance at the local level". But no change is made. Marvin Wallin is elected president; Carleton Anderson of Isanti Local is chosen vice president. Mrs. Thelma Binger becomes secretary-treasurer.

1973: Oct. 27 – Co. delegates urge special low-interest loans be established for young farmers. Special speaker was Markell Gjellstad, youth and family director for Minnesota Farmers Union.

1974: June 11 – Dennis Sjodin of Cambridge reported on new Farmers Union reward signs available to all members and new ones recruited.

Oct. 26 – In response to State President Cy Carpenter's invitation for county Farmers Union to select long-time members who helped build the organization, and to be considered for special recognition at upcoming state convention. Isanti Co. chose the Ralph Johnsons for the honor.

1978: Mar. 31 – Co. president Marvin Wallin reports membership at 170 (Isanti – 73; Cambridge (Rum River) – 62; and, Braham – 35).

1979: Mar. 30 – Franklin Lundquist spoke to county officials about "getting young people into the Farmers Union and active in cooperatives", recorded Secretary Myra Westphal. Her minutes noted Lundquist's long career helping the cooperative movement, and that he felt "Farmers Union is the only one fighting hard for co-ops".

Oct. 5 – Annual convention speaker Roger Vogt explained the difference between "budget fund" (money donated by members and locals to be used for lobbying activity) and Farmers Union's Education Fund (which can be used only for promoting cooperatives through educational and related activities).

1984: April 11 – Dennis Sjodin of Cambridge becomes "acting president" of Minnesota Farmers Union after Cy Carpenter is elected president of National Farmers Union. Sjodin will hold the office temporarily until Willis Eken of Twin Valley takes over Aug. 1. Eken's term was postponed until his legislative chores as majority leader in the Minnesota House of Representatives ends.

Sjodin, a family farmer with his wife Joanne and some of their children, has been assistant to the state president, field representative, director of program development, and president of the Rum River Farmers Union Local for many years.

Dec. 3 – Co. leaders approve suggestion by Joanne Sjodin that advertisements be used to promote a "no" vote in upcoming soybean checkoff referendum, and that telephone calls be made to growers promoting a "no" ballot. The action came after a letter was received from State President Cy Carpenter urging growers to defeat the checkoff measure, according to minutes taken by Marian Eklund.

1985: At county convention Dennis Sjodin explained the recent sale of Minnesota Farmers Union headquarters building in St. Paul. It was, Sjodin said, "a good time to sell, space wasn't needed, and rental property was available".

Sjodin also discussed the farm economy, reporting, "It's even worse than many realize – a few years ago farmers were encouraged to increase, expand, get more efficient, and now, they claim they are poor managers... "

THE "CORE" – THE FARMERS UNION LOCAL

At the core of Farmers Union is the local organization, the grassroots group that is the heartbeat of the entire movement, the unit which links the member to the county, state and national segments. It is the "action unit", where things begin to happen.

This is the story of one local, Becker Co.'s Sugar Bush Local #171

National Farmers Union "Leadership Achievement Award" presented to Sugar Bush Local, Becker Co., Minnesota, for "increased organizational activity and membership growth... "

(Signed) Leland H. Swenson, President, David E. Carter, Secretary, 1991.

"I don't think we've ever missed (holding) a monthly meeting... "

Blanche Rousu Niemi, Becker Co. Youth Leader, Charter member, Sugar Bush Local No. 171.

It wasn't exactly a landslide election of a president when the new Farmers Union local was organized in Becker Co., Minnesota, on a Saturday night in June, 1944. Minutes of the meeting show only seven people voted. And the man they elected president, Joseph Werre of Ogema, declined the office.

Nevertheless, it was the first official written record of activity for Sugar Bush Local No. 171, and a document carefully preserved in a musty, wrinkled and tattered official "Farmers Union Secretary-Treasurer Record Book" commonly used in that generation when the movement was relatively young.

This part of Farmers Union history was jotted in ink in the official record book by Emma Rousu of Ogema who was "unanimously elected" that June Saturday night when the birth of Local #171 took place in what was then "Finn Hall" – a rural community center built by the Finnish Society in the Strawberry Lake area.

Emma Rousu was to be an active member/leader for a half-century in Sugar Bush, and her daughter Blanche was to be a local and county youth leader for nearly 40 years as this history is being written. In addition, 47 members of the one family – children, grandchildren and great-grandchildren – achieved the Farmers Union Torchbearer award. That's almost half of all Torchbearers (97) originating from Sugar Bush Local for the years 1944-1997.

State Farmers Union President Einar Kuivinen of New York Mills spoke at Sugar Bush's second meeting (on July 12, 1944) and told the group that farmers were pretty much in the same economic position as organized labor.

At only its sixth meeting, the members voted to buy the Finn Hall facility which became known as the "Farmers Union Hall". It was used for dances, pie socials, basketball for junior members, and other activities as well as for Farmers Union meetings, and occasionally was made available to other groups, like 4-H.

In June, 1946, a letter from Finland was read at the monthly meeting. It thanked Sugar Bush for its shipment of "relief clothes" sent to that country. The Farmers Union local had shipped 145 pounds of clothing and $30 in cash to Finnish Relief, Inc., three months earlier. Minutes show that many of the residents in Becker Co. boasted Finnish heritage.

In 1947, members elected George Fitch president, with Bill Koski vice president. Sugar Bush sent two juniors to state Farmers Union camp at Detroit Lakes, and Glenn Talbott of North Dakota spoke at Becker Co. Farmers Union picnic. Sugar Bush used to send 20 to 30 youths to summer camp each year. And the local established a day-camp program for three groups – seniors, reserves and pre-school.

Issues discussed at the meetings in the post-World War II era pertained to world peace, rural electrification, corporate farming, separation of Farm Bureau and Extension, compulsory military service, farm programs and such. Letters outlining how members felt were mailed to members of Congress, state legislators, and even to President Truman.

There was a "problem" noted in the Sugar Bush minutes concerning providing a delegate to state convention in 1957 – the Farmers Union convention dates conflicted with the Minnesota deer hunting season.

The Sugar Bush records indicate members often contacted government officials concerning programs, policies and proposals. At monthly meetings, letters from officials were often read – letters from people like U.S. Senators McCarthy and Humphrey of Minnesota, and U.S. Agriculture Secretary Orville L. Freeman. These letters happen to pertained to proposed increases in dairy price supports.

Senator Hubert Humphrey wrote to Ale Rousu of Callaway in Mar., 1954: "I am going to keep right on fighting against the (Eisenhower) Administration's legislative program on farm prices and against (Agriculture) Secretary Benson's actions in cutting (government price) supports... "

In May, 1955, U.S. Rep. Coya Knutson (D-Minn) wrote to Emma L. Rousu of Callaway:

"Yesterday, after several days of debate... the 90% rigid parity farm bill was passed by the House... we can now only wait and see what the Senate will do... the problem of the farmer is critical... "

Wrote Eugene J. McCarthy in part to Emma L. Rousu of Callaway in Mar., 1963: "I share your concern about the welfare of dairy farmers... (we have) scheduled hearings on bills for a dairy program... "

Secretary Freeman wrote to George A. Rousu of Callaway in July, 1966: "Senators McCarthy and Mondale have introduced a bill which would authorize a temporary program involving production adjustment payments to (dairy) producers... the possible use of direct payments... certainly merits careful study...

288

direct payments have proved to be a very useful tool in some of the other commodity programs... "

Sen. Walter F. Mondale in Aug., 1970, wrote to Phyllis St. Claire of Callaway about the 1970 farm bill: "Since farmers have been caught in a terrible cost-price squeeze in recent years – one that has made our farmers the chief victims of both recession and inflation – an improved farm bill is essential... I am doing all I can to improve the Senate version so it more adequately meets the needs of Minnesota farmers... "

Ale and Emma Rousu were original members of the Sugar Bush Local (which got its name from the Sugar Bush Lakes and from Sugar Bush Township. The Rousu couple had 15 children, including Paul, who earned the Farmers Union Torchbearer award and Blanche Rousu-Niemi, the wife of John Niemi who has been Becker Co. Farmers Union chairman for nearly 20 years.

The third generation of Farmers Union members is represented by Kenneth and Donna Niemi, who farm nearby. Three of their children (Brooks, Christopher and Monica) are Farmers Union Torchbearer award winners.

Over the years, Blanche Niemi recalls, several families have been dedicated members of the Sugar Bush Local – the Rousus, the Niemis, the Koskis, the Bakkas, the Latvalas, the Soybrings, and several others.

Things were "different" years ago, Blanche Niemi commented in a 1997 interview. "Back then where we lived there were a lot of tiny, little farms. And there was enthusiasm that those early members had for participating and working toward the goals of Farmers Union – all of them were very dedicated... maybe society has changed, maybe there's too many distractions, people busy every day of the week... whatever, the younger generation definitely hasn't been as involved... "

Today, Sugar Bush's old Farmers Union Hall now is a community-township "Town Hall" facility. However, when the hall was sold, Farmers Union retained the right to use the facility for its functions. The president of Sugar Bush is Marge Schouville, the mother of a pair of Torchbearer-winning sons (Dennis and Tom). Vice president is Nydia LeFebvre (all of her seven children took part in the Farmers Union youth program).

A LOOK AT COTTONWOOD CO.

"Prior to 1949 there was only a small membership in the (Cottonwood) county, but through the tireless efforts of George Schwandt of Germantown Township, the membership was increased several hundred and the first Local (No. 264) was organized in his home township... "

Elnora Swenson, Revere, MN.

From the records of Elnora Swenson of Rosehill Farmers Union local

Much credit for the rapid growth in Farmers Union membership in Cottonwood Co. is given to George Schwandt of Germantown Township because membership had been quite limited prior to 1949 until Schwandt began what have been described as "tireless efforts" to recruit new farmers.

The first Farmers Union local was Germantown (No. 264). Its members elected Russel Schwandt of Jeffers as the local's first president. However, membership in this local eventually was transferred to the third local organized, Jeffers No. 266 and U. D. Fisk became its first president.

Amo Township (Local No. 265) was the second to be organized in Cottonwood Co., with Albert Piper elected president (Amo later was transferred to Windom Flaxters local in Great Bend Township). Adolph B. Ness was the first president of Ann Local No. 302 organized in Ann Township on May 29, 1950. On Sept. 23, 1950, Storden Highwater Local was organized. A month later when its charter was presented, membership was 125 and Kermit Polk was chosen president.

Another 1950 charter was earned by Westbrook Local No. 305, with membership originally in four townships (Westbrook, of course, but also Rosehill Township, Dovary Township, and Des Moines River Township). Art Kuehl of Westbrook became its president. But by a year later, membership had grown to the point where the decision was made to split into two locals, Rosehill Local No. 331 in Cottonwood Co. and Des Moines Local (taking the Des Moines River and Dovary membership) in Murray Co. Elder Engler was the first president of Rosehill Local.

When Cottonwood Co. Farmers Union was established in Dec., 1949, Russel Schwandt, the son of county organizer George Schwandt, became the unit's first president. Others elected first county officers included U. D. Fisk, vice president, and Albert Piper, secretary-treasurer. However, about a year later, the county was reorganized at a meeting held in Jeffers. Greg Wagner became president, with Harry Anderson the vice president. A. J. Nelson was elected secretary-treasurer, but was unable to accept this office. Mrs. Clayton Thomas then was chosen for the post by the county board of directors. Other members of the board included George Mann (who would become a director of Farmers Union Grain Terminal Association), Art Kuehl, Ernest Johnson and E. E. Hocke.

In 1951 the Mountain Lake Local entered the Farmers Union picture, with P. P. Lohring elected president. This addition meant Cottonwood Co. was virtually covered by active Farmers Union locals, with the county organization holding monthly meetings.

However, George Schwandt, the spark-plug organizer in that county, didn't live to see it. He died in Sept., 1950. His son, Russel, boasted some of his father's traits, and was destined to become a highly successful state Farmers Union organizer. Then he was state secretary of Minnesota Farmers Union from 1952-55. After an unsuccessful bid for the state presidency against the popular Edwin Christianson of Gully, Schwandt went to work for Farmers Union Grain Terminal Association.

Then, in 1960, Russel Schwandt lost to incumbent Ancher Nelson in a Congressional race for Minnesota's Second District. However, he then was appointed state commissioner of agriculture by Gov. Karl Rolvaag.

At the county convention in Oct., 1951, Harry Anderson of Walnut Grove was elected president. George Mann of Windom was selected vice president while Mrs. Clayton Thomas continued as secretary-treasurer. Resolutions adopted by Cottonwood Co. delegates included a call for "the divorce of the Co. Extension Service from the Farm Bureau" and for farmers to receive "100 percent of parity". The speaker was Gordon Roth, director of public relations for GTA.

Then in 1952, the premier agricultural figure of the time – M.W. Thatcher, general manager of GTA – was the speaker at the county spring conference. He spoke to a crowd of more than a thousand who packed the Windom Armory for the meeting, and emphasized why he personally and the Farmers Union in particular sought "100 percent of parity" prices for farm commodities.

"Parity is a perfectly reasonable goal for grain farmers, as well as for all farmers," Thatcher said that night. "It means parity of income, parity of opportunity, parity in the ability to finance decent living standards... "

Then, at its Aug., 1952, county meeting was held in Westbrook Park where a candidate for Governor of Minnesota (and future U.S. Secretary of Agriculture in the Kennedy-Johnson Administrations), Orville L. Freeman, shared the podium with Farmers Union President Edwin Christianson.

Another major activity for members included the sponsored bus tours to the Twin Cities and regional cooperatives there, as well as to Washington, D.C.

At the 1952 county convention, president Harry Anderson asked that he not be re-elected to that office. George Mann of Windom was nominated, but declined because he was a member of the board of directors of GTA and its policy that board members cannot also hold office in any county or local organizations simultaneously. Mann also was one of 11 appointed by Secretary of Agriculture Ezra Taft Benson to a national feed grain advisory committee, and chairman of the Minnesota House of Representatives agriculture committee when he was a member of the State Legislature.

After Mann declined, Clifford Swenson of Westbrook was elected, with Claude Meade of Windom named vice president and Mrs. Clayton Thomas again re-elected unanimously. Other county offices: Educational director – Elmer Deutschmann. Legislative director – Harry Anderson. Organization director – Lloyd Giersdorf. Junior leader – Mrs. Phyllis Brandt. Recreation director – Cleo Hoffman. Publicity directors – Mrs. Clarine Takle and Mrs. Stella Piper.

The speaker at the county convention was the editor of the Farmers Union *Herald* who spoke about his recent trip to Europe.

THE SWENSONS OF COTTONWOOD CO.

THE TORCHBEARER PLEDGE
I accept my responsibility as a torchbearer of the Farmers Union, and I pledge myself to bear that torch with clean hands and a courageous heart.

I unite with my fellow Senior Youth in a pledge to work unceasingly for the cooperative movement – to practice tolerance and brotherhood – to keep my torch ever one of those which shall light the way in a warless world.

A "PREMIER TORCHBEARER FAMILY" – THE SWENSONS
The ceremony saluting the 1975 Minnesota Farmers Union class of Torchbearers and Youth Leaders featured an address by Leland Swenson, then of Huron, SD He was the secretary-treasurer of the South Dakota Farmers Union at the time, and destined to become the state president a few years later and then, the organization's national president in 1988.

Leland Swenson's speech that night was, "The Torchbearer". It was a topic he knew well, having attained that distinct honor and recognition a decade earlier himself when he was active as a "senior youth" participant in his Minnesota local (Westbrook) in Cottonwood Co.

In fact, it could be that the Swensons of Cottonwood Co., Minnesota, are the "premier Torchbearer Family" in the state. That's because Lee is one of six children, who all attained the Farmers Union Torchbearer Award, raised by Elnora and the late Clifford Swenson. This ceremony was part of the state convention held in the Prom Center in St. Paul in November, the traditional month for the Minnesota Farmers Union's annual meeting. Twenty-three young people from 10 counties became "Torchbearers" in the candlelight ceremony, receiving their awards from Cy Carpenter, the state president then and a leader who would become the national president nine years later.

Figure 103 Lee Swenson & Parents Clifford & Elnora Swenson

The reading, "Hold High the Torch", was delivered by Margaret Pribyl of Maple Lake, the Farmers Union youth leader. The ceremony closed with the group singing of the Farmers Union's "Hail Our Union" which ends with these words: "Hold the ranks – forever onward, treading paths of heroes dead, who have blazed the path before us, let us follow where they led."

Also sung that night was a song that was written to the tune of "Battle Hymn of the Republic" but which featured the lyrics authored by Ted Townsend and which focused on matters of current concern:

"The land without the people is a land without a soul,
"And the people without land become a mob without a goal,
"But the people on the land have made a nation strong and whole,
"A nation proud and free."

CHORUS:
"Keep the country in the family, keep the family in the country,
"Keep the country in the family, for a nation strong and free."

"The foreign corporations would monopolize the soil,
"Disinheriting the farmers from the fruits of honest toil.
"For their profit they propose to take, to plunder and to spoil
"Our nation strong and free."

"We have sweated in the summer sun, we've fought the winter gale,
"We have licked the drought and dust storm and the grasshoppers and hail,
"And we're not about to quit the farm, we're not about to fail
"Our nation strong and free."

A HALF-DOZEN "TORCHBEARERS" FROM A FAMILY

The family of Clifford and Elnora Swenson of Cottonwood Co. may have set a Farmers Union record: all six of their children achieved that prestigious award.

The first to become Torchbearers were the two oldest, Donald (born in Aug., 1937 and now living in Beaver Dam, WI) and Curtis (born in Feb., 1939 and now living in Lakeville), who both earned the honor in 1957.

The other Swenson Torchbearers in order: Inez (born in Dec., 1940 and now living in Fond du Lac, WI), Kenneth (born in June, 1942 and now living in Marshfield, WI), Kathryn Swenson Erickson (born in Oct., 1945 and now living in Thief River Falls), and Leland (born Nov. 25, 1947 and now living in Evergreen Co., Colorado).

Inez in the spring of 1960 was appointed a delegate to participate in The White House Conference on Education in Washington, D.C.

The last of the six Swensons to achieve Torchbearer status, Lee Swenson, was to advance upward through the Farmers Union ranks to its loftiest post – the national presidency.

The Swenson family was extremely active in Farmers Union, and operated a diversified farm in southwestern Minnesota where they raised crops, livestock and had a dairy herd. Leland Swenson was elected to the National Senior Youth Advisory Council in 1965. He had attended Mankato State College and served in both the U.S. National Guard and the U.S. Air Force. He worked for the local Cenex cooperative in Windom, MN, and operated a sprayer-fertilizer business before entering the Air Force.

In 1971, Lee Swenson was appointed secretary-treasurer of South Dakota Farmers Union. Then, seven years later he was named field coordinator for the National Farmers Union at its headquarters in Denver. He coordinated the successful "Save Our Co-op" campaign. In 1980 he returned to South Dakota as field services director. A year later he was elected president of South Dakota Farmers Union and Farmers Union Service Association.

In 1984 he coordinated the founding of the Farm Alliance in South Dakota, bringing together farm and church groups to seek better farm and tax policies. And a year later he organized a "rural crisis rally" in which 8,000 people visited the state capitol. Then, in 1988, Leland H. Swenson would succeed a fellow-Minnesotan, Cy Carpenter, as National Farmers Union's tenth president.

ABOUT CLIFFORD AND ELNORA SWENSON...

Clifford Swenson, who was born and raised in Cottonwood Co., accompanied a brother-in-law to a Rosehill Local meeting in the fall of 1948. Not only did he join Farmers Union that night, but he returned home with the job of secretary-treasurer. The local had over 100 family members then, and each had to be send cards by mail notifying them of Farmers Union meetings and activities.

Then, Clifford became a state convention delegate. And later, he was elected president of the county Farmers Union organization in 1952 – an office he held through 1957. He was "a firm believer in Farmers Union and its family farm policy, and a strong believer in cooperatives", recalled his widow, Elnora Swenson.

Elnora Swenson, who grew up in Cottonwood Co., was county Farmers Union secretary for a number of years. She is best-known for her work in education, having been the chief organizer of efforts to establish youth groups in 1952 when 14 young people attended Farmers Union camp. She was then elected to the state Education Council in 1955, a post she would keep for 20 years – a period of unmatched growth for Farmers Union in Minnesota. She resigned the position only because she and her husband were moving to Colorado.

In retrospect, Elnora Swenson recalls a wide range of activities – taking part in educational conferences, helping with summer youth camp chores (including the year when she was a cook and took young Leland with her while her husband cared for the other five at home), attending the inauguration ball of Governor Freeman in 1955, being a national convention delegate, and Farmers Union group trips, bus tours, fly-ins, and such.

She also remembers how her Rosehill Local put on several plays for the community as money-raisers, and received support from the public. One of those who really enjoyed this activity was Clifford Swenson, who invariably played the lead role. He also organized one-day, Twin Cities tours featuring the two big farm cooperatives, Farmers Union Grain Terminal Association (the luncheon stop for the group) and Farmers Union Central Exchange (coffee and pie were served to the visitors there).

Clifford Swenson was an active membership worker, winning a trip to the national convention (which he gave to a friend when his schedule didn't permit him to attend). "He enjoyed doing membership work, and talking to people about Farmers Union and what it was doing for farm people," Elnora said. Then, when elected to the Agricultural Stabilization and Conservation committee, Clifford had to give up his Farmers Union county presidency.

A highlight came in Nov., 1953, at the state convention, when the Clifford Swenson Family received the Minnesota Farmers Union Farm Family award – an honor based on activities in the community and other factors. The Swensons, indeed, were kept busy with church, 4-H Club work, Future Farmers of America (FFA) involvement and other activities, including Farmers Union. All six young Swensons attended the Farmers Union All-States Camp in Colorado – Donald and Curtis in 1957, Inez in 1959, Kenneth in 1960, Kathryn in 1962, and Leland in 1965.

Here's a brief look at each of the Swenson offspring:

Donald – He went to England in a farm youth exchange program in 1957, and received his Torchbearer award in Dec., 1958, after his return. He is a U.S. Army veteran, and now manages a civic arena in Beaver Dam, WI.

Curtis – He was elected to the Farmers Union National Junior Advisory Council in 1957, the same year he became a Torchbearer. The following year he was a speaker at Central Exchange's 27th annual meeting. He is a Lakeville teacher, and also is a U.S. Army veteran.

Inez – She was invited by President Eisenhower to participate in the Golden Anniversary White House Conference on Children and Youth in 1960, an event begun by President Theodore Roosevelt. She worked on the Farmers Union camp staff several years, and has been a teacher and librarian in Fond du Lac, WI.

Kenneth – He received his Torchbearer's award in 1960, and became a counselor at the Bailey Education Center. After farming for a while, and working in Windom, he attended college and became a teacher in the Marshfield, WI, schools for more than 30 years.

Kathryn – She worked at Minnesota Farmers Union state camps several years, and became a counselor at All-States Camp in Colorado for a few years. She attended the All-States Camp as a Torchbearer honoree in 1962. She works for an electronic company in Thief River Falls.

Leland – An All-States camper in 1965, the year he became a Torchbearer, he was elected to the national advisory council the following year. His career has involved numerous state and national Farmers Union positions before he was elected National Farmers Union president in 1988. He has been re-elected unanimously in 1990, 1992, 1994, and 1996.

Clifford and Elnora Swenson bought and operated a small grocery store in Revere after their farm sale in 1967. They moved to Colorado in 1973 to work as caretakers at Farmers Union Educational Center in Bailey Co. After retiring in 1982, they moved to South Dakota where Clifford again became active, helping with Farmers Union membership work in Iowa, Kansas and Oregon.

Then, a month after celebrating 50 years of marriage, Clifford died of cancer in Feb., 1987 – a year before his youngest son attained the highest Farmers Union office possible. (When his son disclosed his candidacy, Clifford told him he thought he "was a little too young" for the national office being sought also by the president of Minnesota, Willis Eken, and Rocky Mountain Farmers Union, John Stencel.)

However, Elnora Swenson was in the Albuquerque convention audience when their son, Leland, was elected national president.

Later, Elnora Swenson moved back to Cottonwood Co., to Westbrook and her Farmers Union roots of so long ago. And, her story doesn't end with her homecoming. She was elected secretary-treasurer of Rosehill Local – a job Clifford accepted 45 years earlier – and a post she still holds as this is being written.

THREE "TORCHBEARERS" HONORED IN 1943

At the second state convention of the new Minnesota Farmers Union in Alexandria in 1943, National Farmers Union President James G. Patton presented the awards to three junior members recognized as "Torchbearers". Patton told the group, "I present you with this badge, the highest honor and symbol to be bestowed on you by the National Farmers Union."

Honored were Monophaye Hagen of Crookston, Ellsworth Smogaard of Madison, and Jaroslav Kruta of Gatszke. In an interview more than 50 years later, Smogaard recalled getting the award. "There was no national convention that year because of the war (World War II)," he said. "We (Torchbearers) all got a U.S. War Bond." He earned the award the same year he graduated from high school.

Smogaard reported he "just sort of grew up with Farmers Union", mainly because his dad was involved early in the original organization. He remembered the first Minnesota Farmers Union president, John Erp, who came from the southwestern part of Smogaard's county (Lac Qui Parle Co.). "He (Erp) had the state office in his home near Canby, and was pretty determined everything had to go his way, and it didn't go too long before it broke up," he said about the early leader.

Then, when the new organization was getting started, Smogaard became involved in junior class work which statewide then was directed by Junice Dalen of Madison. Farmers Union grew gradually, he recalled, and in the mid-1960s had over 1,200 members active in nine locals in the county. "There were a lot of good local activities back then," Smogaard, a member of Lakeside Local, recalled. "But we had a lot of people in rural areas. Now, we've got a lot of senior citizens who are sort of out of it now, and we've got the big farmers who don't seem to have time to go to meetings."

He was county secretary of Farmers Union for a number of years and the Lac Qui Parle Co. chairman, a post he had to resign in 1974 when he was elected to the Minnesota House of Representatives where he served two terms. Among others who served with distinction in this charter county: Lowell Miller, Milo Hanson and Harold Windingstad. Each was a member recruiter, county president and member of the state board. In addition, Windingstad was the longest serving congressional district president the Minnesota DFL. Hanson served for many years as state soil and water conservation director.

Looking back, Smogaard felt M.W. Thatcher, the long-time general manager of Farmers Union Grain Terminal Association was one of the most influential persons he had known. "We have had a lot of good leadership over the years. But you have to give a lot of credit to Bill Thatcher. At that time things were going great. He kept the board (of directors) in order, lived by the rules and gave good education funds to Farmers Union. Thatcher helped Farmers Union a lot. You could always get him to come to meetings, and educational funds meant an awful lot to him. This is why we're hurting now because our co-ops don't have the membership and all of them wanted to drop the Farmers Union name. This has hurt us."

Smogaard recalled "strong leadership" in the state Farmers Union, mentioning Edwin Christianson and his immediate successor, Cy Carpenter. He was a member of the Minnesota Farmers Union executive committee when Christianson, the state president for more than 20 years, suffered a disabling stroke that left him speechless in Nov., 1971, and Carpenter was selected to take over.

Smogaard reflected, things "are different – years ago Farmers Union was more effective when everybody belonged." He added, "Now, farmers got to belong to their own thing – they got to belong to corn growers, the wheat growers, they don't realize we all got the same thing to sell – our time, and that if we worked together, we'd get some place... ".

THE FARMERS UNION EDUCATION COUNCIL, 1942-72

In 1942 the Education Council of Minnesota Farmers Union was first organized. It was to be used more extensively later, during the years when Lura Reimnitz was the director and record growth was taking place in the organization.

For years, members of the Education Council were the camp staff as well as those who provided materials and direction to working with junior and senior youth, and the Torchbearer Awards Program.

The state council was discontinued after Lura's retirement in 1972, with senior youth beginning to assume camping staff assignments along with other responsibilities including conferences.

Records show these members and years of service to the council:

Freda Eisert, Euclid (1943-47). Ione Kleven, Appleton (1943-44, 1947-48). Lulu Pearson, Roseau (1943-33). Laura Wolff, Montevideo (1947-48). Ruth Lorenzen, Lake Park (1943-44, 1947-48). Evelyn Jacobson, Lake Lillian (1943-4, 1947-48).

Mable Mosbeck, Red Lake Falls (1947). Mrs. Clarence Anderson (address unknown) (1947-48). Alice VanDyke, Bejou (1945-48). Frances Wolfe, Bemidji (1945, 1950). Lura Reimnitz, Kennedy (1945, part-time). Mrs. Alfred Johnson, Underwood (1945). Fred Harris, Spicer (1946).

Laura Long, Ortonville (1947-48). Nolla Audette, Red Lake Falls (1949). Verna Owens, Crookston (1949). Myra Kruschke, Ortonville (1949). Alvera Harris, Spicer (1949). Mrs. Arley Miller, Cottonwood (1949). Jean Pastir, Red Lake Falls (1950-58).

Mrs. Ed Tuveng, Fosston (1951-61). Mrs. Art Gjervold, Moorhead (1950). Mrs. Lemuel Swenson, Spicer (1950). Audrey Jerve, Clarkfield (1950). Mrs. Clayton Thomas, Windom (1951-54). Mildred Seelig, Barrott (1951). Mrs. Rufus Lundquist, Parkers Prairie (1952-57). Mrs. Milville Kleberg, Willmar (1952-68).

Elnora Swenson, Westbrook (1955-72). Luella Jacobson, Hitterdal (1958-72). Alice Grove, Roosevelt (1962-72). Bessie Klose, Atwater (1965-72).

Figure 104 Clint Hess

CLINT HESS – "DESTINATION WASHINGTON"

A man of varied talents was Clinton V. Hess of Minneapolis. For nearly a decade, beginning in 1955, he was state secretary of Minnesota Farmers Union. Earlier Hess had been assistant to State Secretary Russel Schwandt. One of his favorite projects was the "Fly-In" to Washington, D.C., where Farmers Union members participated in legislative activities.

To commemorate the Fly-In, Hess produced a special documentary on 16mm movie film titled, "Destination Washington". This video captured the excitement surrounding a trip to the nation's capitol and the opportunity to visit with key Congressional and agricultural leaders.

Then, in June, 1964, Hess headed to Washington to accept the post as director of international projects for National Farmers Union. His job was to coordinate Farmers Union programs operated through the U.S. Agency for International Development in South America, Africa and Asia. These projects included farm exchanges that once brought 81 farm and ranch leaders from South America to Midwest states.

Hess' promotion came after 13 years on the Minnesota state Farmers Union staff where he represented the organization on various rural and civic committees, including being an officer on the town and country commission of the Minnesota Conference of the Methodist Church.

Figure 105 Archie Baumann

Succeeding Clint Hess as state secretary of Minnesota Farmers Union was the multi-talented Archie Baumann, who then had been on the state and field staff for 10 years. In addition to his regular staff chores, that included working with groups like CROP (Christian Rural Overseas Program) and Future Farmers of America, Baumann sketched editorial cartoons almost regularly for Farmers Union publications, including Minnesota Agriculture.

His flair for drawing is matched by his knowledge of politics, issues and concerns that form the main message Baumann seeks to communicate. As the old saying goes, "One picture is worth a thousand words", in Baumann's case, a lone cartoon can speak louder than a dozen well-prepared editorials.

A PAIR OF FARMERS UNION WORKERS... THE GRASDALENS

In Nov., 1985, Marion Grasdalen of Albert Lea accepted a special Minnesota Farmers Union award for her late husband Ray, a dedicated Farmers Union member and worker for many years. The plaque presented at the state convention recognized Mr. Grasdalen's organizational work for the farm organization.

Both Grasdalens were active in Farmers Union. Ray for years was area field supervisor for a four-county southern Minnesota area – Freeborn, Mower, Steele and Waseca. However, his skills were often sought by state officials when encountering "a problem" elsewhere. "Ray was a good trouble-shooter," explained F. B. Daniel, long-time state staff member. "If there was a problem that needed solving, he often was the one asked to look into it. And, he never let us down."

Ray and Marion Grasdalen began life together as newlyweds in the depths of the Great Depression – 1932, when, she recalled, "Hogs sold for 2 ½ cents a pound, butterfat for 17 cents a pound, and eggs for six cents a dozen." Those early, tough financial years, she said many years later, made the farm couple more appreciative of the better times that followed.

Ray Grasdalen was a doorman at the Minnesota House of Representatives, a local assessor, census-taker, and officer in his Lutheran Church and its

Brotherhood. Both he and his wife held offices in their Farmers Union Local, Riceland, for many years, with Marion a junior leader for some time.

STATE VICE PRESIDENTS...

Thirteen have held the office of Minnesota Farmers Union vice president, including William Nystrom of Worthington who had the longest tenure – 21 years, from 1951 to 1972, when he declined to seek re-election.

Nystrom was a long-time Nobles Co. farmer, orchardist and one-term member of the Minnesota House of Representatives from the 11th district in 1934-36 sessions. He was born on Nov. 16, 1902, in Indian Lake Township where he was to spend his entire life, and was only 32 years old when elected to the state legislature. Nystrom married Violet Pratt on Jan. 4, 1937. He died Dec. 23, 1989 shortly after his 87th birthday. Survivors included three daughters.

Nystrom's unmatched tenure as Farmers Union state vice president corresponds with the years Edwin Christianson of Gully held the state presidency.

Second-longest tenure is held by Vere Vollmers of Wheaton, who became a special vice president in 1974 and who was elected state vice president in 1978. Vollmers' Farmers Union career includes achieving the Torchbearer award, a seven-year stint as president of his local, and then chairman of the Traverse Co. Farmers Union. He also served as a member of the National Farmers Union resolutions committee.

MINNESOTA FARMERS UNION EXECUTIVE COMMITTEE... 1997

Chairman – Merlyn Hubin, Westbrook. Vice chair – Orion Kyllo, Pine Island.

Secretary – Bessie Klose, Atwater. Vice president of young farmers – Randy Schmiesing, Chokio; Darrel Mosel, Gaylord; Jeanne Wertish, Renville. Other members – Dave Johnson, Fergus Falls; Gary Gregerson, Badger; Roger Vogt, Palisade; Dennis Sjodin (Minnesota Farmers Union vice president), Cambridge.

State Board of Directors (county presidents):

Aitkin – Roger Vogt. Becker – Roger Schaefer. Beltrami – Adrian DeVries. Benton – Bud Lubbesmeyer. Big Stone – Bud DeNeui. Blue Earth – Francis Bach. Brown – Fred Tauer. Carlton– David Borchardt. Carver – Ken Kirchenwitz. Cass – Dave Butcher. Chippewa – Lyle Koenen. Chisago – Roger Eklund. Clay – Larry Jacobson. Clearwater – Richard Moen.

Cottonwood – Merlyn Hubin. Crow Wing – John Borden. Dakota – Orin Legare. Dodge – Wava Larson. Douglas – Marvin Jensen. Faribault – Palmer Langsev. Fillmore – Eunice Biel. Freeborn – Donald Gooden. Goodhue – Orion Kyllo. Grant – Bob Klaasen.

Houston – Maynard Welscher. Hubbard – Jim Gildersleeve. Isanti – Alice Jaloszynski.

Jackson – George Paulson. Kanabec – John Ripka.

Kandiyohi – Bessie Klose. Kittson – Hugh Hunt. Lac Qui Parle – Harold Windingstad.

Lake of the Woods – Grant Slick. LeSueur – Rebecca Carson. Lincoln – Gary Van Overbeke.

Lyon – Nels Myhre. Mahnomen – Jean Nelson. Marshall – Steve Sparby. Martin – At Large.

300

McLeod – Leonard Pikal. Meeker – Glensdale Rumsey. Mille Lacs – Bob Williams.

Morrison – Mike Kliber. Mower – Lorene Ingvalson. Murray – Martin Hoekman. Nicollet – Bruce Hulke. Nobles – Joe Landhuis. Norman – Robert Hoekstra. Olmsted – Mike Clemens. Otter Tail East – Jim Adamietz. Otter Tail West – Dave Johnson. Pennington – Korydon Chervestad. Pine – Bob Pigeon. Pipestone – Marvin Conrad.

Polk – Conrad Zak. Red Lake – Steve Linder. Redwood – Richard Berg. Renville – Gary Wertish. Rice – Gene Werner. Rock – Ron Rentschler. Roseau – Gary Gregerson. Scott – Gerald Williams. Sibley – Lowell Grams. Stearns – Curt Wegner. Steele – Duane Hortop. Stevens – Randy Schmiesing. Swift – Judith Anderson.

Todd – Herman Childberg. Traverse – Alan Peterson. Wabasha – Mike Wobbe. Wadena – David Gilster. Waseca – Jim Byron. Washington – Bill Herzfeld. Watonwan – Steve Rumsdahl. Wilkin – Robert Roach. Winona – George Brown. Wright – Kevin LeVoir. Yellow Medicine – Morris Behrman.

13... GREEN VIEW – A FARMERS UNION SUBSIDIARY...

"We have not yet recorded the history of Minnesota Farmers Union... If we fail to do it, I'm afraid its efforts to preserve family farm agriculture will eventually go unnoticed and unappreciated. We can't let that happen... "

Don Knutson, Executive Director, Green View, Inc., Nov., 1995.

Figure 106 Green View contract signing, 1969. L to R: Wm Heuer, Ed Christianson, Dale Wreisner, Richard Braun, Percy Hagen

How best do you measure the success, or even progress, of a program? Do you use numbers of dollars, or projects, or people? What kind of yardstick reflects true value?

When it comes to Green View, Inc., of Minnesota, Executive Director Don Knutson has his own method. Simply put, Knutson reported at the 1996 state convention of Minnesota Farmers Union, it was the best year yet for the organization.

Records were set in several meaningful categories – number of workers (600 on the payroll of Green View, Inc., with 800 workers participating), the number of work sites (200 locations across the state), and the size of the budget (project revenues topped $4 million for the first time, only three years after they first reached $3 million).

Yet, impressive as these statistics may be, Knutson added in his 1996 convention speech in Minneapolis: "Our real success of Green View, however, is clearly not solely economic. The real success can be measured through the efforts of hundreds of people across Minnesota who provide exceptional work to the Minnesota Department of Transportation, to the Minnesota Department of Natural Resources, and to several other government entities under (Green View) contract."

Here are the details of Green View with its staff of only six persons:

Green View's biggest contract is with the Minnesota Department of Transportation, and involves 77 highway rest area locations.

There is at least one Green View work site in nearly every Minnesota county, with a state total of 200 work locations statewide.

The Green View program growth is largely due to new and expanding contracts, particularly with the Minnesota Department of Natural Resources (forestry and parks work experience programs).

A report on Green View is an annual event at the Farmers Union state convention because this organization is a subsidiary of Minnesota Farmers Union. And, Knutson always notes the important role this organization played in the origin of Green View, Inc.

"Farmers Union members should take a great deal of pride in the role this organization played to create Green View in 1969," he noted in his annual report 27 years after Green View was formed. "We have enjoyed a very successful year due in large measure to the dedication and hard work of hundreds of individuals across Minnesota."

Then he added, "While we continue to look for new contract opportunities, our primary focus is the employment of low income, elderly Minnesota residents."

THE BIRTH OF GREEN VIEW

Providing responsible, paying jobs for older, limited income residents in Minnesota was the goal when a group met in the late 1960s to discuss ways to meet the needs of a growing population of low-income, elderly residents. Taking part in that discussion were high-level state officials, representatives of the labor movement, and the Minnesota Farmers Union. Out of that discussion emerged an embryonic Green View, Inc., organization.

One of those present was Cy Carpenter; one of Carpenter's deep-seated interests over the years was working with other groups, including organized labor, the clergy, and others in seeking ways to better serve rural people and communities. He felt strongly that alliances with other councils, associations, and organizations could be beneficial for rural America.

Back in 1969, when Green View began, it was a new, untested, idea. That first year it achieved a lone contract, with the Minnesota Department of Highways. Twenty-six years later, it still has only one contract with the Minnesota DOT. However, it also has seven contracts with the Minnesota DNR, six with county governments (Stearns, Winona and Wright counties are good examples), and two with Minnesota cities.

Total revenue generated that first year totaled slightly over $100,000, with 55 Green View program workers at 10 highway rest areas. Contrast those small numbers with the Green View of 26 years later – 600 workers, contracts totaling in excess of $4 million, and 200 job sites (76 highway rest areas, 100 parks and forestry sites, and a couple dozen small contract locations).

From time to time, Knutson said, "We still hear someone asking the question, 'What's Green View got to do with Farmers Union?' My response to that question is, 'A lot.' There's a historical connection, there's a participation reason, there's a credibility factor, and, a financial relationship."

Green View is nearly one-half as old as Minnesota Farmers Union, and the organization is the source of many Green View employees, who bring to their jobs the dedication and work ethic of farm people. At the same time, Knutson sees "Green View bringing to Farmers Union an additional level of credibility which money can't buy, credibility delivered every day through the actions and deeds of Green View workers across the state".

BACKGROUND OF GREEN VIEW

The minutes of the executive committee of Minnesota Farmers Union and an early report prepared by Green View, Inc., provide historical background of the organization. The concept of this program had its inception with the origin of Green Thumb, a federal program authorized under the Older Americans Act to help provide employment for senior aged women and men living on small, fixed incomes in rural communities.

Green Thumb started in Minnesota in Feb., 1966. Seventy people were employed that year in Ottertail, Todd and Wadena counties. In 1967 sufficient funding came to expand Green Thumb into 10 counties, with 210 workers. In 1972 there were 303 "Green Thumbers" in 21 counties.

In keeping with the Farmers Union tradition of helping people and building communities, when the Minnesota Department of Transportation announced it would build a number of all-season rest stops along freeways in Minnesota, Farmers Union promptly asked who would staff these facilities. Since this was not yet established, Farmers Union asked for a contractual agreement whereby Farmers Union would recruit and manage the operation of these facilities, employing low income older workers. This required legislation to modify the regulations for DOT on outside contracting and the establishment of a non-profit entity to assume separate and arms-length operation from Farmers Union in accounting and auditing. Necessary legislation was quickly passed, with strong support from labor, and Green View was established as a non-profit Minnesota corporation.

MAJOR DEVELOPMENTS RELATED TO GREEN VIEW IN 1969

First, the organization held its first official meeting at the Rumble law firm office May 12, and articles of incorporation were approved and filed. Minnesota Farmers Union President Edwin Christianson became president of Green View, Inc., and state vice president William Nystrom was elected to that position with the new subsidiary. Gale Haukos was secretary-treasurer and Cy Carpenter was named assistant secretary, with authorization to sign all necessary papers and act on behalf of Haukos. The remaining board members were identical to the Farmers Union state executive committee. Green View's designated bank was First National Bank of Wadena.

Second, that same day in 1969 a new branch of Green Thumb was created, and set up expressly for women. It was called Green Light. It hired 45 female employees that initial year, with Lucille Pfeffer of Bertha hired as its director. By the beginning of 1973, Green Light reported 49 workers in seven Minnesota counties. Green Light was merged into Green Thumb in 1971. In 1990 there were 600 women employed part-time.

THE ORIGIN OF THE NAME "GREEN THUMB"

In 1970 Dr. Blue A. Carstenson, national director of Farmers Union Green Thumb, Inc., of Washington, D.C., spoke at the annual Minnesota Farmers Union state convention in St. Paul.

"Give the poor a chance for a job, and to earn their way out of poverty with respect and dignity," he told the delegates. Green Thumb, he added, offers "a workable and effective alternative to public welfare".

Minnesota Green Thumb Director Paul Anderson credits Dr. Carstenson with helping found and name this organization. Writing in Minnesota Agriculture, Anderson said Green Thumb "grew out of the vision of three Farmers Union leaders – Jim Patton, Red Johnson and Dr. Carstenson. National Farmers Union President Patton formed the National Policy Committee on Pockets of Poverty in 1963, and then urged President Johnson to "declare war on poverty". Then Patton hired Dr. Carstenson to work with the Johnson Administration to develop community action programs which focused on the needs of impoverished rural areas where there was a need to help farmers who had retired and needed to supplement low retirement incomes as well as to remain active and feel productive members of society.

Dr. Carstenson is said to have commented to Ladybird Johnson, who advocated a highway beautification effort, that "the National Farmers Union proposes to take the green thumbs of the poor, older, and retired farmers and put them to work to beautify our nation's highways."

Green Thumb and Green Light, Dr. Carstenson said, have served as a pilot program of the U.S. Department of Labor. These jobs have helped people cope with what he said was a serious lack of nutrition among the nation's older citizens. "We need a new program of food and nutrition... including meals-on-wheels, senior citizens' lunch programs, and other nutritional efforts to help older Americans who are not getting enough to eat," Dr. Carstenson said.

He added that innovative programs like Green Thumb and Green Light, which employ older and retired low-income workers in jobs dealing with ecology, conservation and beautification of the environment, can help stem the migration from rural America and help meet the problems facing the 20 million Americans who are past age 65.

The first contracts were signed in Dec., 1965, to hire 70 men each in four states – Minnesota, New Jersey, Oregon, and Arkansas. Minnesota Farmers Union president Edwin Christianson took a strong interest in the program, and Todd,

Wadena and Otter Tail were the first three counties in the state to participate. The main activities were highway beautification, tree plantings and park development.

Alec Olson of Spicer, former Minnesota Lieutenant Governor, Congressman and long-time Farmers Union member, was national administrator of Green Thumb from 1982-88.

The first Green Thumb director in Minnesota was Percy Hagen, who directed the program from its inception in 1965 to 1978 when he became the first state director of Green View, Inc.

In Minnesota, the organization Green View, Inc., was a direct outgrowth of the state's successful Green Thumb program. In 1967 the Minnesota Department of Highways (now DOT) indicated it would like to hire senior citizens like the Green Thumb workers to be in charge of the network of rest areas being developed as a part of the Interstate Highway System as well as at locations on major highways in the state. However, at the same time, the highway officials let it be known that such a plan would not be possible unless legislation was enacted providing for contractual arrangements permitting the state agency to contract for this work.

After legislation was enacted the following year (1968), the executive committee of Minnesota Farmers Union took steps to create a new, non-profit corporation called Green View, Inc., with a mission of promoting employment among needy, older Minnesotans.

In July, 1969, Minnesota Farmers Union president Edwin Christianson signed a contract with state highway officials which marked the beginning of Green View, Inc., a new non-profit subsidiary of Farmers Union. Christianson said the new organization would supply personnel for the staffing of rest areas on designated freeways and primary highways across the state. Christianson said that the executive committee of Minnesota Farmers Union would be the board of directors for Green View, Inc.

In 1971, when Edwin Christianson served as chairman of the National Green Thumb advisory committee, he participated in several Green Thumb events, including the dedication of a plaque commemorating the work of Green Thumbers in developing a Covered Bridge Park in Zumbrota. Christianson also visited Bemidji State Park facilities where Green Thumb workers prepared park signs and did other woodworking projects. At Walker that year, Green Thumb workers constructed a bark-covered wigwam and wildlife display for the city museum. The state director of the program then was the original director, Percy Hagen.

Few then could envision how this new program – Green View, Inc. – originating from the federal Green Thumb program would grow. Where there were 40 men working at seven rest areas in 1969, three years later, Green View, Inc., was responsible for 18 rest areas with 92 men employed at rest areas along Interstates 94, 35, and 90, as well as along primary roads near Frazee, Elk River and Anoka, north of Stillwater on U.S. 95 and near Garrison on U.S. 169 (the latter two were seasonal rest areas).

The 1972 report noted how a contract with the forestry division of the State Conservation Commission (now DNR) provided seasonal jobs for 14 workers who maintain camping locations in state forests near Zimmerman, Bemidji, Deer River, Hibbing, Virginia, Orr, Tower, Grand Marais, and Hovland. These workers cut wood for fireplaces, tidied up camping areas, hauled out rubbish, and did general maintenance work.

The pay in 1972 was $2 an hour, with a limit of 32 hours per week. Foremen at rest areas were paid monthly ($348 to $375), depending on whether they supervised one or two highway rest areas. One-half of the workers were in their 60s, while nearly 40 percent were 70 years of age or older. Only 10 percent of the workers fell into the 55 to 60 years age group.

There are, however, some inherent problems with such an aged work crew, based on the first Green View, Inc., report for its initial year (July 1, 1969, through June 30, 1970): "July 3 was the opening date for the Kettle River Rest Area along I-35... we had no Green Thumb program there, so we found it necessary to rely on the (Minnesota) State Employment Service as well as the Farmers Union to locate men within commuting distance who were eligible for work... this held true for the General Andrews Rest Area located a few miles north on the opposite side of I-35... of 11 men originally hired at these two places, one is deceased, and one quit because of illness... " At another location, Lake Hansel in Douglas Co., four men were hired. "But of this original group, one died, one quit to take a better paying job, and one had to quit because he had earned the maximum allowed by Social Security... "

In 1973 Minnesota Farmers Union President Cy Carpenter received a letter on Green View, Inc., letterhead which listed its office in Wadena, MN. The writer, William Heuer, a farmer and former state legislator from Hewitt, reported new contracts totaling nearly a half-million dollars for 1973-74. "Of this amount, 82.3 percent was allocated for worker wages, 3.2 percent for administrative salaries, and the balance (14.5 percent) covered Social Security, Workmen's Compensation, liability insurance, mileage expenses for workers and administrators, per diem expenses, uniforms for the workers, and office expense... " There were 12 working on DNR projects, 110 at Class I rest areas and 14 at Class II rest areas.

In 1974 there was a threat that the Nixon Administration was attempting to phase out Green Thumb nationally, according to Executive Committee minutes for Feb. 19. President Carpenter urged keeping Green View under a Farmers Union program then reported to have about 250 workers employed seasonally. In July Farmers Union legislative director David Velde was named Green View state director, to succeed William Heuer who would retire on Dec. 1, 1974. Velde will continue his legislative duties.

A 1975 letter to Carpenter signed by Larry Roe, Green View, Inc., administrative assistant to David Velde, provided an update. There were now Green View workers operating 24 Class. Rest areas and 10 Class II rest areas, Roe reported. In addition, Green View had a dozen workers assigned to nine locations for the DNR.

The Roe report also noted "a problem" at the Albert Lea I-35 rest area where there had been an abnormally high turn-over of workers. An investigation indicated a misrepresentation of duties given by the local Manpower Services office which led applicants to believe they would be "official greeters" at the I-35 Rest Areas and simply stand around visiting with tourists rather than the real chores having to do with daily maintenance and care of the facility.

In 1976, the bank account was shifted from Wadena to the Drovers State Bank of South St. Paul, and the state Green View office moved from Wadena to Minnesota Farmers Union's new offices at 1717 University Avenue. In 1977

Dennis Sjodin was named manager to replace David Velde on Green View's staff. And in June, 1978, Green View had two full-time staff members, Percy Hagen and Larry Roe. Hagen later was named director.

Twenty-some years later, Green View Executive Director Don Knutson reports the completion of rest area evaluations on all 55 Class I sites in Minnesota, the development and coordination with DOT of a rest area landscaping pilot project for selected areas where the goal is to improve overall rest area appearance, expanded contract coverage in the cities of Inver Grove Heights and Roseville and in Winona and Stearns counties. Also Green View published an employee newsletter, called View Points, in 1996.

In his 1995 report, Don Knutson noted:

A goal of providing financial support to cover the costs of caretakers at Farmers Union Park and Campground at Lake Sarah in 1996. This would be over and above the existing commitment of Green View, Inc., to Minnesota Farmers Union.

Providing financial assistance to prepare and publish of a history of Minnesota Farmers Union so that "its efforts to preserve family farm agriculture will not go unnoticed and unappreciated".

Knutson in particular paid tribute to Dennis Forsell, Green View's senior staff member and director of field operations, as "a major factor" in the growth of the organization. But Knutson also noted the "teamwork effort" of the other staff as well – Joe Kausner (field supervisor), Carolyn Atz (personnel director), Jody Meyer (administrative assistant), and Nancy Beimert (accounting).

Dennis Forsell, Knutson pointed out, has been very important in the life of Farmers Union and Green View. Forsell joined the state staff of MFU in 1972 as director of organization. The Green View project was only three years old, and because of a small working budget and the fact that the highway rest areas where it operated were located far apart, the MFU field staff helped monitor things.

"We helped interview new employee-applicants when we happened to be in the area," recalled Forsell about those early years. In 1978 Percy Hagen resigned as director of Green Thumb to become the first full-time executive director of Green View. Hagen is credited with establishing strict, uniform statewide standards for maintenance of the rest area facilities. When he retired in 1991, Don Knutson was his successor. Knutson had been assistant to the president of Minnesota Farmers Union.

Knutson's technical skills and his experience in accounting and working with computers coupled with the assistance of accountant Nancy Beimert, upgraded Green View's bookkeeping system so it was efficiently more compatible with that used by the State of Minnesota. This expedited necessary paperwork and transmissions between Green View and state workers and improved the working relationship.

Further, Forsell recalled, Knutson's leadership resulted in "a much higher level of expectation of services and responsibility by crew leaders". Also, he added, "Field service spends much more time in training and working with each crew leader."

Such things, Forsell said, "have combined to give Green View a highly respected image throughout the state". He feels, as a long time employee, "Green View has become one of the most productive and highly-respected programs in the state, and will continue to be a credit to both Farmers Union and the State of Minnesota for many years to come."

GREEN THUMB STATE DIRECTORS:

Percy Hagen, Wadena 1965-78
Paul Anderson, Wadena 1978 to present.

GREEN VIEW STATE EXECUTIVE DIRECTORS:

Wm Heuer, 1969 – 1974
Dave Velde, 1974 – 1977
Dennis Sjodin, 1978 - 1978
Percy Hagen, 1978 - 1991.
Don Knutson, 1991 to present.

14 ...ANOTHER FACET: INSURANCE

(Much credit goes to Floyd C. Borghorst, regional manager, Minnesota Farmers Union Insurance, 1954-1989, and to Paul E. Huff and Raymond F. Novak, authors of "A Rich Heritage of Community", published in 1995.)

Farmers Union's "Farm Neighbor" Award Goes To Floyd Borghorst
News item, Dec. 8, 1988, *Minnesota Agriculture.*

Figure 107 Floyd Borghorst

Minnesota Farmers Union agency regional manager Floyd Borghorst has been presented 1988 MFU "Farm Neighbor" Award. Borghorst has been manager of Farmers Union Insurances in Minnesota for nearly 35 years. The award was presented at the annual Farmers Union state convention in St. Paul.

Borghorst, a graduate of the University of South Dakota, joined National Farmers Union Insurance Companies in 1951 as a district manager. Since then he has attained designation as chartered underwriter for property, casualty and life insurance. He was honored for his long and distinguished career with Farmers Union and its insurance affiliates. Since Floyd Borghorst jointed the organization, Minnesota insurance premiums have grown from $154,000 annually to more than $15 million.

Long-time National Farmers Union President James G. Patton, who was born in 1902 – the year the organization was founded, is the originator of Farmers Union life insurance services. In 1937 a life insurance operation was established in Colorado. Patton was 35 years old, and destined to become the seventh president of the non-profit Farmers' Educational and Co-Operative Union of America, or as it was better known, "Farmers Union", in 1940.

Patton had helped Farmers Union create the "Farmers Union Mutual Life Association" in 1932. It was an assessment burial company, and Patton became its president. In 1934 he became secretary of the Colorado Farmers Union as well.

Marketing insurance was far from new in 1937. Even earlier, beginning in 1912, state Farmers Unions in seven states (Arkansas, Colorado, Kansas, Montana, Nebraska, Oklahoma, and Washington) had created mutual property insurance companies to provide needed property coverages for their family farmer-members. North Dakota founded its mutual company many years later, in 1944, according to "A Rich Heritage of Community", a history of the National Farmers Union Insurance Companies, authored by Paul E. Huff and Raymond F. Novak and published in 1995.

Despite the widespread economic problems during the difficult depression years of the early 1930s, Patton invited C.E. Huff, then 55 and who had been national president from 1928-1930, to head up the effort to establish a national life insurance company. The home office would be located in Denver.

However, before the Farmers Union life insurance could be made available nationally, it had to be reorganized on a broader scale and be highly promoted, according to a historical review of Farmers Union Life by Floyd C. Borghorst, the regional manager of Farmers Union Insurance in Minnesota for 35 years.

Why establish a life insurance company in 1937, in the midst of the Great Depression? James Frederickson, the Minnesota Farmers Union agency director, provided this answer in a 1997 interview. "In order to stabilize and even increase its membership during these very tough financial times (the 1930s) in rural America and to supplement its dues income, National Farmers Union needed to provide a (insurance) product or service that could offer a product or service that could be attractive to its members and their neighbors," Frederickson said. "Also, many Farmers Union people had experience in running insurance companies."

However, an entirely new company had to be set up. But it could not be established under Colorado mutual benefit laws which were seen as too narrow for a national program. So the new company was established under "fraternal laws". This translated into no less than 500 applications of $1,000 or more. It didn't take long to meet those requirements. On April 20, 1938, National Farmers Union Security Association was born – a name later changed in 1945 to National Farmers Union Life Insurance Company.

The birthplace was a small northeastern Colorado town (Peetz) where 72 applicants took physical exams in one day to qualify for the first Farmers Union life policies. Then in 1941 the new company received a fortuitous opportunity – providing term life insurance policies for all borrowers of the Farm Security Administration in seven states. The FSA agency had been established during the Roosevelt Administration to provide operating funds after banks had failed during the early depression years and farmers had trouble getting loans even to plant a crop. This agency eventually became the Farmers Home Administration and in the 1990s was known briefly as Rural Economic Community Development, and later became "Rural Development – USDA". Then, in 1952, C.E. Huff, the general manager of Farmers Union's insurance companies nationally, recommended that

312

dues billings be added to the renewal premiums of Farmers Union policy holders – a suggestion that was enacted and continues today.

It should be pointed out that the insurance companies were initially organized as "a service" to members of Farmers Union. These companies had a couple of significant fringe benefits:

Agents could provide good public relations by promoting Farmers Union philosophy as they contacted farmers about their insurance needs.

Insurance could provide financial help to both national and state organizations. (The original allocation was 2 percent of all new and renewal premium given to state units as "Service fees" while one percent would be paid to the national organization. Then in 1958 the state share was reduced to ½ percent and the national set at ¾ percent.)

It was for these same reasons that insurance operations were begun in Minnesota. Over the years service fees to both the state and National Farmers Union have totaled in excess of $5 million.

Because the Farmers Union's Property and Casualty Company, which was chartered in 1945, lacked a Minnesota license, members were able to get auto insurance from a company known as "American Farmers". The first insurance checks for this protection were issued in the fall of 1946 in Minnesota. Oscar Haugo of Waubon was the first to get a commissions check (for $5.50 in Sept., 1946). Others who received auto insurance commission checks shortly later were A. M. Larson of Climax, George Magnuson of Fosston, Randolph Windingstad of Dawson, and Arnold Ackerman of Willmar.

Erwin Reimnitz, brother-in-law of Minnesota's long-time educational director Lura Reimnitz, joined the insurance operation in Oct., 1946. He was active as an agent until he retired in 1963. Also in 1946, Bud Johnson of Montevideo was appointed to build the insurance program in Minnesota. He was succeeded by Jack Witt of New York Mills.

The first Farmers Union Insurance service to be offered to members in Minnesota was the health program in 1947. Agents including Randolph Windingstad, Alton Isaacs and Oscar Nordine helped establish the first county health groups were. Isaacs was the 1975 "agent of the year" while Nordine was given a special plaque in 1976 for his many years of service. Windingstad and Isaacs worked in the insurance program for more than 40 years.

In 1949 the insurance activity in Minnesota had not progressed as much as some had hoped. Consequently, the state board of directors asked the National Farmers Union to take over. Arthur R. Tisthammer, a native of Nebraska, was asked to be state manager. In five years state premium increased sharply, from just under $72,000 to nearly $880,000.

This remarkable feat under the Tisthammer regime won for him a regional, six-state insurance post in 1954. However, less than a year later he left the insurance companies to become an assistant to the Minnesota Farmers Union President Edwin Christianson.

The number of Farmers Union agents grew to nearly one hundred, with many of these new agents farmers supplementing their income selling insurance part-time. One of these part-timers was Robert Bergland of Roseau, who later became a Congressman, then, U.S. Secretary of Agriculture and finally, the executive director of the National Rural Electric Cooperative Association. Bergland was an agent from 1950 to 1953. Another part-time insurance agent was Robert Wiseth of Goodridge, son of former state Farmers Union president Roy E. Wiseth.

It was in Feb., 1951, that Tisthammer hired a recent graduate of the University of South Dakota who would play an important role in Minnesota Farmers Insurance for nearly 40 years. He was Floyd Borghorst, a South Dakota native.

In 1951 the agents first began writing Farm Liability policies in Minnesota. There wasn't much demand then, since lawsuits were few and far between and judgements generally in those days were small.

By 1953 there were 11 insurance districts in operation. District managers were: Herman Anderson, Baudette; Oscar Nordine, Lancaster; Don Ogaard, Ada; Julius Fossum, Kragness; Walter Grothe, Benson; Louis Eichorst, Montevideo; Cecil Hofteig, Cottonwood; Milo Johnson, Ihlen; Ivan Wyum, Hills; Elmer Deutschmann, Windom; and, G. H. Featherstone, Hastings. Floyd Borghorst, who had been a district manager in northern Minnesota, was named state life supervisor in Mar., 1953. In Sept., 1954, Vivian Otto was hired to manage the office for the insurance operation with additional responsibility for accounting and agent records for the state. She served in this capacity until retirement in 1991.

In Nov., 1954, Floyd Borghorst hired Alec Olson of Pennock to assist with Life Insurance production in areas not assigned to district managers. Later he became District VI manager in southwestern Minnesota. In 1962 he resigned to run for Congress and was elected to the U.S. House of Representatives, and later headed up Green Thumb, Inc., nationally where older, low-income persons are employed.

In 1955 crop hail insurance was made available to members. Crop hail was the ninth insurance line offered — the others were auto, farm liability, life, cargo, fire and extended coverage, personal comprehensive liability, accident and health, and animal mortality.

Concerned about farm safety and in particular the farm machinery hazard on highways and public roads, Farmers Union in 1954 came up with a simple, effective and inexpensive program – providing free three-inch reflectorized tape that could be affixed to farm machinery. This tape was provided until 1960 when Minnesota law required white and amber lights plus red reflectors on such equipment moving on public roadways. Then in 1968 the now familiar triangular "slow moving vehicle" emblem was required by law for vehicles moving 25 miles per hour or less.

Jorgen Fog, who had worked from 1955 to 1958 with the Minnesota insurance operation, left a family insurance business in Lisbon, ND, to manage the Farmers Union's property operation. It was during his tenure that Minnesota developed the picture program where photos were required on new and renewal policies. When the property insurance operation was returned to the home office in 1961, Jorgen Fog became a district manager – a post he was to hold for more than 30 years.

By 1958 premium income in the Minnesota insurance program grew to more than $2 million, compared to $671,000 in 1953. This growth enabled the transfer of the state Farmers Union insurance operations from the national office to the Minnesota Farmers Union on July 1, 1958, and Farmers Union Agency, Inc., was created.

Leo Klinnert of New York Mills became a Farmers Union Insurance agent in 1959. He also was the president of the East Otter Tail Co. Farmers Union president. His son-in-law, Craig Palan, took over the insurance account in 1986.

In short, the decade of the 1950s was a period Floyd Borghorst described as a time of "remarkable growth" for the Farmers Union insurance activity. Here is a summary of annual gross premium by year: 1950 – $154,379; 1951 – $267,535; 1952 – $455,158; 1953 – $670,976; 1954 – $879,209; 1955 – $1,206,341; 1956 – $1,287,546; 1957 – $1,693,529; 1958 – $2,074,659; 1959 – $3,228,734.

This increase in premium related to membership. There were 17,000 members in the Farmers Union in Minnesota in 1950; by 1954, the state membership had grown to 30,000. By 1958, there were 41,068 farm families belonging to the organization.

Borghorst gives much credit for the success of the insurance operation to Raymond Novak, the president and general manager of National Farmers Union Insurance Companies. Novak worked in North Dakota's insurance program for years, and was state manager for 10 years before heading up the national insurance activity for Farmers Union. Novak was general manager from 1960-1969, and then when the title was changed, he became "president" of the insurance companies for the balance of his career, 1969-1984. He was the eldest son of Aton Novak of Alexander, ND His father was an early Farmers Union activist, and a member of the board of directors of the Farmers Union Livestock Commission Co. in the 1930s-1940s.

When he retired in 1989, Floyd Borghorst described those early insurance days in an article in Minnesota Agriculture. He was a district manager based in Thief River Falls in the early 1950s when part of his duties included organizing Farmers Union locals. Borghorst would line up 20 members, then call state president Edwin Christianson or an organizer to charter a new local.

This worked out well for Borghorst, because he needed 10 members to write a group hospital insurance plan. In one 16-week period he established 13 hospital groups. And, all told, he figured signing up 250 to 300 Farmers Union members in 10 locals.

In the 1950s, he recalled, there were about 120 part-time insurance agents. This changed, though, as the business expanded and more types of policies became available and more complicated and specialized.

In 1963, the Minnesota insurance operation was sold to National Farmers Union Service Corp. because of slower than anticipated growth in the state. It was felt that Minnesota, a major "cooperative state" with two to three times the farm numbers compared to the Dakotas, should have seen its insurance business growing far more rapidly. However, the business was among the first to feel the brunt of the recognized constant exodus from rural America which at times accelerated in relation to the status of the farm economy. Jorgen Fog became a district manager – a post he held for more than 30 years prior to his 1992 retirement.

Then later, a major financial problem emerged and the National Farmers Union had to sell a significant portion (90 percent) of its insurance business to the Baldwin Piano Company that later became Baldwin United. Its president, Morley Thompson, was a proponent of "bigger is better", and urged the national insurance companies to invest more money in individual states for growth. Minnesota was one of the target states, and received large subsidies during the 1970s. which sparked a premium increase from $4.1 million to nearly $20 million in 10 years. The number of new agents increased from 62 to 76, with more agents able to meet the new "Mile High" Club qualifications. This meant those agents won trips to places like San Francisco (1970), Nassau (1973), Mexico City (1976), Lake Tahoe (1977), and Hawaii (1978).

A "special accounts" department was established in 1965 to provide a multi-line package to certain concerns, including rural electric, telephone and cable cooperatives as well as major or regional cooperatives. By the end of 1977 Farmers Union had written $5.3 million worth of new premium for organizations, including: Farmers Union Central Exchange (Cenex), $1,648,000; the Farmers Union Grain Terminal Association, $565,000; Green Thumb, $243,000; and, local cooperatives and Farmers Union Marketing, $88,000.

About that same time Minnesota manager Floyd Borghorst met with the Iowa Farmers Union, and outlined plans to expand insurance activity in that neighboring state. Iowa was split into three districts and 32 agents were hired. Then in 1981 the Iowa operation reverted to direct control out of the home office in Denver. But, because of the "farm crisis" of the 1980s, the national organization had to withdraw from five states in 1987. Iowa was one of those states. But good growth continued in Minnesota – except for one thing: special account losses were mounting.

Then, by 1982, in the throes of the worst depression since the 1930s, losses mounted to the point that a premium increase was ordered. Then the GTA policy was transferred, and $2.5 million in premium lost. Also, Cenex shifted a portion of its insurance business away from Farmers Union causing the group accident and health division to lose two-thirds of its income. In addition, workers compensation premium was dwarfed by its claims increase. The bottom line was that by 1984, special accounts premium income had declined by a disturbing amount – $5 million in Minnesota.

In 1986 Stanley Moore of North Dakota became president and CEO of National Farmers Union Insurance Companies. One of his first objectives was to shift the Farmers Union Agency of Minnesota out from under the ownership of the National Farmers Union Service Corp. to Minnesota Farmers Union. The offering price of $100,000 was to be paid in yearly installments of $10,000. However, the state board of directors declined, and on Dec. 31, 1986, the Insurance Acquisition Corporation (IAC) assumed ownership. "It has proved to be an excellent investment for I.A.C.," noted Borghorst in his recounting of the transaction.

And, in a somewhat ironic turn of events, Baldwin United, the corporation which acquired most of the Farmers Union insurance interests earlier, found itself in serious financial difficulty.

And Farmers Union, by establishing the IAC and with support and help from members and cooperatives, was able to buy back the share of its insurance business sold to Baldwin United.

THE FARMERS UNION CLAIMS DEPARTMENT

Prior to 1955 local adjusters operated directly under the home office insurance claims department in Denver, CO, while working out of their offices in Thief River Falls, Paynesville, and Sanborn – the areas where most of the Farmers

Union premium was being written then. But then Archie Dawes became the first state claims manager and worked out of the new state claims office.

Dawes was succeeded by a North Dakota attorney, William Murray, in 1957. Murray, in turn, was succeeded in 1962 by Joe Scherman who was the state claims manager for 18 years. In 1986 Jim Sullivan, a Wisconsin claims official, became the Minnesota assistant claims manager. A year later, after the claims operation for six states (Kentucky, Illinois, Indiana, Ohio, Washington, and Wisconsin) was placed under the supervision of the Minnesota office, Sullivan was put in charge, a position he still holds today.

THE "MILE HIGH" CLUB

Over the years various incentive programs have been developed and are still used to motivate insurance agents and boost morale. In 1963 the "Mile High Club" was founded as a vehicle to recognize and reward top premium producers within Farmers Union insurance companies.

Then, later the leaders of the "Mile High Club" organized another group, the "President's Council". The top Farmers Union producer annually becomes president of the Council.

Over the years there have been 24 "Mile High" sales contests, with winners earning trips to vacation jaunts in places ranging from Hawaii and Mexico to cruises in the Caribbean as well as trips to Orlando and Washington, D.C.

Minnesota agents have participated in these contest and insurance campaigns, with a dozen earning trips to Vancouver, British Columbia, Canada, in 1997.

"If the farmers don't make it, nobody else does."
Sam Genereux, Thief River Falls agent.

Sam Genereux, one of the Farmers Union's most successful career insurance salesmen, had a strong feeling about the importance of a good farm economy. If farmers aren't doing good, neither will anybody else in rural Minnesota, he was known to say.

However, in his case, there developed a special relationship which in 1976, for the second consecutive time, the Thief River Falls insurance agent was named "President of the President's Council". That's the loftiest honor possible in the Farmers Union insurance arena.

That title carried with it the distinction that Genereux was the No. 1 salesman in the National Farmers Union Insurance Companies network for two years in a row. As such he was the leader of the ninth "Mile High Club" where qualifying agents won trips to places like Vancouver, Canada, where the annual awards meeting was held in 1975. And then again Genereux won a trip to the tenth "Mile High" gathering held in Mexico City the following year.

Genereux operated out of his office in Thief River Falls, a city of then about 9,000 located in northwestern Minnesota about 60 miles south of the Canadian border and about 45 miles east of Grand Forks, ND The city serves a farming and urban area that in the mid-1970s represented an estimated 20,000 people in total.

In an interview for The Farmers Union Agent publication, Genereux expressed his philosophy about rural America and Thief River Falls in particular: "It is a very prosperous town, but if the farmers don't make it, nobody else does."

Red Lake Co. then had "104 percent" of the farmers signed up as Farmers Union members in that county, with Genereux explaining that this total includes some members who live in town but own farms or have farming interests. Nearby Pennington Co. then boasted 88 percent Farmers Union membership.

Sam Genereux was born in 1922 in Terre Bonne, an all-French community not far from Thief River Falls. He became a member of a U.S. Army Ranger Regiment during World War II, enlisting shortly after completing his high school studies. He made three landings (Sicily, Italy and southern France) and spent 42 days pinned down by German fire on the Anzio Beachhead. In all, Genereux won seven medals and decorations without being injured. "I could run faster than the bullets," he joked about his military career.

Later, though, as a farmer, he wasn't to be quite as lucky – he lost his right arm in an accident on his farm. Then, in 1958, after working as a truck driver, farmer, and salesman, he was asked by Minnesota Farmers Union state insurance manager Floyd Borghorst to join his insurance team.

However, Genereux was to eventually end the fruitful Farmers Union relationship, and began a business of his own and competed against his former employer – a step which naturally provoked many of his former Farmers Union associates.

BORGHORST RETIRES; FREDERICKSON TAKES OVER

On Dec. 31, 1988, Floyd Borghorst retired, ending a Farmers Union insurance career that had begun 37 years earlier, and relocated in Florida. His achievements included developing a full-time agency force generating over $20 million of annual premium and profits for the agency owners, Farmers Union Insurance Acquisition Corp.

Figure 108 James D. Frederickson

His successor was James D. Frederickson who had been first a life supervisor, and then a district manager in central Minnesota since 1976. Frederickson grew up in a Farmers Union family in west central Minnesota. His father was president of both Swift Co. Farmers Union and his local. Raised near Murdock, and a cousin of David Frederickson, president of Minnesota Farmers Union since 1991, Jim

318

Frederickson participated in Farmers Union camps and other youth activities. Jim Frederickson is married, and he and his wife have two sons.

THE INSURANCE PROGRAM TODAY

The objective outlined by Farmers Union veteran leader Stanley Moore of North Dakota to return Minnesota's insurance operations to Minnesota Farmers Union was realized on Jan. 1, 1995. Jim Frederickson credits three people who played strategic roles in this development – Stanley Moore, the president and CEO of National Farmers Union Insurance Companies; David Frederickson, president of Minnesota Farmers Union; and David Velde of Alexandria, legal counsel for Farmers Union.

Since this transfer, the Minnesota agency has continued to grow and prosper, according to Jim Frederickson. It has 40 full-time agents and eight employees who produce over $23 million in premium. In addition to its property and casualty insurance products, the agency also has a brokerage department which allows it to provide insurance needs not available under Farmers Union. Further, the Minnesota agency offers a full line of financial service products such as life insurance and annuities, health, disability and long term care, retirement and estate planning and mutual funds.

The business activity of the insurance operation contributes to the financial security of Minnesota Farmers Union. For example, in the two years the agency has been under control of Minnesota Farmers Union, it has generated over $235,000 in service fees and profits – a good investment for the organization, Frederickson points out. "Farmers Union Agency will continue to grow and prosper," he added. "Its agents, managers and employees are committed to provide service, not only to Farmers Union members, but all of rural Minnesota and those who support it."

"BEHIND OUR SUCCESS... "

Several individuals merit credit for the success of the Farmers Union insurance operation in Minnesota, including those mentioned earlier in this chapter, Jim Frederickson pointed out in 1997. Despite the chance of missing deserving persons, he listed the following for special mention:

Elmer Deutschmann, who began his career with Farmers Union in 1951 and who has developed the Windom account to one of the larger operations in Minnesota. He has qualified for several "Mile High" awards and continues to represent Farmers Union at this writing.

Alton Isaacs, a veteran insurance worker and who boasts the longest active career – 48 years. He has made the Montevideo agency one of the top units in the state. Vivian Otto served the agency as office manager for 38 years. Steve Hedeen was a district manager in the southern part of the state for 32 years. His son, Dan Hedeen, is the agent in Goodhue and has qualified for several "Mile High" contests.

George Tengwall not only developed a sizeable account in Willmar and qualified for many "Mile High" awards, but has had four sons enter the business as well.

For 38 years Jorgen Fog, now retired, was a key insurance staff member – a district manager and the state manager of property insurance operations. Kenneth

Jones was a district manager for over 20 years. The "manager of the year" title was won by district manager Rodney Allebach in 1992, and by Jon Hedman, a 15-year employee, in 1993. Hedman has been a Moorhead agent, company education director and district manager for southern Minnesota.

David Kompelien is an eight-year veteran as agent and district manager in eastern and central Minnesota. His son Steve Kompelien is an agent in Owatonna, and has qualified for the "Mile High Club".

Three generations are represented by the Schuster Family. Albert Schuster (1951-66) developed the Morris area. He was succeeded by his son, Dean Schuster (1963-93), a "Mile High" winner and President's Council award-winner. The third family member is Tom Schuster (1993-present), the son of Dean Schuster. He also has qualified for those same insurance honors won several times by his father.

The late Maurice Melbo developed the St. Paul office agency prior to his death in 1994. He was a former Farmers Union bus driver who joined the agency in 1967.

Other notables: Dave McCollum, Mahnomen, 1978-present. Ron Hanson, Rochester, 1989-present, a "Mile High" qualifier. Mark Sagvold, Moose Lake, 1983-present. Dan Sjostrand, Hallock, 1978-present, "Mile High" member. Ron Nerstad, Spring Grove 1978-present.

Also, Deb Breberg, Dawson, 1991-present, a "Mile High" qualifier who succeeded Randolph Windingstad, one of the agency's first agents in 1947. Joyce Presley-Johnson, Crookston, 1990-present, also a "Mile High" clubber.

And, Dennis Fjeld of Moorhead, another "Mile High" agent, 1989-present. Don Noetzelman, Alexandria, 1967-present, also a "Mile High" winner. Bob Pampusch of Elk River, 1978-present, a "Mile High" qualifier, as is Ron Solheim of Fergus Falls, 1975-present, who also is a member of "President's Council". Joe Griesch of Little Falls, 1979-present, is another "Mile High" member.

A father-daughter insurance team belongs to Mervin Eischens of Canby, 1951-present, and his daughter Joyce Eischens who will succeed her dad.

Also, Clarence Anderson, Cokato, 1963-present, and Jerry Sullivan, Redwood Falls, 1965-present, another "Mile High" member, as is Ken Lindsay, St. Paul, 1985-present, and Helen Johannes, Willmar, 1991-present.

And, Jeri Nesland, New York Mills, 1997-present; Rodney Mathsen, Ada, 1990-present; Tom Hollingsworth, Benson, 1993-present; Jerry Wisuri, Menahga, 1992-present; and, Delbert Audette, Thief River Falls, 1992-present.

More "Mile High" winners are David Henke, Hutchinson, 1977-present; Steve Cattnach, Luverne, 1981-present; Sue Christen and Barb Sweep, both of Fosston, 1995-present; and, Paul Johannes, Olivia, 1992-present.

Also, Don Mackey, St. James, 1982-present; Dennis Klocow, Marshall, 1993-present; Gloria Heinen, Princeton, 1979-present; Steve Dostal, Roseau, 1997-present; Eric Frederickson, Dayton, 1994-present; Alan Neurer, Bemidji, 1994-present; and Wade Motter, Farmington, 1992-present.

PREDICTION – "WE'LL BE AROUND A LONG TIME... "

Effective in 1987, the State of Minnesota required agents in the state to complete 20 hours of "continuing education" annually in order to retain their insurance licenses. This educational program coupled with Farmers Union's statewide district meetings for agents has helped their agents to be recognized for their broad knowledge of insurance, according to Borghorst in his history authored at the close of his 34-year career.

"We have a group of fine agents," Borghorst wrote. "We work with Farmers Union – an organization dedicated to helping not only farmers but all people in America. The Farmers Union will be around a long time into the future, and Farmers Union (insurance) agents will be there, too, ready to serve... "

At this writing, despite economic fluctuations, droughts and floods and even the farm credit crunch of the 1980s, Borghorst's forecast appears right on target.

ANOTHER FARMERS UNION COOPERATIVE – FUMPA

When the Equity Cooperative Exchange tried to rent space in the Livestock Exchange Building in South St. Paul in 1916, they were turned down cold. That's because back then cooperatives represented change, and a threat to the business-as-usual atmosphere at the yards.

However, Equity leaders were determined to have a livestock market presence. And they solved the space problem simply by building Equity's own facility right across the street from the Exchange – a location where the farmer-owned market outlet did business for 54 years.

Some of this history was related by long-time marketing official Ed Wieland in a news feature published in *Minnesota Agriculture* in July , 1986. Wieland told how Farmers Union Marketing and Processing Association, "FUMPA", represents what were two organizations originally – Farmers Union Marketing Association and Farmers Union Processing Association.

One of the early managers of Farmers Union Marketing was Marv Evanson and Milton Holtan was manager of the Farmers Union Processing Association in Redwood Falls. Following World War II, Glenn Long, just returning from service, was hired to manage the marketing association with the idea of renewed activity. Long was a very active and progressive individual. His long time association with Ed Christianson and his strong support for Farmers Union assured continuous cooperation between the two. Long's enthusiasm drew needed attention and support for the marketing association, but as both Swift and Armour packing plants left So. St. Paul and more and more livestock was sold direct or in the country, the inevitable decline of So. St. Paul stockyards as a major livestock market continued and accelerated. Facing the inevitable, the two Farmers Union cooperatives were merged to become Farmers Union Marketing and Processing. The headquarters were established at Redwood Falls and Milton Holtan was named general manager. Glenn Long became manager of the Union Stockyards Company, a private company, and moved to St. Joseph, MO., the company headquarters. He later died unexpectedly of a heart attack. It was Long who originally hired Cy Carpenter to work northwest Minnesota to recruit livestock for So. St. Paul and Fargo markets and to work with livestock truckers. In harmony with Ed Christianson, Carpenter's duties also included working with local and county Farmers Union units.

The merger of these two organizations in 1967 – the year Wieland resigned as secretary of the Farmers Union in North Dakota and joined the consolidated Farmers Union companies as director of research and development.

The modern-day organization can be traced back to 1929 when North Dakota Farmers Union chartered a cooperative under the name, "Farmers Union Livestock Commission Company". It had operations in Minnesota as well as in its home state. Then in 1947, after Farmers Union purchased Central Bi-Products of Redwood Falls, MN, the cooperative entered the agricultural processing business.

Wieland became assistant manager of FUMPA in 1972, and five years later became the co-op's general manager. In 1978 he was named president. In 1986, Wieland reported that the organization had 5,600 patrons in five states (Minnesota, North and South Dakota, Montana and Wisconsin) and 225 employees.

The co-op had offices in South St. Paul, and did business at West Fargo's terminal market. Its processing division then operated two rendering plants in Minnesota and one in South Dakota. Two protein blending plants located at North Redwood and Long Prairie, provided standard meat and bone meal as well as special rations for some customers.

The difficult economic situation for farmers in the mid-1980s was something that also was felt in the marketing and processing arenas as well, Wieland reported. The co-op had to begin charging for picking up waste items at packing plants, and it had to close its West Fargo livestock office because of smaller livestock volume marketed at that facility. However, the co-op made arrangements for livestock sales to be handled at the North Dakota market.

SUPPORT FROM FUMPA FOR CO-OP EDUCATION

In 1989 the largest Farmers Union educational fund support in Minnesota came not from the source one might expect. Not from Cenex, or Harvest States Cooperatives, the two super-co-ops based in the Twin Cities. Instead the biggest check came from Farmers Union Marketing and Processing Association (FUMPA), the five-state cooperative based in Redwood Falls, MN.

FUMPA's contribution was $169,054 – or $28,000 more than was contributed by another cooperative Farmers Union helped create, Cenex. While this cooperative doesn't get the attention given to other interstate cooperatives, it occupies an important agribusiness niche.

"FUMPA remains true to the Farmers Union and the cooperative philosophy," said Willis Eken, president of Minnesota Farmers Union during a tour of the co-op's headquarters in 1990. "It's a cooperative our members should support."

Gordon Scherbing, president and CEO of FUMPA, reported that the co-op then had 6,530 members in Minnesota (which had the biggest membership – 4,157), North Dakota, South Dakota, Wisconsin and Montana. Its processing division contracts with independent haulers that make pickups of dead livestock, bones and tallow, poultry processing byproducts and beef packing plant byproducts. These items are trucked to plants in Long Prairie, North Redwood or St. Cloud where the items are converted into "value-added" products utilized in the production of a variety of items ranging from pet food and mink feed ingredients to industrial tallow and hides. In Jan., 1989, FUMPA's long link with the South St. Paul terminal livestock market came to an end, and it closed its office there after making arrangements with a competitor to handle animals marketed by Farmers Union patrons.

At the National Farmers Union convention in Des Moines in 1992, Scherbing was a featured speaker. FUMPA, he told the group, is the only cooperative rendering facility in the nation. In the 1991-92 year, he reported, the cooperative earned $35 million in sales, and returned about $200,000 in educational funds to affiliated Farmers Union organizations. Part of the success hinged on its production of highly-nutritious livestock feed and pet foods, both profitable activities.

The cooperative, Scherbing explained, is basically a rendering operation and feed manufacturer that offers the service of dead animal pickups. This involves a perishable liability which must be disposed of in an ecologically sound and legal manner. Producing livestock feed and pet foods represents another service and benefit for farmers.

And, less than 18 months later, on Aug, 8, 1993, Gordon Scherbing, the man who had been at the helm of the growing and successful Farmers Union Marketing and Processing cooperative died of a sudden heart attack. The Foley native was only 53.

Gordon Scherbing started working at the cooperative as a wintertime temporary worker. But he never left, and years later became its president. At his death, the cooperative employed 237 at plants in Redwood Falls, Long Prairie and St. Cloud where dead livestock is rendered and pet food is manufactured. The cooperative has two divisions – Central Bi-Products (which processes the dead livestock) and Commodity Trading Co. (which handles the sales of certain by-products and pet food ingredients). However, despite the "marketing" in its formal name, and the heritage of livestock sales activities for members at major terminals, FUMPA no longer had a live animal marketing division.

At the time of Scherbing's death, Leo Fogarty, a member of the board of directors from Belle Plaine, described him as "One I think understood the cooperative side of things real well. He knew there was an inter-relationship that is important." Commented Marv Jensen, another director: "Gordy was very comfortable working with Farmers Union and the relationship between the two. He took actions that helped strengthen that relationship."

But in addition to the good relationship between the cooperative and Farmers Union, Gordon Scherbing also left behind a progressive, innovative and strong organization, one that represented a profitable and competitive farmer-owned enterprise.

THE ROLE OF WOMEN IN FARMERS UNION

"There are many, many ingredients involved in the rise and fall of good organizations... one of these is the non-action of farm women... the true partners in the family farm, integrated into the farming organization, into positions of decision-making, and power... "

Marion Fogarty, LeSueur Co., Member, State Board of Directors, MFU.

Figure 109 Marion Fogarty

THE PATTERN FOR WOMEN: CERTAIN JOBS ONLY

There's been one thing in farming circles that has really bothered Marion Fogarty, a LeSueur Co. farmer – the way farm women were received – and recognized – in most of those farm organizations. To Fogarty, Farmers Union was just like the rest of them – women were welcomed, of course, with "Mother of the Year" awards, spouses' luncheons, buffets, sessions on home crafts, style shows, trips to antique malls, how to make flower arrangements and so on.

At the local and county levels, they could be secretaries – efficient and knowledgeable. Or, youth directors. That was a source from which to instill the beliefs and benefits of Farmers Union.

Even the meeting or convention name tags, Fogarty witnessed, year after year, and, meeting after meeting, provided really a no-name identity for women. They were only "Mrs. Joe Blow", or "Mrs. John Jones". That's all. That was enough identity.

And, always, no real important stuff. No coming to, or putting hands-on grip with what brought the organizations together – no direct leadership role. Women were delegates, though, at state and national conventions – hence they did have a vote on policies and resolutions. Beyond that, though, is where women were noticeably absent. Their choices were limited. Mainly just secretaries – very efficient here, and maybe youth director.

She also became irked when speakers or farm writers referred to a spouse as a "farm wife". Women, she would quickly point out, "do not walk down the aisle and marry a farm". Her suggestion was to better describe them for what they really are – spouses, yes, but better yet, "partners", "co-farm managers", etc., and never ever use the erroneous terminology of "farm wives" again.

To Fogarty, limiting the role of women to what for years had been the traditional roles, was irritating. And when Cy Carpenter's Farmers Union came up with a "Good Guys Award", she felt that designation was just too much. Fogarty banged on Cy's presidential desk in St. Paul, complaining the new Minnesota Farmers Union award was blatantly "sexist". (It didn't matter, as she learned later, that the initial honoree was a nun from Wisconsin and that the "Good Guy" moniker didn't exactly fit the gender.)

But Fogarty's influence was felt. And the "Good Guy" title was changed. The next time it was the Farmers Union's "Good Neighbor Award". Carpenter, however, feels Farmers Union was "out in front in recognizing and encouraging women to participate". Also, he counters, "there was reluctance of farm women themselves to be more involved".

Over the years Fogarty has worked for such changes, putting women more into the farm organization mix and into the agricultural arena as the role they really play – "farm partners".

She recalled changing the Farmers Union's "Ladies Luncheons" from the traditional good meal, be entertained with a speaker or slide-show program of some exotic place most of us will never get to visit, and the like. Fogarty carefully selected speakers on topics like the "International Women's Conference", "The Economic Status of Women", and so on.

But then came financial pressured cutbacks, and the cancellation of the women's luncheon program.

THEN A NEW IDEA, "CITY COUSINS"

Then as a volunteer, Fogarty became involved in a letter-writing campaign with a significantly different twist. It was called, "Letters to Congress Through Our City Cousins", and involved contacting Farmers Union women in their county, and asking them to write to their relatives (a form letter was provided to assist participants) anywhere in the nation and asking them, in turn to contact their local representatives in Congress to support the federal farm bill proposals. This was a legislative program instigated by state Farmers Union president Cy Carpenter, and Fogarty was one of his volunteer leaders at the state level.

"There really was no yardstick by which to measure the results of this," Fogarty related. She'd drive from her home in Belle Plaine to the state Farmers Union office in St. Paul many days and would get on the telephone and ask members across Minnesota to take part by lobbying their "city cousins" to lobby Congress. It was, she indicated, a novel, effective campaign.

As director of a "Humanities Grant" program obtained through Minnesota Farmers Union, she arranged town meetings in small communities around the state – again working through local Farmers Union women – to present sessions focusing on Ignatius Donnelly, a turn of the century Populist farm leader and a central figure in the attempt to politicize the Minnesota State Grange. Donnelly was a powerful orator, and had been a force in the anti-slavery movement, a three-term Republican Congressman and a Populist candidate for U.S. Vice President, and a constant critic of the old political parties.

SHE RESEARCHED OTHER FARM GROUPS

"In my travels, I did my own interviews to find out what made organizations tick," Fogarty related. "I went to the newest kid on the farm block group, the Farm Wife meeting in convention in Milwaukee. I did my own research, of women who belonged to commodity groups, or no farm group, Farm Bureau women – why, in Farm Bureau, women didn't rate a vote.

"I found a lot of sponsors were agribusinesses, seed companies, farm machinery, etc. They had no interest in issues like helping the family farm; they were interested only in selling goods. I felt it was time to promote farm women."

She remembers being frustrated at finding "these women to be articulate, knowledgeable, and creative, looking for a way to help keep their family farm and their family's way of life". She recognized "the flash growth of the commodity groups and their checkoffs, seeing the seed companies put up lots of money to run ads in all of the local weekly papers promoting checkoffs, and the land-grant colleges getting money from agribusiness to do the research".

Fogarty felt this represented a roadblock against those working for the family farm. Lobbying in Washington, D.C., she reports "seeing the soybean growers saying they could do it on their own and keep government out of their business", and then later, when their prices dropped, responding with something like "we do need support money, but without any strings attached, like limiting the acreage we can plant".

AN EXPERIMENT WITH A PETITION

In short, Fogarty feels, "There are many, many ingredients to the rise and fall of good organizations. One of these is non-action, or non-brain storming, to get the farm women, true partners in the family farm integrated into the organization, and into positions of decision-making and power."

It's going to be a huge chore changing things, and the traditions and attitudes reflected by farm organizations and rural communities which have prevailed for years.

"Once, when I was talking to a group of co-op patrons at their annual dinner-meeting, I passed around a petition to be signed. I was seeking support for blocking the plans by trying to prevent an electric company from building a coal-powered generating plant on prime agricultural land. All of the men signed, of course.

"When I asked the women why they had not signed, they replied that they didn't think their signatures were important. Not as long as their husbands signed."

That experiment alone communicated clearly to Marion Fogarty that, indeed, her quest for change is going to take a little more time than she had hoped. And a lot more work.

"OPPORTUNITIES (FOR WOMEN) WERE LIMITED... "
Mildred K. Stoltz, Education Director, Montana Farmers Union, 1930s-1950.

One of the most influential persons in a Farmers Union leadership role is the state education director. Three who earned accolades were Gladys Talbott Edwards of National Farmers Union, Lura Reimnitz of Minnesota and Mildred K. Stoltz of Montana. They were contemporaries who exhibited a rare caliber of dedication blended with professionalism.

Edwards was the daughter of the C.C. Talbotts of North Dakota and a brother of Glenn Talbott, who succeeded his famous father as president of the Farmers Union in that key Farmers State. C.C. Talbott was a member of the famous Northwest Organizing Committee which in the late 1920s helped spread the movement into several states, including Minnesota.

Reimnitz and her husband Oskar were life long Farmers Union loyalists and staff workers for Minnesota Farmers Union for about 30 years. She played a key role in developing educational, camping and awards programs for young members of Farmers Union families, including the operation of the Torchbearer Awards recognition program (See Chapter 3).

Mildred Stoltz for about 20 years was in charge of Montana's education program from the late 1930s to 1950. While her career brought plaudits for her work, she wrote in a 1947 column that she felt that leadership opportunities for women in Farmers Union "were limited", according to historian William C. Pratt in his essay presented at the Northern Great Plains History Conference held in Mankato on Oct. 3, 1991.

Observed Pratt: "While Mildred Stoltz was never a board member, she was a key figure in Montana leadership... I realize that education work is often considered part of the traditional nurturing role performed by women... my point is that education was deemed an important part of the Union's program, and that some women, particularly in the Montana organization, had more influence than one might have assumed given the fact that men held most of the leadership posts... "

A LOOK AT SECRETARY BESSIE KLOSE...

Minnesota Farmers Union Secretary Bessie Klose is proud of the fact that her family represents four generations of active membership in the farm organization covering a period of almost a half-century.

In her case, she was quite young when she attended Farmers Union meetings with her parents, Arthur J. and Myrtle Johnson of the Gennesse Local. And she is a relative of John Bosch, an early Minnesota Farmers Union member who became one of the leaders of the Farm Holiday movement in the early 1930s. Bosch was her mother's uncle. And his father, J. B. Bosch, was her mother's grandfather and Bessie's great-grandfather.

Bosch was the brother-in-law of Franklin Clough of Lake Lillian, who was president of Minnesota Farmers Union in the mid-1940s. Clough had married Bosch's sister Minnie, making him a great uncle of Bessie Klose. Only a young teenagers at the time, Klose recalls little about either of the pioneer Farmers Union leaders.

As a high school senior, she held the office of education director for her local. She worked membership, including one year when she teamed with her father and they won a trip to Washington, D.C. Bessie declined so her father could make the award trip. Later, Bessie was to win two "Silver Star" awards presented at the National Farmers Union convention. Those honors are given to those who sign up 25 new members in a year.

Bessie participated as a junior member for about eight years, climaxing her early Farmers Union years as a member of the Torchbearer Class of 1954 when she was 17. She remembers attending many state camps beginning in 1944 (when she was eight years old), and attending Farmers Union's "All-States Camp" when these sessions were held at the old CCC Camp in the Red Rock region near Denver.

Her son Jeffrey, who also achieved the Torchbearer award as well, and also attended "All-States Camp" in 1977 near Bailey where the event has been held since the late 1950s. And his daughter, Nicole Klose, became the third generation to attend All-States when she participated in 1997.

She has held several local offices over the years, as her Farmers Union local (Gennesse) has divided (forming Kandy Local) and then later merged into East Central. And she has been Kandiyohi Co. secretary and then became county president.

Bessie Klose was first elected state Farmers Union secretary in 1982, and has continued at that post since – longer than any of her predecessors. In addition she has been a member of the board of directors of Green View, Inc., the Farmers Union Services Corp., and a member for years of the state Educational Council Advisory group. She also was a member of the building committee when Farmers Union purchased its state office building at 1717 University in St. Paul.

And she was a delegate to the Farmers Union Marketing and Processing Association convention for many years "back when most state organizations were still only electing men".

Bessie Klose recalls traveling to Washington, D.C., by bus in 1964 and taking part in the first "Ladies Fly-In" jaunt in 1966 – something she later did many times, including 1995 when she was a group leader. Further, she has been a Minnesota delegate to National Farmers Union conventions several years.

About her long, active career with the organization, Bessie Klose said she has enjoyed all aspects, from the camps and picnics as a youth to taking part in state and national meetings and hearing speakers like the late Hubert H. Humphrey. "When you sit across a table at a (Farmers Union) picnic from someone as prominent as Senator Humphrey, it's an experience you never forget," she commented. "Farmers Union offers us opportunities like that which we otherwise might not have had."

Bessie Klose has her reasons for a lifetime of activity to the organization. "I believe in the principles of Farmers Union. At the same time I have enjoyed a lot of the work, being on the education advisory committee, working up to the county presidency and then the (state) executive committee... I feel like I have something to give to the organization, and that it has something in return for us... "

WOMEN ON STATE BOARD...

As of this writing, there are seven women serving as county presidents, and automatically are members of the board of directors of Minnesota Farmers Union.

The list includes State Secretary Bessie Klose of Atwater, the president of Kandiyohi Co. Others are: Wava Larson, Hayfield Local, Kason (Dodge); Eunice Biel, Chosen Valley Local (Fillmore); Alice Jaloszynski, Isanti Local (Isanti); Rebecca Carson, Montgomery (LeSueur); Lorene Ingvalson, Blooming Prairie, Grand Meadow Local (Mower); and, Judith Anderson, Benson, Clontarf-Tara Local (Swift).

15... EPILOGUE ... A "THANK YOU"...

Behind them, the pioneers leave many things – monuments, buildings, structures, institutions, memories, records, footprints in the pages of time. For them there were the times of successes, and of shortfalls, of targets achieved, and goals attained, but also disappointments, perhaps regrets, the sad, anguished feeling that more should have been done, or attempted, in the years that so swiftly pass by.

Could any of those immigrant pioneers from the Old Country have foreseen the day when their bankrupt grain marketing cooperative of the 1920s might emerge from its financial shambles and within 75 years be dispatching to the world 1.6 billion bushels of grain from the fertile states of the Upper Midwest and the Northwest?

Could any of those early Americans envisioned the struggling farm cooperative they helped create in the depression era of the 1930s becoming a $2-billion a year enterprise and one of the nation's biggest corporations 70-some years later?

Do the generations of today recognize how such things just don't happen? Harvest States of the '90s stands as testament to the contributions, dedication and perseverance of those who chartered a course and then navigated through troubled waters when things like drought, embargoes, changing farm programs and policies, and uncertainty were periodic companions and confronters.

Is the Minnesota farmer from Ottertail Co. knowledgeable about the roots of Cenex, and how it steadily grew in a constant competitive battle for its share of the agribusiness market? How does an embryonic experiment in cooperative marketing spawned in a depression environment in 1931 become a multi-billion-dollar operation two generations later?

Do the members of the Farmers Union local in Yellow Medicine Co. understand how both of those highly-successful regional cooperatives share a common heritage – a partnership that dates back to their origin when Farmers Union played a major role in their birth? A special relationship when "Farmers Union" was a part of their corporate name.

Is a Jackson Co. cooperative patron of today aware of the leadership and effort made in the early 1920s to organize farm people into a pair of pioneering Farmers Union "central exchanges" in Lakefield and Jackson to purchase things like twine, peaches, coal in bulk for members?

Does the Farmers Union member in Cottonwood Co. know that the Minnesota organization produced a pair of presidents of the National Farmers Union consecutively? That from 1984 through today the national president of this important farm organization has been a native of Minnesota?

Is it important to know something about how men like M.W. Thatcher fought for the right of farmers to be represented in the commodity exchanges of the nation's Grain Belt? That Minnesota leaders of the stature of Edwin Christianson of Gully battled for farm people and rural communities in the state legislature and the nation's capitol?

Or that Cy Carpenter of Sauk Centre became a staunch advocate for a family-type agriculture and a steadfast foe of greedy, corporate farm entities?

Or that two former Minnesota legislators, Willis Eken of Twin Valley and Dave Frederickson of Murdock, would serve as state presidents of Minnesota Farmers Union?

And does the Farmers Union member of today know that former Congressmen Bob Bergland and Alec Olson were part-time and full-time employees of Minnesota Farmers Union at one time or another? That Bob Bergland is a former U.S. Secretary of Agriculture (1977-81) who fathered the concept of "the farmer-held grain reserve" where producers stored (and received storage payments from the government) rather than commercial concerns and agribusinesses?

As Don Knutson, executive director of Green View, Inc., so clearly stated in his annual report at the 54th Minnesota Farmers Union convention: "If we fail to record the history of Minnesota Farmers Union, I'm afraid its efforts to preserve family farm agriculture will eventually go unnoticed and unappreciated." Then Knutson added, "We can't afford to let that happen."

Don Knutson's message that November day appealed for a comprehensive history to definitely be written and published, that such a record might in some way express appreciation to those who preceded us here and put in place and established so many things on which future generations might grow and build.

In a way, a history is recognition of an indebtedness to those who passed this way before us. A sense of gratitude and thankfulness that much of what our predecessors did promised to spare us some of the trials and tribulations with which they had to deal. Perhaps what they did for us in their time in their own way is a lesson for us, like a reminder that we really owe as much or more to our children and our children's children.

Those pioneers blazed the trail for us to follow, a route to a better, more promising future, just like what they saw ahead when they and others crossed the ocean, moved westward to establish a new home on the frontier in the Upper Midwest – a place for dreaming of a better time, a brighter day and a future worthy of the next generation.

FARMERS UNION: "WE WERE ONE BIG 'FAMILY'... "

Perhaps Elnora Swenson of Westbrook summed up best what has happened in her time, to her generation, and to Farmers Union, in a thoughtful note sent along with some biographical and historical information about her family and its relationship to the organization.

"I have tried to write down some things for you... seems like it's so long ago, and when you are by yourself (her husband and partner in Farmers Union), it is hard to remember things, but I hope this will help you," she wrote.

"We all worked hard in Farmers Union, but enjoyed every bit of it. And there were many people in Cottonwood Co. that put lots of effort to have an active group – the Hubins, the Andersons, the Deutschmanns, the Clayton Thomas family, the Claude Meade family, and many, many more. We were one big 'family' – and still are – those of us who are left...

"But the younger generation seem to have a different attitude than we did. There's much more competition for time... "

THE CHALLENGE... CAPTURING THE PAST

It is not an easy task, trying to sum up and capture in words and within the confines of a few pages the significant happenings and far-reaching accomplishments of three or four generations. How does an author in a solitary chapter catch the full scope of visionary leaders like a Bill Thatcher or a Ed Christianson or a Cy Carpenter? How can someone who never knew M.W. Thatcher successfully capture the breadth of such an inimitable agricultural cooperative pioneer?

Hopefully what is written here, and the words on these pages, will communicate for all time and for all generations some of the things they did, present an image of what kind of a person some of these people were, and are. At the same time, we strived to reflect accurately something of their world – the challenges, the battles, the victories, and, in some cases, the defeats.

And most important, we trust that what was compiled through personal interviews, through research and study in books, the library and archives, provides in some measure proper credit due so many.

To do less would be unforgivable, and an injustice – not only to them and their contemporaries, but as well to those who some day may come this way.

The Editors.

ACKNOWLEDGMENTS:

CY CARPENTER

Figure 110 Cy Carpenter

No one was more determined to make sure a history of Minnesota Farmers Union was put down on these pages than the "old dean" of the organization, Cy Carpenter of Stearns Co. His enthusiasm for the history project has been untiring and admirable.

Cy's contributions to this book are many. First and foremost, he is one of the truly legendary figures featured in this account – as president of both Minnesota and National Farmers Union in a leadership career spanning 40-some years. Secondly, Cy Carpenter was a resource, providing historical leads to pursue, a provider of confirmation of historical happenings, and an interpreter of many things that happened to provide a better understanding of what and who were involved.

Finally, Cy Carpenter helped with the editing of the manuscript, making suggestions, finding errors in fact, or raising questions that future readers might possibly have about what was written. As always, Cy was tolerant, and understanding, and encouraging as these pages were filled, and a history of the organization he loves took shape. He really wanted this book to happen. While he may be the first to say our history project may not be perfect, or as complete as some might like, or that we didn't recognize all who should be a part of this record, at the same time, the author is confident no one could be more proud of this book than Cy Carpenter. Truly, with his strong feelings for farmers and farm people, he long wanted such a record to be a reality.

F. B. DANIEL ON COOPERATIVES

"Cooperatives do not do things for people... But we have known people to do some really remarkable things for themselves... through cooperatives."
F. B. Daniel, Cooperative Specialist, Minnesota Farmers Union, 1986.

Figure 111 F. B. Daniel

Brief Biography of F. B. Daniel
Born: 12-17-18, Ray, ND.
Education: Agricultural Education Degree, North Dakota State University.
Employment: Farmer until 1958, field representative and manager of member relations for Farmers Union Central Exchange, 1958-72; executive director, Iowa Farmers Union, 1973-74; staff, Minnesota Farmers Union, 1974-88.
Military Service: U.S. Navy, World War II.
Community Service: Member, Minnesota Vocational Education advisory council; Member, Minnesota Board of Vocational Technical Education; member, University of Minnesota Institute of Agriculture, Forestry and Home Economics

Advisory Council; delegate, National Farm Coalition Conference, 1970-72; delegate, World Farm Congress, London, 1982; member, trade missions to Egypt (1980) and Bulgaria (1983).

Honors: Special National Farmers Union award, 1986; John Carroll Award (Minnesota Vocational Association); honorary American Farmer Degree (National FFA); Award of Merit (Region III, American Vocational Association); Meritorious Service Award, Farmers Union Milk Marketing Association; Exemplary service award, Minnesota Technical College System.

Married: Mary Rose Vogel on 8-23-43. Children: F. Bernard, James, Gregory, Margaret, Ann, Rosemarie, Nancy.

"In the history of Farmers Union and service to farm families and Cooperatives, no one has served with more commitment and personal contribution than F. B. Daniel... he is a model... "
Cy Carpenter, NFU President, NFU Ring presentation, 1986.

This history of Minnesota Farmers Union would not have happened if it hadn't been for the dedicated, conscientious labors of F. B. Daniel. Time and time again F. B. came to the rescue with old, weathered and fragile documents yellowed with age and clippings or papers faded, torn, and dog-eared with use, to provide a tidbit of history, an unknown fact, or a verification of an event, action or decision, or a lead to be followed or checked out.

F. B. truly is a walking, talking Farmers Union encyclopedia, a data base of information with a recall worthy of a computer chip. But more than that, F. B. is a highly-dedicated disciple of the farm cooperative movement, a family farm advocate with little patience or respect for those who threaten family agriculture of a type he has long championed as a farmer himself, a farm organization leader and staff employee, and, above all, a person deeply committed to Farmers Union. In short, F. B. Daniel is a product of much of what is written about in these pages – like Cy Carpenter, and others, he experienced, worked, or lived through much of the times and things which we acknowledge in this book.

And perhaps, a wiser, older F. B. Daniel – like others – would like to relive history, knowing what is known now, turn back the clock, revisit history, and work to shape it differently than the way things progressed or turned out. The author is sure that's the case.

Time and time again the past two years, F. B. displayed undue patience with the author who was ignorant of so much of the Minnesota farm history, and the skills of a seasoned editor in reviewing chapter outlines and drafts, suggesting changes that might make the final product more substantive and worthwhile to future generations who otherwise might not learn of the times and of the people.

As Minnesota Farmers Union President Willis Eken said at F. B.'s official "retirement" recognition in 1988: "His central course has always been Farmers Union ... and will always be... F. B. has served and built as a member, as a local officer, as an employee of Farmers Union Central Exchange (Cenex), as an employee of National Farmers Union, as executive director of Iowa Farmers Union, and for a decade and a half (since 1975), an employee of Minnesota Farmers Union.

"But the areas of employment or assignment do not tell the whole story. F. B. Daniel has been a contributing part of every aspect of Farmers Union, from speaking at meetings to writing policy to helping with state and national convention, and all the chores in between, including that of song leader... his record is full, and long, and rich with accomplishment... "

Two years earlier F. B. Daniel received a special award presented by National Farmers Union President Cy Carpenter. The honor came in the form of a NFU ring, presented only for the second time in history (the first such award was presented to Paul Huff, retired secretary of the national organization).

"In the history of Farmers Union and service to farm families and cooperatives, no one has served with more commitment and personal contribution than F. B. Daniel," Carpenter said in presenting the NFU ring. "He is a model of the hearty and hard work of what Farmers Union and this ring stand for."

Carpenter told how the board of directors of the National Farmers Union have authorized the casting of the official ring, directing that it be presented only to persons chosen by the board who have made exceptional contributions to the organization. The board voted unanimously to present the award to F. B. Daniel.

IN F. B.'S OWN WORDS... .

During October Cooperative Month, 1986, F. B. Daniel authored a column printed on the front page of Minnesota Agriculture. "Many of the best things of life are eventually taken for granted," F. B. wrote. "Included are the rather remarkable gains made, for all farmers, by those farmers who organized and developed farmer-cooperatives."

He went on to say that "not many of today's grain farmers know that before the successes of regional grain marketing cooperatives in the late 1920s, farmers were barred from participating in grain trading in any of the nation's grain exchanges". There was little choice of markets for most grain growers, between local elevators owned by or affiliated with one of several private grain commission firms.

"Not until Farmers Union members, acting through Farmers Union Terminal Association (the predecessor to Farmers Union Grain Terminal Association and what today is Harvest States Cooperatives), received a favorable ruling from the United States Supreme Court were farmers permitted to buy a seat on the Minneapolis Grain Exchange," F. B. pointed out.

Another major development came when a North Dakota agricultural professor teamed up with grain cooperatives and confirmed what many growers had suspected – buyers were not paying what they should for the high-quality grain raised in the Upper Midwest, yet were getting a premium from the flour mills.

Later, farmers sought better methods and compensation for other products – milk and livestock. Again, cooperatives helped in two ways – in supplying what was needed to grow crops and produce milk and meat animals, and then by providing a marketing structure.

When farm machinery manufacturers balked at producing modern machines and equipment, farmers through their cooperatives took action. One product was the "Co-op Tractor" in 1935 – a revolutionary machine boasting rubber tires, self-starter, electric lights, and moved at road speed.

"For many years in many, many communities, there has been a competitive price across the street primarily because there is a strong competing cooperative on this side of the street," F. B. said. "Cooperatives do not do things for people. But we have known people to do some really remarkable things for themselves — through cooperatives."

MILTON D. HAKEL, SR.

"Milt Hakel was among the Farmers Union 'Giants'...
Leland Swenson, president, National Farmers Union, Oct., 1996.

Figure 112 Milt Hakel

The news career of renown Farmers Union editor Milton D. Hakel, Sr., began in rather ordinary circumstances, working as a "printer's devil" and reporter for the Silver Lake (MN) *Leader* during the depression era of the 1930s. In those early newspaper days at Silver Lake and later at the Brownton *Bulletin* and the Hennepin Co. *Review*, he was to become a journalistic jack-of-all-trades — reporter, editor, publisher and owner. Newspapering proved to be a family affair, with Hakel's wife Emily handling yeoman chores in assisting her husband publish his community newspaper.

In the process Hakel developed a rare talent of writing about complex and controversial issues, and explaining matters in easy-to-understand language. His skill caught the attention of Hubert H. Humphrey who became a close friend, and who often sought Hakel's assistance as a researcher and speech writer. It was to become a special relationship where Hakel as a fine researcher and writer worked closely with Humphrey who was the consummate politician and an eloquent speaker during his long career in Minnesota, in the U.S. Senate, the U.S. Vice Presidency, and the Democratic presidential contender in 1968.

In 1952 Hakel was hired by Minnesota Farmers Union President Edwin Christianson who later described him as "the greatest editor in the farm field". For 20 years Hakel was editor of the Farmers Union's newspaper in Minnesota, and the original editor of *Minnesota Agriculture* when it was founded in Jan., 1957. Then in 1975 he became Washington Editor for National Farmers Union and the chief author and editor of the often-quoted weekly *Washington Newsletter*.

His career with Farmers Union spanned a 40-year period, with Hakel continuing to write and editor even after his "official" retirement in 1983. After his death in Oct., 1996, National Farmers Union President Leland Swenson commented, "Milt Hakel was among the Farmers Union 'giants'."

Hakel showed an extraordinary interest in Farmers Union history, and for many years authored a column of news items focusing on the organization, its early leaders, and some of the more significant events in its past.

Reflecting the high regard held for Milton D. Hakel is the coveted agricultural communications award bestowed by National Farmers Union and named in honor of the Minnesota country editor who earned the description of "a giant" and "the greatest" in his farm-writing field.

FARMERS UNION STAFF

The staff of Minnesota Farmers Union is to be thanked for its assistance, for providing temporary office space, for helping locate information and photos from the archives, for its patience and understanding as we researched records and files, and for assistance in verifying things like dates, events, personnel involved, and so on.

In particular, the author/researchers are indebted to Dave Frederickson, MFU president, Vice President and long-time staff employee Dennis Sjodin, and others.

CREDIT TO FARMERS UNION *HERALD*, *MINNESOTA AGRICULTURE*

Much credit for many of the historical items and information in this book must go to the Farmers Union *Herald* and *Minnesota Agriculture*, the official publications of the Farmers Union in Minnesota and the trade area covered by the two major cooperatives, Farmers Union Grain Terminal Association and Farmers Union Central Exchange.

The Farmers Union *Herald* was a splendid source for the researchers/writers of this history. This publication, founded in 1927 when it became the successor to two earlier cooperative publications, was published by Farmers Union Publishing Company. That company was owned by Farmers Union Grain Terminal Association and Farmers Union Central Exchange. These regional cooperatives purchased *Herald* prescriptions for each patron of affiliated farmer-co-ops, which in the early 1960s totaled a quarter of a million farm families in the Upper Midwest states.

The Herald was published twice a month, with its goal to keep farm people informed about cooperatives and other farm news and developments of general interest to co-op patrons that might not be covered in other publications they might receive. Its pages and news columns provided a wealth of authoritative information not available from other sources.

In 1962, the officers and directors of Farmers Union Publishing Company were Thomas H. Steichen, president (general manager of Farmers Union Central Exchange); M.W. Thatcher, secretary-treasurer (general manager of Farmers Union Grain Terminal Association); Glenn A. Long, vice president (Farmers Union Marketing and Processing Association); and directors Emil Loriks (South Dakota), Glen Coutts (Montana), and Fred Seibel (North Dakota).

ADVICE FROM TONY T. DECHANT

When this history project was first envisioned, F. B. Daniel knocked on some doors for help. One of those who responded was the late Tony T. Dechant, past-president and secretary of the National Farmers Union.

"You will have to give some thought as to what kind of a history you want to put together," Tony advised. "There were some turbulent times. MFU membership politics were at times run or influenced by GTA, Cenex, the Marketing group as well as by members determined not to have any of the above dominate.

"And, there were some unpleasant moments within MFU. For example, I remember well a time when MFU was not only broke but many thousands of dollars in debt. NFU, GTA and Cenex put up a big chunk of money, and I was put in charge until the next (state) convention.

"I drove to Minnesota to reduce staff. The then president got drunk. I ended up reducing the staff from 16 people to six that Friday afternoon. It was not one of the better assignments. There were some mad and disappointed people... "

Tony Dechant ended his note with the observation that "there are not many of us old Farmers Union codgers left – good luck!".

GREEN VIEW, INC.

All of the best laid plans could go for naught without the wherewithal to make things happen. So lastly we acknowledge the support morally and financially of the officers, board members and staffs of Green View, Inc., a subsidiary of Minnesota Farmers Union, and the Minnesota Farmers Union Foundation, and express sincerely our gratitude for the opportunity to chronicle the men and women of the farm movement, and the organizations they created and built for family farmers.

16... MINNESOTA FARMERS UNION TIMETABLE

The Year and the Event. Historical tidbits about Farmers Union, its people, and the times:

Sept. 1, 1902 – The Farmers Union originates in Point, Tex., and sets goals including "profitable and uniform prices" for farmers and working "to bring farming up to the standards of business and industry enterprises".

Dec. 5, 1905 – A convention at Texarkana, Tex., establishes a "national" Farmers Union; It had been a regional organization up to that time.

Nov. 10, 1908 – First stock is sold in Equity Cooperative Exchange, the forerunner of Farmers Union cooperatives in the upper Midwest. (It sought a place on grain exchanges for cooperatives in 1910, and pioneered in cooperative livestock marketing on terminal markets in 1916.)

Feb. 9, 1909 – President Theodore Roosevelt's Country Life Commission makes its report. National Farmers Union President Charles S. Barrett is a member of the commission.

Jan. 1, 1913 – U.S. Parcel Post begins, something Farmers Union wanted for years. NFU presented petitions and resolutions to Congress in 1910 representing 3 million farm people in 39 states asking that this be done.

June 30, 1914– Bureau of Crop Estimates established in U.S. Department of Agriculture, something Farmers Union had urged since 1912.

April 20, 1915 – W. B. Evans of Bismarck, ND, begins Farmers Union organizing efforts in Minnesota. Organizational work had begun in North Dakota three years earlier, but that state was not yet chartered.

July 17, 1916 – Federal Farm Loan Act creates network of 12 regional Federal Land Banks and National Farm Loan Associations.

Autumn, 1918 – First Farmers Union in Minnesota, Rost Center Local, is organized, according to charter member Charley F. Wendel of Lakefield.

Jan. 28, 1920 – National Farmers Union organizes Grange, Farm Bureau, cotton and milk groups support for what in 1922 becomes historic cooperative law (Capper-Volstead Act of 1922). The bill had been drafted by NFU President Charles Barrett.

Jan., 1921 – Farmers Union locals in Wright Co., Minnesota, form a Farmers Union Shipping Association. In August on the first anniversary of Dewey Local, members raised funds to establish a Wright Co. Farmers Union Store.

Aug. 15, 1921 – Packers and Stockyards Act is passed, climaxing a four-year campaign by Farmers Union and the National Board of Farm Organizations to break monopoly held by five major meatpackers who process 86 percent of the nation's meat, own railroads, stockyards and cold storage facilities.

Feb. 18, 1922 – The "Co-op Bill of Rights", the Capper-Volstead Act, becomes law. This is something Farmers Union sought for 16 years. A similar bill was vetoed by President Taft in 1913. Passage was aided by support of National Board of Farm Organizations.

Early 1922 – Jackson Co., Minnesota, boasts 27 local unions and 2,000 members. Five locals (Silver King, Des Moines Valley, Star, Hummer, and Banner) hold joint meeting to order carloads of flour, sugar, fruit, and machinery.

Mar. 4, 1923 – President Harding signs "filled milk" law prohibiting interstate shipment of milk adulterated with non-dairy fats, a measure asked by Farmers Union and National Milk Producers Federation.

Agricultural Credits Act creating the "Intermediate Credit Banks" system becomes law. Farmers Union advocated new system to help provide short-term loans in wake of economic troubles for farmers in post-World War I period.

Mar. 10, 1923 – Equity Cooperative Exchange of St. Paul, MN, enters voluntary receivership; W. M. Thatcher is named manager of the bankrupt cooperative.

Sept., 1923 – Farmers' Union Exchange is organized in Jackson, MN, and $6,000 raised in five weeks to launch the enterprise. Fred Hample is president and Charley Wendel is secretary.

Mar. 3, 1924 – Farmers' Union Exchange in Jackson opens for business, with Louis S. Bezdicek as its manager.

Dec. 21-23, 1925 – The Corn Belt Committee, representing 23 agricultural groups and organizations, meets in Des Moines, IA., at meeting organized by the National Farmers Union. NFU President Charles Barrett is committee chairman and A. W. Ricker, then head of Producers Alliance (which would merge with Farmers Union in 1927, and later editor of the Farmers Union *Herald*) is secretary.

1926 – The Northwest Organizing Committee (Myron W. "Bill" Thatcher, C.C. Talbott, and A. W. Ricker) begins re-organizational work in North Dakota, Minnesota, Montana, and Wisconsin. They received $500 from National Farmers Union to organize; Thatcher sought this relationship to help establish a better foundation for his cooperative.

The Farmers Union would link together cooperative and legislative goals, and carry on its education and promotion programs with 5 percent of the savings of the cooperatives. It becomes a mutually advantageous relationship. The Equity Cooperative Exchange becomes the Farmers Union Terminal Association.

July 2, 1926 – Cooperative Marketing Act passed, and provides for new Cooperative Marketing Division in U.S. Department of Agriculture.

1927 – Farmers Union Exchange, a wholesale supply cooperative subsidiary of Farmers Union Terminal Association, begins operation.

Feb. 12, 1927 – The Farmers Union-supported "Import Milk Act" which bans import of milk or cream unfit for human consumption become law.

Feb. 25, 1927 – President Coolidge vetoes McNary-Haugen two-price, cost of production farm support system supported by Farmers Union. Legislation would have provided for two prices (domestic and foreign) for farm commodities.

1927-28 – Something like 50,000 wheat growers submit samples for testing, using small baking powder tin cans. The free tests to determine the protein content of the grain results in premiums that boost the market price by as much as 42 cents a bushel.

1928 – C.E. Huff of Kansas is elected National Farmers Union president.

1929 – Farmers Union membership in about a dozen Minnesota counties is sufficient for a state charter from the National Farmers Union. John C. Erp of Canby from Lac Qui Parle Co. (which has 596 member-families) is state president. Representatives of 60 cooperatives form a new regional, North Pacific Grain Growers, Inc., at a meeting in Lewiston, ID. Fifty-four years later it would merge with Farmers Union Grain Terminal Association of St. Paul.

April, 1930 – Brown's Valley Local near Ortonville with 93 members is the biggest local in Minnesota Farmers Union.

Nov. 18-21, 1930 – Minnesota Farmers Union hosts 26th annual National Farmers Union convention in St. Paul. It is a stormy meeting and John A. Simpson of Oklahoma wins NFU presidency in contest with incumbent C.E. Huff.

Jan. 3, 1931 – Farmers Union Central Exchange is formally incorporated; it is successor to the Farmers Union Exchange which has operated for four years.

1933 – John C. Erp of Canby, the first Minnesota Farmers Union president, is seated on the NFU board of directors.

Feb. 23, 1933 – Farmers Union Central Exchange joins group to build a manufacturing plant for farm inputs and supplies. National Cooperatives is the result, with main plant located in Albert Lea, MN

May 12, 1933 – The 1933 Agricultural Adjustment Act and the Emergency Farm Mortgage Act are voted, along with the Federal Emergency Relief Act. Farmers Union favored the first two.

May 27, 1933 – The Farm Credit Administration is created by another FDR executive order. Its mission is to address credit situation regarded by Farmers Union as one of the worst problems confronting farmers.

Oct. 17, 1933 – President Roosevelt issues executive order establishing Commodity Credit Corp. to make loans to farmers and enable them to store their commodities while waiting for better markets. Such loans were advocated for years by Farmers Union, and North Dakota's crop loan system provided model for historic legislation.

1934 – John A. Simpson dies, and E. H. Everson of South Dakota becomes national president.

Jan. 21, 1934 – Federal Farm Mortgage Corp. established. Farmers Union sought this to help farmers facing farm foreclosure action.

April 30, 1935 – President Roosevelt issues executive order establishing the "Resettlement Administration", the predecessor to the Farm Security Administration and later the Farmers Home Administration.

May 1, 1935 – First "oil-blending" plant in South St. Paul established by Farmers Union Central Exchange.

May 11, 1935 – President Roosevelt issues another executive order. This one creates the Rural Electrification Administration. Farmers Union urges cooperatives take responsibility to implement the program once Congress approves the plan.

June 29,1935 – Bankhead-Jones Act is approved, providing for broadening farm research, developing new crops and uses, and improved livestock breeds.

Feb. 29, 1936 – Legislation creating soil conservation districts and Agricultural Conservation Program (ACP) is enacted. Farmers Union and other groups supported action after U.S. Supreme Court rules part of 1933 Agricultural Adjustment Act unconstitutional.

1937 – John Vesecky of Kansas is elected president of the National Farmers Union.

June 3, 1937 – Authority for Federal milk marketing order goes into effect as President Roosevelt signs Agricultural Marketing Agreement Act.

July 22, 1937 – Bankhead-Jones Farm Tenant Act becomes law. It resulted from recommendations of Farm Tenancy Commission of which M.W. Thatcher was a member. Under this act Federal Government can buy land and sell it to disadvantaged families up to 100 percent of value in 40-year, 3 percent loans.

Dec. 29, 1937 – M.W. Thatcher visits President Roosevelt about the problems of farmers, and lack of funds for cooperatives. Visit results in FDR memo note to Agriculture Secretary Henry A. Wallace directing him "to try to work out" a plan to permit farmers to gain control of grain-marketing facilities and to build local cooperatives. This led to Farmers Union Grain Terminal Association in June, 1938.

April 9, 1938 – Tensions in Minnesota Farmers Union related to Farm Holiday movement cause state membership loss. National Farmers Union suspends, and then revokes state charter.

June 31, 1938 – Farmers Union Terminal Association changes name to Farmers Union Grain Terminal Association, with M.W. Thatcher in charge of reorganization. Thatcher was instrumental in gaining changes where Bank for Cooperatives could provide loans to regional cooperatives, and where farmers could borrow government funds to buy stock in both local and regional cooperatives. The GTA office is located at 1923 University Avenue in St. Paul.

July 15, 1938 – Farmers Union Grain Terminal Association is incorporated, replacing Farmers Union Terminal Association that originally was Equity Cooperative Exchange and earlier was part of President Hoover's Farmers National Grain Corporation.

Feb., 1939 – M.W. Thatcher brings together officials of nine regional cooperatives, forming the National Federation of Grain Cooperatives. He becomes its perennial president. Thatcher also becomes legislative representative for National Farmers Union.

1940 – James G. Patton of the Colorado Farmers Union becomes national president. He will hold that office for 26 years.

1940 – Farmers Union *Herald* editor A. W. Ricker, M.W. Thatcher of Farmers Union Grain Terminal Association, and Glenn Talbott of North Dakota plan a meeting to be held at Moorhead to plan the rechartering of Minnesota Farmers Union.

April, 1940 – A "war of words" broke out between the two chief cooperative figures, M.W. Thatcher and Emil A. Syftestad, in the form of sharply critical letters the two wrote to each other but also mailed to key Farmers Union and regional cooperative officials. The first volley was fired by Thatcher who wanted to merge the grain marketing and distribution cooperatives into one giant organizations. Syftestad and others opposed the consolidation of the two Farmers Union cooperatives, the Grain Terminal Association and the Central Exchange. The exchange of letters was the forerunner of political struggles and tension ahead.

April 27, 1940 – More than 20,000 from 24 states pack St. Paul City Auditorium for a mass meeting focusing on "farm credit crisis" called by M.W. Thatcher of GTA. Action by Congress later lowers interest rates from 5.5 to 3.5 percent. Meeting also is a promotional, political event staged in part to support Henry A. Wallace for the U.S. Vice Presidency.

July, 1941 – Reorganization effort in Minnesota is excellent, with 70 local unions chartered. National Farmers Union President James G. Patton asks Mrs. Ione Kleven of Appleton to head up a Minnesota youth education committee.

1941 – The most modern grain elevator in the world is built at Superior, WI, by the GTA. It can house 4.5 million bushels, and links the northwest grain farmers with the world market via the Great Lakes-St. Lawrence Seaway.

June 15, 1942 – Durum wheat mill at Rush City, MN, acquired by Farmers Union Grain Terminal Association. It is first time a cooperative begins milling activity, and sets the stage to initiate oilseed processing and feed production .

Nov. 10, 1942 – Minnesota Farmers Union applies for a state charter at convention in Fergus Falls. A state organization had been chartered by the National Farmers Union in 1929, but had become fragmented in the late 1930s, with charter revoked in April, 1938. One of incorporators is Edwin Christianson of Gully, MN, who would become a legendary Farmers Union and cooperative leader. Dues set at $3.50 per family.

Dec. 30, 1942 – National Farmers Union issues state charter to Minnesota. Einar Kuivinen of New York Mills is elected president, with E. L. Smith of Montevideo vice president. Irene Paulson of New York Mills is state secretary-treasurer, and Junice Dalen, Education Director. State office is located in New York Mills. Edwin Christianson of Gully elected to state board of directors.

M.W. Thatcher and Glenn Talbott sit in on board meeting, offering "full cooperation", with Kuivinen responding that "Farmers Union intends to cooperate with regionals".

Resolution calling for 5 percent educational funds from cooperatives, with 2.5 percent going to counties. Vote to waive Farmers Union dues for members in military service.

Jan. 2, 1943 – Articles of Incorporation of Minnesota Farmers Union filed in Otter Tail Co. M.W. Thatcher starts radio program on four-state network, publicly advocates "Food For Freedom" effort in post-war years to use American "food-for-peace".

Mar. 1, 1943 – Farmers Union Central Exchange buys oil refinery in Montana.

May 1, 1943 – Farmers Union Grain Terminal Association acquires 57-year-old St. Anthony and Dakota Elevator Co. which owned 135 elevators and 38 lumber yards. Acquisition prompts competitors and cooperative critics into forming National Tax Equality Association. This group will spend millions trying to convince people that farm cooperatives are a severe business threat, tax-evaders, and "un-American", and attempt to eliminate tax laws favorable to cooperatives, and to reduce federal farm program activity.

May 18, 1943 – National Farmers Union President James G. Patton outlines his plan for a "food and agriculture organization" in the proposed United Nations at Hot Springs, Va., meeting.

Aug. 2, 1943 – Farmers Union Central Exchange purchases oil refinery at Billings, MT and helps organize National Cooperative Refinery Association at

McPherson, Kans., and assumes a one-third interest in what is then the world's largest cooperative oil refinery.

Nov., 1943 – Second state convention held in Alexandria. Membership – 2,550. Delegates elect B. Franklin Clough of Lake Lillian state vice president.

May, 1944 – A "Warning" is printed in The Farmers Union *Herald* that "agents of the Farmers Union Life Insurance Co. Of Iowa" are "using the prestige of Farmers Union's name to solicit insurance business and securing agents. This company, the notice said, "no longer has any connection with the Farmers Union... has no cooperative characteristics and makes no contribution to Farmers Union".

June 22, 1944 – Legislation creating G.I. Education and training programs becomes law.

1945 – Tony T. Dechant becomes national secretary of the Farmers Union.

1945 – Minnesota Farmers Union President Einar Kuivinen of New York Mills declines re-election at fourth convention in Willmar. State vice president B. Franklin Clough of Kandiyohi Co. is elected, and moves state office to Willmar where it will remain until 1951. Edwin Christianson of Gully elected state vice president. Dues increased from $3.50 to $5 per year. Membership – 3,316.

1946 – Millions of bushels of "Mercy Wheat" shipments are donated by farmer-members of Farmers Union cooperative elevators for shipment to war-ravaged Europe. It is a humanitarian effort that receives national publicity.

April 3, 1946 – Farmers Union Central Exchange opens first L-P (liquid petroleum) bulk plant at Glendive, MT, marking a new important service for farmers.

May 3, 1946 – First producing oil well for Farmers Union Central Exchange begins production in Cat Creek area north of Billings, MT

Sept. 9, 1946 – Farmers Union Central Exchange is co-founder of Central Farmers Fertilizer Co. of Chicago.

Nov., 1946 – Einar Kuivinen of New York Mills is elected president of the Minnesota Farmers Union. (He had been president 1943-45.) Edwin Christianson re-elected state vice president. Delegates vote 25 cent dues checkoff for building fund. Membership increases to 6,729. By-laws changed to make all county presidents members of the state board of directors.

April 3, 1947 – M.W. Thatcher and Farmers Union leaders help organize a "Mercy Wheat Campaign" rally at Climax, MN, to encourage support for world food relief. Farmers are told to "scrape their grain bins clean" to augment food supplies. A featured speaker was New York City Mayor LaGuardia who had been appointed by President Roosevelt to head up efforts to help combat hunger in post-war Europe.

Feb. 14, 1947 – The right of farmers to market grain cooperatively on terminal markets is upheld by Minnesota Supreme Court in case involving Farmers Union Grain Terminal Association which had protested grain trade discriminatory practices.

Oct. 30, 1947 – General Agreement on Tariffs and Trade (GATT) is signed, supported by Farmers Union which wants to expand world trade in an orderly manner.

April 3, 1948 – President Truman signs Foreign Assistance Act. This Farmers Union-supported act leads to "Marshall Plan" to help war-ravaged countries, and its success paves the way for later "Food For Peace" program.

May 28, 1948 – Charles F. Brannan is confirmed as U.S. Secretary of Agriculture, and later becomes advocate for Farmers Union proposal of direct support payments to farmers.

June 28, 1948 – Commodity Credit Corp. Is created by law sought by Farmers Union. Its mission is to stabilize, support and protect farm income and prices while facilitating the orderly disposal of agricultural commodities.

Dec. 14, 1948 – National Farmers President James G. Patton tells annual meeting of Farmers Union Grain Terminal Association of plans to organize member-trips to Washington, D.C., to influence legislation.

1948 – M.W. Thatcher receives Farmers Union "meritorious service award".

Jan. 15, 1949 – First Farmers Union member bus trip to visit Congressmen and government officials in Washington, D.C.; GTA general manager M.W. Thatcher meets with President Truman in The White House.

Feb. 24, 1949 – New U.S. Senator from Minnesota, Hubert H. Humphrey, co-sponsors rural telephone administration bill as part of Rural Electric Administration. It is legislation sought by Farmers Union.

Oct. 28, 1949 – Farmers Union officials on hand as President Truman signs Rural Telephone Act.

Nov. 9, 1949 – United Nations Relief and Rehabilitation Agency is established, a measure urged by Farmers Union as best for post-war recovery and to combat hunger.

Nov., 1949 – Roy E. Wiseth of Goodrige in Pennington Co. is elected president of the Minnesota Farmers Union. (Willmar *Daily Tribune* headline of Nov. 14, 1949 proclaimed: "Roy Wiseth Defeats Einar Kuivinen as Liberals Outvote Leftists".) Minutes show vote was: Wiseth – 6,184; Kuivinen – 5,915; Russel Schwandt – 3,171. Wendell Miller of Cottonwood was elected vice-president.

Tony T. Dechant of national staff becomes acting secretary of Minnesota Farmers Union and will supervise office and operations. Arthur Tisthammer of Denver takes over FU insurance operation.

By-laws changed to establish Farmers Union districts for the purpose of electing a member of the state board of directors from each district to be a member of state executive committee.

First Farmers Union bus is purchased, with Sidney Olson of Willmar hired as driver. Initial bus trip to Washington, D.C., is made.

Jan. 1, 1950 – Lura Reimnitz takes over state Farmers Union education department duties. She is destined to become one of the best-known, deeply dedicated professional staff members and historian.

April, 1950 – Farmers Union *Herald* reports 10 new locals formed in the first three months of the year, or more than were organized in all of 1949.

Ralph Ingerson, the "creator" of the famous Co-op Tractor dies. Burial in Maple Plain, his boyhood home. He also negotiated first oil refinery contracts for local oil co-ops in 1930.

Sept. 7, 1950 – New Hampshire Senator Styles Bridges makes speech on Senate floor smearing National Farmers Union, its leaders, and cooperatives as "communistic". His comments are rebuked immediately by other members of the U.S. Senate.

Summer, 1950 – A 5.7-million-bushel addition is made to Lake Superior elevator by Farmers Union Grain Terminal Association. It is first of three expansions there; the other improvements were made in 1954 and 1958.

Nov. 11 1950 – Edwin Christianson of Polk Co. is elected president of Minnesota Farmers Union in contest with Roy Wiseth at state convention in Willmar. Minutes show vote was: Christianson – 12,071; Wiseth – 6,845. William Nystrom is only vice presidential candidate, gets 18,445 votes. Two-year terms of office approved.

Minnesota has membership of 12,413, with steady growth since re-chartering in 1942. (State membership in 1943 was 2,558). Family dues are set at $5 per year, first increase since rechartering.

1951 – Minnesota Farmers Union moves its state office to St. Paul to more closely monitor state legislation. Purchase of second bus announced at tenth annual state meeting.

1952 – Edwin Christianson arranges for a Minnesota Farmers Union special news section to appear in Farmers Union *Herald*. (Later a Minnesota edition is published; then shortly later, *Minnesota Agriculture* is created.)

Maurice "Mel" Melbo of Gully is asked by Christianson to be bus driver for the Minnesota Farmers Union. (Later, in 1967, he becomes insurance agent. Then, only two years later, he is named Minnesota's insurance "Agent of the Year".)

1953 – Emil A. Syfestad, receives Farmers Union "meritorious service" award. For 26 years he was general manager of the Farmers Union Central Exchange, and directed the cooperative from an indebted organization in 1932 and less than a million in sales to a successful firm with $68 million in sales.

1953 – Fund-raising for first permanent home-headquarters for Minnesota Farmers Union begins, with share of membership dues ear-marked for that purpose.

1953 – Farmers Union Exposition Building erected at Minnesota State Fairgrounds by Minnesota Farmers Union and the cooperatives near a major entrance and on major thoroughfare used by the public.

April 10, 1953 – Minnesota Farmers Union President Edwin Christianson praises passage of "Extension Separation" Bill by state legislature. He calls it "a victory" after 16-year effort with The Grange to end Farm Bureau-Extension close relationship.

July 10, 1954 – Public Law 480 becomes law, opening the door to provide food aid to developing countries. Farmers Union testimony was titled, "Use Food to Win the Peace".

Jan. 1, 1955 – Farmers are covered by Social Security, something NFU advocated as early as 1936.

1955 – Long-time Farmers Union editor and advocate A. W. Ricker receives National Farmers Union "meritorious service" award posthumously after death on Feb. 11.

Sept. 14, 1955 – Ceremony held for construction of $1.5 million FUCE office center at 1185 North Concord Street.

May 28, 1956 – Legislation is passed that creates "the soil-bank" program as well as a commission to seek increased industrial utilization of agricultural products.

July 2, 1956 – Minnesota Congresswoman Coya Knutson writes to Harold Johnson of Greenbush thanking him for his idea of federal student loans, and enclosed a copy of her bill providing for the government student loan program. Johnson was a Farmers Union member and a district supervisor for the organization when he suggested the idea that the government provide loans to students just like it does to REA and other agencies.

Jan., 1957 – First edition of weekly *Minnesota Agriculture* is published by Minnesota Farmers Union.

Feb. 24, 1957 – Farmers Union Central Exchange moves into new headquarters office building in South St. Paul.

June 28, 1957 – Farmers Union Central Exchange opens its first bulk fertilizer blending plant in Trimont, MN. Twenty-five Soils Service Centers will follow.

July 16, 1957 – Long-time Farmers Union Central Exchange General Manager E. A. Syftestad dies. He had been manager for about 25 years, or almost from the beginning of the Exchange. His assistant, Thomas H. Steichen is named successor.

July 22, 1957 – Membership card No. 40,000 issued by Minnesota Farmers Union to Roger and Judy Janzen of Mountain Lake (Cottonwood Co.).

July 21, 1958 – Frank Brooks of the Mankato Local in Blue Earth Co. and his family become the 41,000th member of the Minnesota Farmers Union. It is the tenth consecutive year of record membership, with 40 counties achieving membership goals which required an average gain of 10 percent, according to state MFU administrative assistant Art Tisthammer who heads up field service operations.

Oct. 1, 1958 – Farmers Union Grain Terminal Association takes over the McCabe Company elevators. The firm dates back to 1886 and has elevators and feed plants in Montana (22), North Dakota (15), and South Dakota (4). The 350 McCabe employees are to remain with GTA. Ben McCabe, the president of the Minneapolis-based family business, had been head of the National Tax Equality Association, a arch-enemy of cooperatives. When the sale to GTA was announced, McCabe said, "A cooperative has many advantages over private business. First is customer ownership. The majority of farmers prefer to do business with a cooperative in which they share the ownership."

Jan. 1, 1959 – B. J. "Barney" Malusky becomes the new field director for Farmers Union Grain Terminal Association which completed its 20th year in 1958. He is destined to head GTA in 1976 when M.W. Thatcher retires.

April 24, 1959 – Minnesota Gov. Orville L. Freeman signs farm trucking rights bill supported by Farmers Union.

April, 1959 – A Danish ship is the first ocean-going vessel to arrive at the Farmers Union Grain Terminal Association elevator in Superior. It takes on 430,000 bushels of barley before heading home via the newly-opened St. Lawrence Seaway.

Oct. 11, 1960 – Farmers Union Central Exchange takes full control of Cenex Pipeline, carrying gasoline and fuels more than 420 miles from Laurel, MT, refinery to Minot, ND

June 14, 1961 – National Farmers Union testifies in favor of starting U.S. Peace Corps.

Sept. 26, 1961 – Farmers Union Grain Terminal Association buys Minnesota Linseed Oil Co. of Minneapolis to provide farmers a market for flaxseed.

Nov., 1961 – Minnesota Farmers Union membership peaks at more than 41,000 families, with county organizations in all but three of state's 87 counties. Minnesota trails only Oklahoma and North Dakota in number of members. Membership growth credited to President Christianson and Russel Schwandt of Cottonwood Co., state secretary and organization director.

May 29, 1962 – Assets of Farmers Union Grain Terminal Association top $1 billion.

Sept. 27, 1962 – Farmers Union hails 1962 farm bill which authorizes direct payments to Feed Grain Program participants, calls it "a major advance in farm programs".

Nov., 1962 – Minnesota Farmers Union buys office building at 1275 University Avenue in St. Paul.

1962 – Orville L. Freeman honored by National Farmers Union for "outstanding service to agriculture and world agriculture".

1963 – Special one-day convention held in St. Paul where delegates vote $7.50 annual family membership dues, and 2-year membership for $10 and 5-year membership for $25.

June 12, 1963 – Farmers Union urges pilot food stamp program be made permanent.

July 15, 1964 – Farmers Union Central Exchange listed in directory of Fortune's 500 largest industrial companies for the first time.

Sept. 11, 1964 – Minnesota Farmers Union President Ed Christianson joins national president James G. Patton in urging manufacturing milk price supports be increased to $3.40 a hundredweight.

Nov., 1964 – By-law change on state and national levels to count family memberships to determine number of voting delegates (one delegate per 20 family memberships).

1965 – Hubert Humphrey receives National Farmers Union's award for "outstanding service to world agriculture".

June-Aug., 1965 – First "Legislative Fly-Ins" to Washington, D.C., is organized by Minnesota Farmers Union, with 270 participating in three trips staged to lobby 1965 farm bill. One group included Minnesota Gov. Karl Rolvaag.

Nov. 29, 1965 – National Farmers Union President James G. Patton announces at Minnesota State Convention that he will not seek re-election in 1966. He was in 25th year as NFU president. Special advance, 25-year membership for $100 is voted to supplement other annual, 2-year and 5-year membership options. There are 130 enrolled in 25-year memberships.

Dec. 22, 1965 – "Green Thumb" Program is born as National Farmers Union signs contract with R. Sargent Shriver, director of the Office of Economic Opportunity. Minnesota becomes one of the first states to develop this program emphasizing employment opportunities for low-income senior citizens.

Mar. 1966 – Tony T. Dechant elected National Farmers Union president; Edwin Christianson, Minnesota president, is chosen national vice president at Denver convention.

April, 1966 – Second Farmers Union "Ladies Fly-In" to Washington, D.C., occurs. They lobby for "truth-in-lending" legislation.

June 14, 1966 – Farmers Union Central Exchange becomes full owner of Farmland Industries plants at Pine Bend and Winona to produce ammonium phosphate and other fertilizer ingredients.

July 4, 1966 – Farmers Union Central Exchange opens large distribution center in Inver Grove Heights. Facility houses oil-blending plant and a merchandise warehouse.

Aug. 22, 1966 – The Farmers Union Central Exchange's plant at Pine Bend, MN, begins operation and is the nation's biggest anhydrous ammonia fertilizer plant.

Sept. 19, 1966 – National Farmers Union Vice President Ed Christianson testifies before "Blue Ribbon" Commission on Food and Fiber Policy, and urges future farm programs be "income-oriented" rather than "surplus-oriented".

Oct. 8, 1966 – Farmers Union Central Exchange begins distribution of $7 million in cash – the largest single cash redemption of stock in history of U.S. cooperatives.

Dec. 4, 1967 – Minnesota Farmers Union President Edwin Christianson appoints a Task Force on Corporate Farming.

1968 – Farmers Union Grain Terminal Association is one of several cooperatives investing in new Farmers Export Co. terminal on Mississippi River near New Orleans.

Mar. 19, 1968 – U.S. Secretary of Agriculture Orville L. Freeman announces 90 percent of parity program ($4.28 per hundredweight) for milk at National Farmers Union convention in Minneapolis.

April 1,1968 – First Farmers Union "Young Farmers Fly-In" to Washington, D.C., takes place. The 235 young members urge support for G.I. Education Bill, and other measures to help young farmers.

June 1, 1968 – B. J. Malusky succeeds M.W. Thatcher as general manager of Farmers Union Grain Terminal Association, only the second to head the giant marketing cooperative.

Nov., 1968 – Cyril H. Carpenter of Stearns Co. becomes secretary of the Minnesota Farmers Union. He is destined to later head both state and national organizations.

Dec., 1968 – Minnesota Farmers Union "task force on corporate farming" makes its report. Long-time communications chief and editor Milt Hakel was staff director of this project.

1969 – First contracts are signed by Green View, Inc., a new subsidiary of Minnesota Farmers Union, with state highway officials to provide for workers at rest areas. This program is modeled after Green Thumb, where jobs are filled by low-income senior citizens.

1971 – Led by Minnesota Farmers Union, a "corporate farm reporting bill" is enacted by Minnesota Legislature. Minnesota Farmers Union vice president William Nystrom of Worthington announces his retirement after holding that post for 21 years. Returning home after a National Farmers Union board meeting in Denver, Edwin Christianson suffers severe stroke at Twin Cities airport.

Nov., 1971 – Cyril H. Carpenter of Stearns Co. is elected state vice president and then acting Minnesota Farmers Union president when it is learned that Edwin Christianson will not be able to continue his leadership. Carpenter had been state secretary since 1968.

1972 – Farmers Union leadership calls attention to the failure to provide farm management training classes for veterans returning to farm and entitled to G.I. Bill of Rights benefits. The number of G.I. Classes is increased from only two to 82. Payments to trainees total $5 million annually in Minnesota. Ralph Whiting is hired as consultant to work with farmer-veterans eligible for such training. About 80 farm wives participated in spring Farmers Union Fly-In to Washington, D.C.

July, 1972 – A new symbol, CENEX, is adopted, replacing the long-used Farmers Union Central Exchange identification. However, the old, full name is to be used for legal and formal purposes, officials said.

Ralph Arends of Buffalo, a dedicated Farmers Union fieldman and family agriculture advocate, is buried in family plot in Rock Co. His widow, Luella, donated land from their farm for a park. A monument in his memory and that of Hubert Humphrey was built there. They were friends of Hubert H. Humphrey who lived in Wright Co. where Arends had been county Farmers Union president. Luella also was an active member and field worker.

Aug. 30, 1972 – Farmers Union Central Exchange General Manager Thomas H. Steichen dies of a heart attack. His successor is John McKay who has new title, "President".

Nov., 1972 – Lura Reimnitz is honored with a "This is Your Life" presentation at the 31st Minnesota Farmers Union state convention. She plans to retire at the end of the year following a remarkable career which began Jan. 1, 1950 and centered on her work as state education director and head of the summer camping program. A Farmers Union member since 1942, Lura prepared a comprehensive 30-year report of the organization, including when various locals were chartered. Cy Carpenter proposes establishing a "Farm Family Communication Center" somewhere in Minnesota.

Dec., 1972 – Jewell Haaland of Yellow Medicine Co. is elected chairman of the board of directors of Farmers Union Grain Terminal Association at the 35th annual meeting of the cooperative. Haaland had been a director 15 years.

Summer, 1973 – Marcia Nygaard is hired to conduct numerous discussion meetings with farm wives. These mini-conferences become known as "Sip and Scheme" sessions, and were designed to get more women involved in Farmers Union programs and policies.

1974 – Farmers Union Grain Terminal Association is one of six Midwest regional cooperatives which purchase Agri-Trans, a Mississippi River barge concern.

Oct., 1974 – Oscar Reimnitz reports the death of his wife Lura at their home. The funeral was held in Willmar. She had enjoyed only two years of retirement following a 30-year career on the staff of the Minnesota Farmers Union. At one time Lura Reimnitz was state secretary, but she was best known for her dedication and work as state education director and supervisor of the state camping program.

The Farmers Union *Herald*, first published in March, 1927, gets a new name – *Co-op Country News. The Herald* was successor to The Cooperators' *Herald* published by Equity Co-op Exchange since 1913, and The Farm Market Guide, a house organ of the Producers Alliance from 1924-27. With its new name, the masthead of *Co-op Country News* also featured the logos of the two regional cooperatives which published the *Herald*, GTA (Farmers Union Grain Terminal Association) and Cenex (Farmers Union Central Exchange).

1974 – Milt Hakel represented Farmers Union at the World Food Conference in Rome, Italy. Minnesota Farmers Union works hard for expanded farm training programs under the G.I. Bill. Farmers Union Grain Terminal Association moves deeper into the food business, with the purchase of Kent Foods of Kansas City, MO, and then acquires creamery company in Carthage, MO, the following year.

1975 – After 23 years on the Minnesota Farmers Union staff, editor Milt Hakel joins National Farmers Union staff in Washington, D.C., as editor of its Washington Newsletter.

Nov., 1975 – Cy Carpenter calls for a "Declaration of Entitlement" for farm families in convention speech noting 1976 bicentennial year, says farm families will control the nation's agriculture with their cooperatives and their rural communities.

1976 – Minnesota Farmers Union develops a mammoth hay-lift, transporting 130,000 tons of hay from Montana, Idaho and elsewhere to augment drought-shriveled supplies of dairy farmers and beef cattle producers. Farmers Union locates supplies, arranges for discount rail shipment rates and provides coordination. This program helped farmers maintain herds, stabilized hay prices, and thwarted distress sales of animals because of lack of feed. Two state office workers, Frank and Ethel Krukemeyer, retire after 13 years. They helped with printing and membership.

Dec., 1976 – M.W. Thatcher dies of apparent heart attack at age 93. Hubert Humphrey praises him as the greatest leader in farm cooperative history, and for engineering "the most significant development in American agriculture since the Land-Grant College".

1977 – The Farmers Union begins successful fight to get Veterans Administration to nullify its new policy which denies G.I. Bill of Rights educational programs to farmer-veterans who fail to qualify under minimum net income provisions. Two years later the VA abandons plan, and invites suggestions from Farmers Union on veterans training program availability.

The Jewett & Sherman Co., a food firm that becomes Holsum Foods, is purchased by Farmers Union Grain Terminal Association. It is a supplier of vegetable oil products, salad oil, jams, sauces, syrups, and is nation's biggest importer of olives.

1979 – Cenex sales top $1 billion for the first time.

July, 1980 – The Minneapolis *Tribune* reports that Minnesota Farmers Union President Cy Carpenter supports the proposed merger of three major Upper Midwest farm cooperatives, the Farmers Union Central Exchange (CENEX), Midland Cooperatives, and Farmers Union Grain Terminal Association.

Mar., 1981 – Minnesota state president Cy Carpenter of Sauk Center is elected chairman of nine-member National Farmers Union executive committee.

1982 – B. J. Malusky retires as general manager/president of the Farmers Union Grain Terminal Association. Allen D. Hanson takes over, only the third CEO inn the cooperative's long history. Bessie Klose of Atwater becomes state secretary of the Minnesota Farmers Union.

June 1, 1983 – One of the nation's largest cooperatives is formed by the merger of North Pacific Grain Growers, Inc., and the old Farmers Union Grain Terminal Association. The name selected for the new super-co-op is "Harvest States Cooperatives".

Mar., 1984 – Delegates meeting in New Orleans elect Cyril H. "Cy" Carpenter of Stearns Co., Minnesota, succeeding retiring George W. Stone of Oklahoma as president of the National Farmers Union. Willis R. Eken, a Minnesota legislator, is elected to succeed Carpenter as state Farmers Union president.

1986 – Willis R Eken, president of the Minnesota Farmers Union, is designated vice president for international affairs by the National Farmers Union in the Uruguay Round of trade talks. Eken was re-elected in 1985 and 1987.

1988 – Cy Carpenter, because of age limitation and health, declines to seek another term as national president of the Farmers Union. Minnesota native Leland H. Swenson, who has been president of the South Dakota Farmers Union becomes tenth national president. But before leaving office, Carpenter succeeds in removing mandatory age limit (no one over 65 could seek elective national office). He declines to be first to take advantage of age bylaw change.

1989 – Jim Frederickson succeeds Floyd Borghorst as Farmers Union Agency regional manager. Frederickson began his insurance career in 1976 and grew up in a Farmers Union family near Murdock. Borghorst had worked for Farmers Union Insurance since 1951.

July, 1989 – USDA cooperative specialist Randall E. Torgerson issues a statement noting "A Crisis in Cooperative Education" exists because "many local and regional cooperatives reduced or eliminated educational programs during the difficult economic times of the 1980s". History has shown, Torgerson added, that "continuous education is the lifeblood of cooperation".

Aug. 13, 1989 – Dedication of the Polk Co. Farmers Union Park Memorial on Lake Sarah was held, honoring two men, Max Larson and Fred Gronenger who spearheaded the park development. The original camp, purchased in 1952, covered six acres now is twice as large and features excellent camp facilities.

Nov., 1989 – Willis Eken re-elected to 3rd term state Farmers Union president.

Feb., 1990 – Curt Eischens of Canby is elected to the Cenex board of directors at the cooperative's 59th annual meeting. Eischens, 37, is member of the executive board of Minnesota Farmers Union and president of Yellow Medicine Co. Farmers Union. Elroy Webster of Nicollet is re-elected chairman of the Cenex board.

Sept., 1990 – Former National Farmers Union President Tony T. Dechant, age 75, dies in Denver hospital. His death resulted from a massive stroke suffered July 15. He was a native of Kansas and had been on NFU's staff since 1943. He was national president from 1966 to 1980.

Dec. 15, 1990 – Minnesota Farmers Union moves office from 317 York Avenue, where it had been since 1986, to Interstate Corporate Center on Co. Road D just west of Interstate 35-W.

Jan., 1991 – Harvest States leases Elevator D in St. Paul to gain better Mississippi River shipping access.

May, 1991 – Minnesota Farmers Union President Willis Eken is named chairman of the nine-member National Farmers Union executive committee.

June, 1991 – Green Thumb celebrates its 25th year. Minnesota was one of four pilot states in which the this program was introduced (the others were Arkansas, New Jersey and Oregon). Farmers Union leaders are credited with the idea of employing retired or older persons in state parks, highway rest areas and other public places.

Cenex breaks ground on $30 million refined fuel delivery pipeline linking Minot to Fargo in North Dakota. The current pipeline extends from the cooperative's refinery in Laurel, MT, to Minot. It bought the Montana refinery in 1943.

Aug., 1991 – Willis Eken, Minnesota Farmers Union president since 1984, resigns post to become agricultural advisor to Minnesota 7th District Congressman Collin Peterson. State Senator David Frederickson, 47, (DFL-Murdock) becomes Minnesota president; Vere Vollmers of Wheaton is elected vice president.

Nov., 1991 – Minnesota Farmers Union celebrates 50th anniversary. Theme is "Golden Heritage, Golden Vision". Twenty young adults name "Torchbearers", receive awards from National Farmers Union President Lee Swenson and state president Dave Frederickson.

Seventh Minnesota Youth Advisory Council hold initial meeting in conjunction with state Farmers Union convention. The group was elected at the summer Senior State Camp. Camp participation has increased considerably since the advisory council was established in 1985 to develop new and creative ideas for the MFU camp program.

Feb., 1992 – The first of 10 Mid-Winter Conferences is held at Thief River Falls Feb. 4, marking the revival of the program by Minnesota Farmers Union where topics and issues of general concern receive attention. State president Dave Frederickson and Legislative Services Director Julie Bleyhl are scheduled as conference moderators, with legislators and resource persons taking part. Other conference sites are Mahnomen, New York Mills, Albany, Moose Lake, Norwood, Rushford, Granite Falls, Slayton and Waseca.

July 2, 1992 – Charles F. Brannan, former Truman Administration U.S. Secretary of Agriculture in 1948-53, and general counsel for National Farmers Union, dies at age 88. He authored "The Brannan Plan" where farmers would receive income support payments if the average market price fell below announced support prices for designated commodities. It was never enacted. He was general counsel for NFU from 1953-83.

Aug., 1992 – Senior State Camp elect four youths to first "Executive Council" to work with Youth Advisory Council. Elected were Hollie Larson, Pennock, president; Matt Adamietz, Staples, vice president; Matt Rosedahl, Dawson, secretary, and Lori Jaeger, Mahnomen Co., public relations chair.

Nov., 1992 – National Farmers Union breaks precedent, endorses Bill Clinton for president, first time in its 90-year history it endorsed a presidential candidate.

1992 – Harvest States adds railcar unloading facilities at Winona elevator to increase the terminal's flexibility in handling shipments from members. Joint venture with Continental Grain Co. is finalized for Tacoma, WA, export terminal.

1993 – Cenex sales top $2 billion for the first time.

Mar., 1993 – Two Minnesota Farmers Union members receive "Silver Star" awards at National Farmers Union convention in Sioux Falls for recruiting new members. It was the third time Wendell Bakker of Renville Co. has won the award, and the second for Bessie Klose of Kandiyohi Co. She is the state Farmers Union secretary. Minnesota President Dave Frederickson is elected vice chairman of the National Farmers Union executive committee.

April, 1993 – Lou Anne Kling of Granite Falls is named state director of the Farmers Home Administration. She joined Minnesota Farmers Union after starting to farm in 1970. She and her husband Wayne are the founders of Minnesota's Farm Advocate Program. Michael Dunn, former National Farmers Union legislative director, becomes national administrator of FmHA.

Aug. 8, 1993 – Gordon Scherbing, president and CEO of Farmers Union Marketing and Processing Association, dies of a heart attack. He had worked at the cooperative for nearly 30 years. He had been the chief officer of the co-op in July, 1987, after the retirement of Edwin Wieland.

Nov., 1993 – Darrell Larson of Upsala, president of the Morrison Co. Farmers Union, is elected vice president of Minnesota Farmers Union. He succeeds Vere Vollmers who held that office for 16 years. Larson worked for Farmers Union from 1972 to 1980 as a field representative and field services director. He and his wife Arlene operate a 450-acre farm. Vollmers declined to seek re-election.

Aug., 1994 – U.S. Census Bureau reports the number of farms has fallen to 1.9 million, the fewest since 1950. Farm numbers peaked at 6.8 million in 1935. Here's how number has dropped: 1940 – 6.1 million. 1950 – 5.4 million. 1959 -- 3.7 million. 1969 – 2.7 million. 1978 – 2.3 million. 1987 – 2.1 million. 1992 – 1.9 million.

Feb., 1995 – Norma Hanson of Goodridge was been elected vice president of Minnesota Farmers Union to complete the term of office of Darrell Larson. She has been a member of the state Farmers Union executive committee since 1986. Norma and her husband Lynn farm 880 acres and a herd of registered dairy cattle. She founded, is past-president of Minnesota Agri-Women. Larson resigned to take a position with the Consolidated Farm Service Agency.

Mar., 1995 – National Farmers Union presents its 1995 Meritorious Service Award to Farmers Union and World Agriculture to Bob Bergland at its 93rd national convention in Milwaukee. He was honored for his 30 years as a farm program official, U.S. Secretary of Agriculture, and executive vice president of the National Rural Electric Cooperative Association. Bergland has retired and lives in Roseau where he farmed before beginning government service and where he grew up in a Farmers Union family.

Nov., 1995 – Dennis Sjodin of Cambridge is elected vice president of Minnesota Farmers Union at its 54th annual convention. Sjodin won over incumbent Norma Hanson of Goodridge. Sjodin joined the state Farmers Union staff in 1966 as a part-time field representative and had been assistant to the president under Cy Carpenter. Minnesota Farmers Union President Dave Frederickson of Murdock was unopposed and was unanimously re-elected to his third two-year term.

Aug., 1995 – Minnesota Farmers Union joins the Internet, with an electronic version of Minnesota Agriculture and other information now available to members via computer hookup. Communications Director Pete Takash and Communications Assistant Stephanie Lake provided demonstrations in the Farmers Union Building at the State Fair.

Nov., 1995 – Green View, Inc., presents $10,000 check at Minnesota Farmers Union convention to help finance this Farmers Union history book project.

1996 – Cenex sales hit another new high – $2.7 billion.

Oct., 1996 – Coya Knutson, the only woman elected to Congress from Minnesota, dies at age 84. Known as a champion of family farms, one of her best legislative contributions was federal student loan program – an idea she got from a Farmers Union member and district supervisor (Harold Johnson of Greenbush).

1997 – Two persons with Farmers Union ties (Bob Bergland and Julie Blehyl) are named to the Minnesota Board of Regents. Former Congressman and U.S. Agriculture Secretary Bergland in the early 1950s worked part-time on the state staff, while Blehyl is a former state legislative director.

Feb., 1997 – The death of Jewell Haaland, long-time Yellow Medicine Co. Farmers Union leader is reported. He became chairman of Farmers Union Grain Terminal Association's board of directors in 1972, when he had been a board member 15 years.

PRESIDENTS OF MINNESOTA FARMERS UNION:

1. 1929 – John C. Erp, Canby (1938 – NFU revokes state charter)
2. 1942 – Einar Kuivinen, New York Mills (re-chartered MFU)
3. 1945 – B. Franklin Clough, Kandiyohi Co. elected
4. 1946 – Einar Kuivinen elected for second time
5. 1949 – Roy E. Wiseth of Goodridge
6. 1950 – Edwin Christianson of Gully elected
7. 1971 – Cyril H. Carpenter of Stearns Co. elected
8. 1984 – Willis R. Eken of Twin Valley elected
9. Aug., 1991 – Dennis Sjodin, Cambridge (named acting president after resignation of Willis Eken)
10. Nov., 1991 – David J. Frederickson, Murdock

PRESIDENTS OF THE NATIONAL FARMERS UNION:

James G. Patton, a native of Bazar, Kansas, served 26 years as president of the National Farmers Union – longer than any other man. He was born in 1902, the year when Farmers Union was founded. Elected to the presidency in 1940, Patton was a tall, distinguished-looking Irishman with a strong, commanding voice and style. One of his traits was a passion for work, and an intense interest in the welfare of farm people, both in the United States and internationally.

Patton once created a small controversy when he refused to attend a U.S. State Department banquet in Mexico City in the "black-tie", formal garb required of participants. Such evening wear, Patton held, was the symbol of elitist snobbery and not appropriate for an agricultural conference.

Over the years Patton was at home and equally conversant with people of all walks of life, from the farm and field to the halls of Congress and The White House, where he visited with presidents from Franklin D. Roosevelt and Harry S. Truman to Lyndon B. Johnson.

One of his gifts to Farmers Union was the creation of its national life insurance company, including the fact that it was able to obtain exclusive sales of term life to all Farm Security Administration borrowers in seven states shortly after the life company was organized. This government contract was extremely beneficial to the fledgling insurance firm.

PRESIDENTS OF NATIONAL FARMERS UNION AND YEARS SERVED:

1902 – Newton J. Gresham of Alabama (founder and former organizer of the National Farmers' Alliance and Industrial Union in Texas and other states).
1. 1906 - 1928 – Charles S. Barrett of Georgia.
2. 1928- 1930 – Charles E. Huff of Kansas.
3. 1930 - 1934 – John A. Simpson of Oklahoma.
4. 1934 - 1937 – Ed H. Everson of South Dakota.
5. 1937 - 1940 – John Vesecky of Kansas.
6. 1940 - 1966 – James G. Patton of Colorado.
7. 1966 - 1980 – Tony T. Dechant of Colorado.
8. 1980 - 1984 – George Stone of Oklahoma.
9. 1984 - 1988 – Cyril H. "Cy" Carpenter of Minnesota.
10. 1988 – present – Leland H. Swenson of South Dakota.

HISTORY OF COUNTY CHARTERS

Here are the dates and years of issuance of county Farmers Union charters after the, reorganization in 1942, with the names of first presidents in parenthesis:

Aitkin – Mar. 15, 1956 (Verle Raatz, Aitkin, 1956-60).

Anoka – Nov. 2, 1956 (Martin Gorham, Anoka, 1956-57).

Becker – 1942 (John Okeson, Detroit Lakes, 1942-43).

Beltrami – Nov. 4, 1941 (Ted Brown, Kelliher, 1941-42).

Benton – June 1, 1954 (Henry Brown, Sauk Rapids, 1954-56).

Big Stone –1940 (Fred Griffith, Ortonville, 1940-43).

Blue Earth – Jan. 21, 1953 (G. A. Strobel, Mankato, 1953-57).

Brown – Sept. 28, 1950 (Fritz Buddensick, Springfield, 1950-52).

Carlton – April 5, 1954 (John Walli, Jr., Wright, 1954-55).

Carver – 1954 (Art Lueders, Waconia 1954-55).

Cass – July 7, 1955 (Otto Gravdahl, Pequot Lakes, 1955-58).

Chippewa – 1942 (Original charter, 4-10-28, Nels A. Peterson, 1928-29); (Cecil Wright, Watson, 1942-43).

Chisago – 1954 (Abner Swenson, Chisago City, 1954).

Clay – Jan. 25, 1942 (Charles Gilbery, Glyndon, 1949-43).

Clearwater – April 22, 1944 (Alf Eldevik, Shevlin, 1944-46).

Cottonwood – Dec. 6, 1949 (Russel Schwandt, Sanborn, 1949-50).

Crow Wing – 1957 (William Cory, Crosby, 1957-58).

Dakota – Aug. 11, 1948 (F. L. Gerten, St. Paul, 1948-50).

Dodge – June 8, 1954 (Reuben Hauglum, Kasson, 1954-55).

Douglas – Oct. 13, 1950 (Fred Stahl, Eagle Bend, 1950-51).

Faribault – Nov. 12, 1949 (George Vikingstead, Blue Earth, 1949-50).

Fillmore – May 5, 1955 (Charles Sellman, Mable, 1955-57).

Freeborn – July 25, 1950 (Joe Solberg, Hartland, 1950-54).

Goodhue – Feb. 3, 1954 (Arnold Boraas, West Concord, 1954).

Grant – July 14, 1948 (Lyle Tobolt, Elbow Lake, 1948-53).

Hennepin – April, 1957 (Henry Pauley, St. Bonifacious, 1957-60).

Houston – Mar. 15, 1956 (Nels Gulbranson, Spring Grove, 1956-59).

Hubbard – Sept. 27, 1945 (Original charter, Verner Tangborn, Laporte, 1945-47); (Reorganized, 1955, Lyman Schmidt, Laporte, 1955-56).

Isanti – June 16, 1954 (Alden Anderson, Cambridge, 1954-57).
Itasca – Sept. 30, 1949 (Miles Nelson, Togo, 1949-54).
Jackson – June 9, 1950 (Leo Freking, Heron Lake, 1954-57).
Kanabec – Nov. 3, 1954 (Robert Ripka, Mora, 1954-55).
Kandiyohi – Jan. 17, 1942 (B. Franklin Clough, Sr., Lake Lillian, 1942-45).
Kittson – Nov. 27, 1942 (William Vesey, Donaldson, 1942-45).
Koochiching – May 20, 1949 (Orlyn Kohlhase, Mizpah, 1949-50) .
Lac Qui Parle – 1943 (Bert Trelstad, Dawson, 1943-45).
Lake of Woods – Feb. 29, 1952 (LaVern Wagner, Baudette, 1952-54).
Le Sueur – 1954 (Roy Sohm, Le Center, 1954-56).
Lincoln – (Original charter in 1933, E. A. Lundberg, Ivanhoe, 1933); (Jan., 1953,
    E. A. Lundberg, Ivanhoe, 1953).
Lyon – Jan. 6, 1953 (Vernon Runholt, Marshall, 1953).
Mahnomen – Mar. 31, 1942 (no record).
Marshall – Co. organized Mar. 13, 1943 (No Records, 1943-48); East Marshall –
    Mar. 5, 1948 (Roy E. Wiseth, Goodridge, 1948-49); West Marshall – Mar. 10,
    1948 (Glenn Schuman, Warren, 1948-49).
Martin – Feb. 4, 1953 (Ross Holland, Granda, 1953).
McLeod – Jan. 6, 1954 (Ben Klima, Silver Lake, 1954).
Meeker – May 29, 1952 (Ben Marshall, Grove City, 1952-55).
Mille Lacs – April 11, 1955 (Ed Bergma, Milaca, 1953-55).
Morrison – Mar. 15, 1956 (George Kliber, Little Falls, 1952-63).
Mower – Nov. 24, 1953 (Harold Murphy, Austin, 1953).
Murray – Sept. 15, 1950 (Harry Nelson, Garvin, 1950).
Nicollet – 1957 (Fred Bruns, Nicollet , 1957).
Nobles – Oct. 26, 1943 (Harry Paulson, Worthington, 1943-45).
Norman – Jan. 9, 1943 (George Schneider, Gary, 1943-44).
Olmsted – Mar. 15, 1956 (Jens Anderson, Byron, 1956-57).
Otter Tail – 1941 (Einar Kuivinen, New York Mills) Co. reorganized: E. Otter Tail
    – 1956 (Roy Fletcher, Vergas, 1956-57). W. Otter Tail – 1956 (Ernest Bartels,
    Jr., 1956-66).
Pennington – Oct. 19, 1943 (Jesse Anderson, Goodridge, 1943-44).
Pine – Jan. 28, 1953 (Frank Adams, Hinckley, 1953-54).
Pipestone – April 2, 1952 (Kenneth Zimmerman, Holland, 1952).
Polk – Mar. 20, 1942 (Edwin Christianson, Gully, 1942-43).
Pope – July 2, 1946 (Walter Gilbertson, Glenwood, 1946-48).
Red Lake – 1942 (William Cassavant, Red Lake Falls, 1942-49).
Redwood – Sept. 14, 1950 (Harold Larson, Springfield, 1950-55).
Renville – Nov. 28, 1949 (Gilbert Nelson, Sacred Heart, 1949).
Rice – June 4, 1952 (Clarence Purfeerst, Faribault, 1952-53).
Rock – Sept. 13, 1950 (Merten Petersen, Hardwick, 1950-51).
Roseau – Jan., 1942 (Bernie Brandt, Roseau, 1942-44).
Scott – July 29, 1954 (Frank Zweber, New Market, 1954-56).
Sherburne – Mar. 15, 1956 (Sherman H. Nelson, Princeton, 1956-61).
Sibley – April 4, 1952 (Elmer Tosch, Winthrop, 1952-55).
Stearns – Oct. 7, 1952 (Lionel Reeck, Paynesville, 1952-56).
Steele – April 23, 1953 (Tyndal Evans, Owatonna, 1953, 1956-57).
Stevens – Sept. 18, 1951 (August F. Fluegel, Morris, 1950).

Swift – July 15, 1941 (Theo Frederickson, Murdock, 1942).
Todd – April 13, 1954 (Laurence Strack, Osakis, 1954-55).
Traverse – June 18, 1952 (Carl Lupka, Wheaton, 1952-56).
Wabasha – 1956 (Leo Deming, Plainview, 1956-57).
Wadena – Nov. 11, 1954 (John Niemi, Menagha, 1956-58).
Waseca – Feb. 2, 1954 (Melvin O'Riley, Waseca, 1954-55).
Washington – April 14, 1955 (Arthur Krueger, Stillwater, 1955).
Watonwan – April 5, 1951 (Frank Vogel, St. James, 1951).
Wilkin – Mar. 12, 1946 (Ralph Pehl, Campbell, 1947-56).
Winona – Mar . 15, 1956 (Gerald Kronebusch, Rollingstone, 1956).
Wright – June 10, 1952 (Andrew Rimpy, Annandale, 1952-56).
Yellow Medicine – Jan. 14, 1939 (Sam Gullickson, Hanley Falls, 1937-42).

TIMELINE OF KEY STATE INVOLVEMENT WITH FARMERS UNION:

1902 – Farmers Union is founded in Texas.

1913-1917 – Organizational efforts begin in 12 states: California, Colorado, Idaho, Iowa, Montana, Nebraska, North Dakota, Ohio, South Dakota, Washington, and Wyoming.

1927 – Northwest Organizing Committee begins organization of North Dakota, and three other northern states – Minnesota, Montana, and Wisconsin. (The committee members are to become legendary figures in the Farmers Union Movement (M.W. Thatcher, C.C. Talbott, and A. W. Ricker).

1929 – Minnesota receives state charter, and John C. Erp of Canby is elected president..

1937 – Minnesota's charter is revoked because of friction and disagreements between state and national organization.

1942 – Minnesota is re-chartered, with Einar Kuivinen of New York Mills is elected president and E. L. Smith of Montevideo wins vice presidency.

1945 – B. Franklin Clough, Kandiyohi Co. elected when Kuivinen declines to run.

1946 – Einar Kuivinen is elected president for a second time; Edwin Christianson of Gully is elected vice president.

1949 – Roy E. Wiseth of Goodridge becomes president; Wendell Miller of Warren Co. is vice president.

1950 – Edwin Christianson of Gully becomes state president, with William Nystrom of Worthington his vice president.

1957 – Minnesota Farmers Union membership card 40,000 is issued to Cottonwood farm family (Roger and Joyce Janzen of Mountain Lake).

1971 – Cy Carpenter of Stearns Co. becomes acting state president when Edwin Christianson is stricken with a disabling stroke just before state convention, and then is elected Minnesota president.

1984 – When Cy Carpenter wins the National Farmers Union presidency, and state board of directors select State Legislative Majority Leader Willis Eken of Twin Valley president.

1991 – Willis Eken resigns to become Congressional agricultural advisor. Dennis Sjodin of Cambridge is named acting state president. Another former state legislator, David J. Frederickson of Murdock, then is elected president at state convention. Dennis Sjodin is elected vice president.

360

MINNESOTA STATE OFFICERS BY YEAR ELECTED:

Presidents:
1929 – John C. Erp of Canby
1942 – Einar Kuivinen of New York Mills
1943 – B. Franklin Clough of Lake Lillian
1945 – Einar Kuivinen regains presidency
1949 – Roy E. Wiseth of Goodridge
1950 – Edwin Christianson of Gully.
1971 – Cyril H. Carpenter of Sauk Centre
1984 – Willis Eken of Twin Valley.
1991 – Dennis Sjodin of Cambridge (acting);
1991 – David J. Frederickson of Murdock.

Vice Presidents:
1929 – Hemming S. Nelson
1942 – E. L. Smith of Montevideo
1944 – B. Franklin Clough of Lake Lillian
1946 – Edwin Christianson of Gully
1950 – Wendell Miller of Cottonwood
1951 – William Nystrom of Worthington
1971 – Cy Carpenter, Sauk Center
1973 – Vincent Ritter, Chokio
1974 – Dean Vattauer, Red Lake Falls, 2nd Vice President
1974 – Vere Vollmers, Wheaton, special vice president, Young Farmers
1978 – Vere Vollmers
1993 – Darrell Larson, Royalton
1995 – Norma Hanson, Goodridge
1996 – Dennis Sjodin, Cambridge.

Secretary:*
1929 – Carl Lundberg, Ivanhoe
1942 – Irene Paulson of New York Mills.
1946 – Lura Reimnitz of Willmar.
1947 – Pete Egan.
1948 – S. J. Christopherson of Willmar.
1949 – Tony T. Dechant of Denver, Co.
1950 – Herman Knutson of Warren.
1951 – Russel Schwandt of Sanborn.
1956 – Clint Hess of Minneapolis.
1964 – Archie Baumann of St. Paul.
1968 – Cy Carpenter of Bloomington.
1972 – Robert Rickert of Okabena.
1979 – Marvin Jensen, Evansville.
1982 – Bessie Klose, Atwood.
*State Secretary is now elected by state board of directors.

BIBLIOGRAPHY:

BOOKS:

Farmers' Organizations, Lowell K. Dyson, Greenwood Press, NY, NY, 1986.

Milo Reno, Farmers Union Pioneer, Roland A. White, The Athens Press, Iowa City, 1941.

A Rich Heritage of Community, A History of the National Farmers Union Insurance Co., Paul E. Huff and Raymond E. Novak, National Farmers Union, Denver, CO, 1995.

50 Years – North Dakota Farmers Union, Charles and Joyce Conrad, North Dakota Farmers Union, Jamestown, ND, 1976.

The Case For Farmers, James G. Patton, Public Affairs Press, Washington, D.C., 1959.

Iowans Who Made a Difference, Don Muhm and Virginia Wadsley, Iowa Farm Bureau Federation, West Des Moines, IA, 1996.

New Frontiers, Henry A. Wallace, Reynal & Hitchcock, NY, NY, 1934.

Yearbook of Agriculture, U.S. Department of Agriculture, Washington, D.C., 1927-50.

Red Harvest, The Communist Party & American Farmers, Lowell K. Dyson, University of Nebraska Press, Lincoln, NE, 1982.

To Gather Together, Leo N. Rickertson, Cooperative Printing Association, Minneapolis, MN, 1980.

The Advancement of American Cooperative Enterprise, 1920-1945, Joseph G. Knapp, The Interstate Printers & Publishers, Danville, IL, 1973.

Great American Cooperators, Joseph G. Knapp and Associates, American Institute of Cooperation, Washington. D.C., 1967.

NEWSPAPERS AND MAGAZINES:

The Canby (MN) *Press* & The Canby *News*
The Farmers Union Herald
Minnesota Agriculture
Co-op Country News
Grain Quarterly (National Federation of Grain Cooperatives)
Minneapolis *Star Tribune*
St. Paul *Pioneer Press*
Mankato *Free Press*
Sioux City *Journal*
Des Moines *Register*

GROWTH OF FARMERS UNION IN MINNESOTA

State Charter: Granted Nov. 10, 1942 (Membership – 1,907)

Here is summary of the chartering of Farmers Union locals by year and county in which each local was organized:

1941(1 local)* Wadena – Menagha. * Some locals organized during the John Erp years (1929-38) may not be included as "new", but instead might have had their old charters "continued" in the new state organization.

1942 (4 locals) Dakota – Home office. Kittson – North Branch, Riverside, Springbrook Juniter. Lac Qui Parle – Harmony. Roseau – Triple.

1943 (99 locals): Becker – Atlant Walworth, Audubon, Callaway, Cormorant, Cuba, Erie Detroit, Richwood Holmsville. Beltrami – Cormorant Quiring, Frohn, Helga, Inez, Medicine Lake, Pleasant Valley, Progressive, Shotley. Big Stone – Barry, Prior. Carlton – Blackhoof, Central, Skelton. Chippewa– Crate Havelock, Grace, Rosewood. Clay – Hitterdal, Morken, Oakprot Kragness. Clearwater – Clearbrook, Eddy. East Otter Tail – Heinola, Homestead, Newton. Grant – St. Olaf Pelican. Kandiyohi – Lake Elizabeth, Irving, New London, Thorpe. Kittson – Cannon Hot Shot, Caribou, Donaldson, McKinley. Jackson– Progressive. Lac Qui Parle – Baxter, Camp Release, Lakeside, Lakeshore, Yellow Bank. Mahnomen– Hello, La Garde, Lake Grove, Marsh Creek, Sunshine. Marshall – Benville, Big Valley, Englund, Gatzke (reorganized in 1952), Holt, Little River, March, Randen, Strathcona, Spruce Grove, Viking. Nobles – Indian Lake, Little Rock, Midway Livewire, Reading Star. Norman – Fossum, Perley. Otter Tail – Almora. Pennington – Erie. Polk – Beltrami, Bransvold, Hubbard, Gully, King, Knute, Neby, Sunnyside. Red Lake – Brooks. Roseau – Triple. Lyon – Eidsvold. Pennington – Good Hope, Pioneer, Union. Ramsey – FUGTA. Rock – Beaver Creek, Steen. Swift – Spring Creek. Wadena – Menagha (Reorganized in 1954). Washington – Langdon. Wilkin – Doran Wi. Yellow Medicine – Eidsvold, Friendship Hazel Run, Lisbon, Normanis, Oshkosh, Sandness, Swede Prairie.

1944: (65 locals) Becker – Lake Eunice, Westbury, Wolf Lake, Sugarbush. Beltrami – Grant Valley, Liberty, Northern. Big Stone – Artichoke, Big Stone Odessa, Clinton. Chippewa – Beacon, Leenthrop, Mandt, Sparta, Tunsberg. Clearwater – Popple, Holts Copley, Whispering Pine. Dakota – Inver Grove. Kandiyohi – Arctlander, Colfax, Gennessee, Green Lake Harrison, Lake Lillian, Mamre Dovre, Sunburg-Norway Lake. Kittson – St. Joseph. Lake of the Woods –

Hagen. Mahnomen – Chief Rosedale. Marshall – Bigwoods, Espelie, Moose River. Norman – Home Lake, McDonaldsville, Sundahl, Waukon. Otter Tail – Last Hope, Tri Township. Pennington – Highlanding. Pine – New Pioneer. Polk – Ash Lee, Cloverleaf, Columbia, Gentilly, Queen, Sletten, Winger. Red Lake – Louisville, Oklee, Plummer, Riverside. Rock – Martin, Rosedale. Roseau – Badger, Central, Cloverleaf, Klectzen, North Star, Willow Creek. Swift – Camp Lake, Chippewa, Hayes Community, Lake Hazel, Tornig. Wadena – Sebeka (reorganized in 1954).

1945: (19 locals) Beltrami – Hornet, Roosevelt. Clay – Oakport, Moland. Clearwater– Berner, Leonard. Hubbard – Becida, Henrietta, Nary, Shell Prairie. Kandiyohi – Holland, White Rose. Mahnomen – Bejou. Nobles – Bloom, Dewald Little Rock, Seward Rangers. Polk – Rosebud. Swift –Hegbert.

1946: (16 locals) Beltrami – Nebish. Clay – Uhlen. Dakota – Salem, Nininger. Kandiyohi – Arctander (rechartered). Mahnomen – Victory. Marshall – Ringbo. Murray – Iona. Norman – Gary. Pope – Grandview, Trappers Run, Midway, Pleasant Hill, White Bear Lake. Wadena – Nimrod (reorganized in 1954). West Marshall –Lincoln Park Valley.

1947:(5 locals) Clay – Hitterdal. Nobles – Dewald Little Rock. Otter Tail – Aurdal, Lakeland. Wilkin – Kutzer.

1948: (11 locals) Big Stone – Foster. Dakota – F. U. Livestock. Grant – Elbow Lake. Hubbard – Laporte. Kandiyohi – State Office. Otter Tail – Otto, Scrambler. Pennington – River Valley. Roseau – Haycreek, Southline. Yellow Medicine – Burton.

1949: (39 locals) Beltrami – Four Square. Big Stone – Beardsley. Brown – Stately. Clearwater –Minerva Rice. Cottonwood – Germantown, Jeffers, Windom Flaxters. Faribault – Blue Earth, Elmore, Emerald, Rome, Seeley. Felton – Felton. Grant – Hoffman. Koochiching – Forest Grove, Gemmell, Mizpah. Itasca – Acme, Cloverdale, Big Fork, Effie Busti, Togo. Jackson –Clear Lake. Lac Qui Parle – Baxter, Rosen. Norman – Halstad. Nobles – Elk Center. Redwood – Wanda. Renville – Ericson, Osceola, South Sacred Heart. Swift – Murdock. Traverse – Parnell. Wantonwan – North Central. West Marshall – Oslo. West Otter Tail – Norwegian Grove, Scambler. Wilkin – Kent Wilkin. Yellow Medicine – Sioux Agency.

1950 (67 locals) Brown – Community, Mulligan. Chippewa – Swift Chippewa Line. Clay – Barnesville. Cottonwood – Westbrook, Storden-Highwater. Douglas – Garfield, Lake Region, Moe, Stormy Creek. East Marshall – Thief Lake. Freeborn – Freeborn, Hartland, Manchester. Jackson – Belmont, Bergen, Double H, Rost, Star, Wei Crosse. Lac Qui Parle – Madison Haydenville. Lyon – Russell. Meeker – Grove City, Union Grove. Murray – Green Valley, Iona, Lake Wilson, Midway, Slayton Hustlers, Ellesboro. Nobles – Adrian, Brewster Booster, Grand Prairie, Leota Lismore, Graham Lakes. Pennington – Goodlanding. Polk – Garden, Gentilly Pioneer, Tabor. Redwood – Belview, Lucan, Clements, Renville – Crooks, Kingman. Rock – Hardwick Traders, Jasper, Luverne. Roseau – Roseau River, Benwood. Sibley – Gibbon. Stearns – Paynesville. Stevens – Alberta. Swift – Mayer. Watonwan – South Branch, St. James. Wright – Temperance Corner, West Albion. Yellow Medicine – Stony Run, Tyro.

1951 (22 locals) Clearwater – Shevlin Copley. Clay – Hagen. Cottonwood – Mountain Lake, Rose Hill. Douglas – West Miltona. Grant– Herman Norcross.

Kandiyohi – Kandy. Murray – Des Moines River, Holly. Norman – Hendrum. Polk – Euclid, Fertile, Honest Liberty, Lakeside. Red Lake – Open Range. Redwood – Milroy, Revere. Rice – Northfield Wheeling. Stevens – Morris, Synnes. Traverse – Clifton Valley. West Marshall – Northwest Valley.

1952 (69 locals) Big Stone – Graceville. Blue Earth – Le Ray, Lincoln. Brown – Essig. Carlton – Moose Lake. Dakota – 4 Corners, Vermillion. Douglas – Liberty, Runstone. Faribault – East Prairie, Winnebago. Freeborn – Twin Lakes. Kanabec – Whited Comfort. Koochiching – Rainy River. Lac Qui Parle – Freeland Manfred. Lake of the Woods – Carp, Pitt,.Spooner Boone, Wabanica, Williams. Lincoln – Alta Vista, Ash Lake, Verdi. Lyon – Dudley, Green Valley. Mahnomen – Church Lake, Wild Rice. Martin – Trimont, Welcome. McLeod – Swan Lake. Meeker – Danielson, Greenleaf, Mahnannah, Trailblazers. Nicollet – New Sweden Bernadotte. Norman – Flom. Pine – Askov, Hinckley, South Pine. Pipestone – Cozenovia, Edgerton, F. P. A., Ihlen, G. R. G. Rice – East Prairie, Lonsdale. Redwood – Wabasso. Renville – Osceola. Roseau – Badger Creek, Greenbush. Sibley – Gaylord, Winthrop. St. Louis – Grass Lands. Stearns – Eden Lake, Sauk Centre. Stevens – Chokio, Donnelly, Hancock. Traverse – Dumont. Wantonwan – Antrim, Madelia. Washington – Washington 1. West Marshall – Florian, Happy Corner . West Otter Tail – Orwell. Wright – Albion Center. Yellow Medicine – Wood Lake Posen.

1953: (74 locals) Becker – Wolf Lake. Benton – Gillman, West Benton. Blue Earth – Danville, Mapleton, Vernon Center. Brown – Hanska. Carlton – Cromwell. Chippewa – Granite Falls. Chisago – Dairy Belt. Clay – Rollag. Clearwater – Nora Falk. Douglas – Belle River, Forada, Midway. East Marshall – Columbus, Grygla (reorganized). East Otter Tail – Hillside. Faribault – Wells. Freeborn – Alden Carlston, Big Four, Mansfield. Goodhue – Kenyon, White Rock. Isanti – Braham, Rum River. Jackson – Alpha. Kanabec – Snake River. Kittson – Hallock, Kennedy. Koochiching – Rapid River. Lincoln – Hope, Marble Hansonville. Lyon – Eidsvold, Tracy. Mahnomen – Rainbow. Martin – Tenhassen, Westford. Meeker – Cosmos, Forest City. McLeod – Glencoe, Pennton. Mille Lacs Princeton. Mower – Austin-Windom, L & U, Lyle-Nevada, Waltham-Red Rock. Murray – Currie Booster, Skandia. Nicollet – Courtland, Nicollet. Norman – Good Hope. Otter Tail – Orwell. Pine – North Pine. Redwood – Lamberton. Renville – Hector. Roseau – Roseau River. St. Louis – Floodwood. Steele – Aurora, Havana, Lemond Somerset, M & D, M & M. Todd – Eagle Creek, Sauk River. Wantonwan – Ormsby-Odin. West Marshall – Old Mill. West Otter Tail – A. W., Dora. Wilkin – Rothsay. Yellow Medicine – Canby, Minnesota Falls, Wergeland.

1954 (114 locals) Aitkin – Lawler. Anoka — Rum River Valley. Becker – Evergreen, Frazee, Ogema, Pine Point. Beltrami – Kelliher, Solway. Benton – St. Francis River. Blue Earth – Garden City. Carlton – Blackhoff, Kettle River, Wright. Chippewa – Big Bend, Clara City, Milan. Chisago – Central, Goose Creek. Clay – Hawley. Dakota - Lakeville. Dodge – Concord, Dodge Center, Ellington. Douglas – Branville. East Otter Tail – Folden, Prairie View. Freeborn – Newry, Riceland. Goodhue – Cannon Falls, Red Wing, Wastedo, White Rock, Wanamingo. Hubbard – Garfield Lake. Isanti – Isanti. Itasca – Bigfork Valley. Kanabec – Brunswick, Ogilvie, Pomroy Kroschel. Kandiyohi – Shipstead. Lac Qui Parle – Marietta. Le Sueur – Cleveland, Elysian, Kasota, Kilkenny, LeCenter,

Montgomery. Lincoln – Drammen Diamond Lake, Hendricks, Ivanhoe, Lakestay-Limestone. Lyon – Florence, Marshall, Minneota. Mahnomen – Parity. Martin – Sherburn. McLeod – Acoma, Glencoe. Meeker – Cedar Mills, Dassel, Forest Prairie, Kingston, Litchfield. Mille Lacs – Onamia, Three Corners. Mower – A & M of Adams. Murray – Fulda. Nicollet – Oshawa Traverse. Polk – Woodside. Pope – Lakeview, Terrace. Redwood – Vesta. Renville – Bechyn, Bird Island, Morton. Rice – M & W, Millersburg, Walcott, Wells. Scott – Credit River, Marystown, New Market. Sherburne – Becker, Big Lake, Santiago-Blue Hill. Sibley – Arlington, Green Isle, Shady Lane. Stearns – St. Joseph Co-op, Western Stearns. Steele – Blooming Prairie, Ellendale. Stevens – Horton. Swift – Clontarf-Tara, Kerkhoven, Shible-Hegbert. Todd – Clarissa, Oak Hill. Traverse – Twin Lakes. Wabasha – Mazeppa, Thomastown, Zumbro Valley. Wantonwan – Darfur. Waseca – Janesville, New Richland, Waseca. West Otter Tail – Pelican Rapids, Urbank. Wright – Maple Lake, Monticello, Mud Lake, Stockholm.

1955 (65 locals) Aitkin – Eastwood, Malmo. Anoka – Lincoln. Becker – Two Inlets-Savannah, Osage. Big Stone –Akron. Carlton – Moose Horn. Carver – Bevens Creek, 4 Corners, Waconia. Cass – Backus, Lake Region, Leader, McKinley, Steamboat Lake, Trelipe. East Marshall – Newfolden. East Otter Tail – Henning, Midway, Sturdy Oak, Woodland. Fillmore – Harmony, Root River, Tawney F.U. Grant – Barrett, Wendell. Goodhue – Vasa. Houston – Houston, Spring Grove. Hubbard – Akeley, Shell Prairie, Yaeger. Kandiyohi – Lake Andrew. Le Sueur – Le Sueur. Lincoln – Lake Benton. McLeod – Lester Prairie, Winstad. Morrison – Pierz. Mower – Grand Meadow. Murray – Lake Wilson. lmsted – New Haven, S. E. Olmsted, Salem. Pennington – Silverleaf. Pope – Bangor & Lake Johanna, W.B. M. Redwood – Morgan. Renville – Fairfax, Franklin. Scott – Jordan. Sherburne – West Sherburne. Stearns – Clearwater, Freeport, Twin Haven. Swift – Danvers. Todd –Bartlett, Browerville. Wabasha – Plainview. Wadena – Compton. Washington – Dairy Center, Woodbury. West Otter Tail – Aurdal, Eagle Lake, Erhard, 4 Square. Wilkin – D & W. Winona – Hart, Pleasant Hill, Rollings. Wright – Albertville, Annandale, Buffalo, Howard Lake, Waverly.

1956 (46 locals) Aitkin – Palisade, McGrath. Anoka – Cedar. Beltrami – Becida. Benton – Duelm, Foley. Brown – Sleepy Eye. Carlton – Perch Lake. Carver – Watertown. Dodge – Hayfield. East Marshall – Newfolden. Faribault – Keister. Fillmore – Lanesboro, Wykoff, Chosen Valley, Peterson-Rushford. Freeborn – London Shellrock, Albert Lea, Olivia. Grant – Erdahl. Hennepin – St. Boni; Hamel. Houston – Caledonia. Hubbard – Farris. Jackson – Okabena. Kanabec – Putnam. Morrison – Little Falls, Hillman, Swanville Upsala. Olmsted – Central Olmsted, Chatfield, Rock Dell. Polk – Prairie. Renville – Buffalo Lake, Olivia. Scott – Cedar Lake, Jordan. Sibley – Rush River. Todd – Pine Grove, Swan Lake. Waseca – Otisco Woodville. Winona – Fremont, St. Charles. Wright – Rockford, Delano.

1957 (17 locals) Chisago – Lent Branch. Crow Wing – Cross Lake, Daggett Brook. Dodge – Kasson. Faribault – Ellmore. Fillmore – Preston. Houston – Hokah. Le Sueur – Kilkenny Waterville. Mower – Le Ray. Olmsted – Kalmer. Morrison – Morrill. Pope – Lakeland. Scott – Belle Plaine. Stearns – Richmond. Wabasha – Oak Center. Washington – Northern. Wantonwan – Rosendale.

1958 (11 locals) Aitkin – Aitkin Deerwood. Crow Wing – Pine Center. Douglas – Friendship. Faribault – Bricelyn. Morrison – Randall Flemsberg. Sherburne – Zimmerman. Stearns – Holding, Greenwald. Traverse – Walls. Waseca – Janesville, Iosco.

1959 (6 locals) Clay – S. W. Community. Polk – Fisher Bygland. Wabasha – Lake City, Millville. West Otter Tail – Lakeland. Wilkin – Valley.

1960 (9 locals) Aitkin – Jacobson. Cass – Boy River. East Otter Tail – Mills. Freeborn – Freeman. Hennepin – Eden Prairie. Todd – Grey Eagle. Traverse – Mustinka. Wadena – Peaceful Valley. West Otter Tail – Dalton.

1961 (3 locals) Benton – Watab, Oak Park. Morrison – Shady Brook.

1969 (1 local) Goodhue - Zumbrota.

1973 – (1 local) Mahnomen – Rising Sun.

1976 – (1 local) Pope – Liberty.

1980 (3 locals) Goodhue – Goodhue. Otter Tail – Southwest Otter Tail. Wadena – Vandale-Heartland.

1982 (3 locals) Chisago – Center City, Dairy Belt, Goose Creek.

1983 (2 locals) Hubbard Co. – Straight River. Pine – North Pine.

1985 (1 local) Grant – Barrett-Hoffman.

MINNESOTA FARMERS UNION MEMBERSHIP BY YEAR

1942 – 1,907	1953 – 21,288	1964 – 31,033
1943 – 2,396	1954 – 31,335	1965 – 27,014
1944 – 3,849	1955 – 35,011	1966 – 24,344
1945 – 4,182.	1956 – 39,590	1967 – 25,032
1946 – 4,689	1957 – 40,313*	1968 – 25,394
1947 – 5,336	1958 – 41,103	1969 – 24,012
1948 – 5,386	1959 – 40,159	1970 – 24,021
1949 – 8,685	1960 – 41,101	1971 – 24,090
1950 – 11,890	1961 – 40,080	1972 – 24,062
1951 – 12,403	1962 – 39,102	
1952 – 15,289	1963 – 35,265	

(*On July 22, 1957 Roger and Joyce Janzen of Mountain Lake in Cottonwood Co. became Minnesota Farmers Union member # 40,000.)

Annual dues: 1942 – $3.50. 1950 – $5. 1963 – $7.50/family (also 2-year membership for $10; 5-year membership for $25.) 1965 – 25-year membership available for $25 along with annual, 2-year and 5-year memberships.

INDEX

A

Ackerman, Arnold, 313
Ackerman, Harm, 25
Adamietz, Jim, 301
Adamietz, Matt, 355
Adams, Frank, 359
Adams, MN Lt. Gov Charles, 20
Aelke, Paul, 16, 24
Agency for International Development, 298
Agricultural Adjustment Act, 62, 75, 84, 343
Agricultural Adjustment Administration, 49
Agricultural Committee for Roosevelt (1944), 71
Agricultural Conservation Program (ACP), 343
Agricultural Marketing Agreement Act, 344
akkers, 13
Albert Lea Rest Area, 308
Albrecht, Tom, 23
Albuquerque, 271, 295
Allebach, Clyde, 192
Allebach, Rodney, 320
Alzheimer's Association, 77
Amalgamation, 48
Amalgamation of 1927, 47
American Agriculture Movement, 21, 210, 252
American Farm Bureau, 59
American National Bank of St. Paul, 46
Amo Township, 290
Anderson, A. A., 17
Anderson, Alden, 284, 359
Anderson, Alvin, 233
Anderson, Bruce, 233
Anderson, Carleton, 286
Anderson, Carlton, 206, 284
Anderson, Clarence, 320
Anderson, Glen, 42, 255
Anderson, Gunder, 32

Anderson, Harry, 290
Anderson, Herman, 314
Anderson, J. Edward, 47, 64
Anderson, Jens, 359
Anderson, Jesse, 359
Anderson, John P., 285
Anderson, Judith, 301, 329
Anderson, Martin, 23
Anderson, Mary (Kuivinen), 69
Anderson, Mrs. Clarence, 297
Anderson, Paul, 306
Anderson, Verner, 65, 76, 77
Anderson, Violet, 283, 285
Anderson, Wendell - MN Gov., 206
Andrews, Mark, 98
Ann Local, 290
Ann Township, 290
anti-Communism, 74
anti-corporation farming law, 258
Atz, Carolyn, 309
Audette, Delbert, 320
Audette, Nolla, 297

B

Bach, Francis, 300
Bakkas, 289
Bakker, Wendell, 355
Baldwin Piano Company, 315
Baldwin United, 315
Bank for Cooperatives, 91
Bankhead-Jones Act, 343
Bankhead-Jones Farm Tenant Act, 85, 344
bankruptcy (federal) laws amended, 62
Banner local, 22
Barrett, Charles, 32, 44, 47
Bartels, Ernest Jr., 359
Batavia (Iowa) Farmers Union band, 60
Battle of the Tin Cans, 88
bauernhof, 13
Baumann, Archie, 151, 261, 268, 285, 299
Beaver Dam, WI, 293, 294
Beckmann, Arnel, 284

Behrman, Morris, 301
Beimert, Nancy, 309
Benda, William, 23
Bender, William, 24
Benson, Ezra Taft - U. S. Agriculture
  Secretary,, 141
Berg, Richard, 301
Bergenheier, Frank, 231
Bergland, Bob, 89, 142, 200, 247, 248,
  249, 251, 252, 264, 313, 332, 356
Bergland, Helen Grahn, 248
Bergland, Selmer "Sam", 89, 248
Bergma, Ed, 359
Bergman, Algot, 17
Bergstrom, Carl, 286
Bernard & Johnson, 193
Bernard, William, 193
Bertels Building, 24
Bezdicek, Louis S., 23
Biel, Eunice, 300, 329
Binger, Mrs. Thelma, 286
Bleyhl, Julie, 255, 355
Blight, Jim, 206
Block, John R., 260
Block, John R. , Reagan Sec. of Ag: Cy
  Carpenter calls for his resignation,
  220
Blodgett, A. J., 16
Blodgett, G. H., 33
Blomgren, Doris, 284
Blomgren, Ralph, 286
Bodeu, Mrs. Elmer, 285
boer, 13
Bohemian Accordian Band, 25
Boraas, Arnold, 358
Borchardt, David, 300
Borden, John, 300
Borden, Winston, 217
Borghorst, Floyd, 311, 318, 354
Bosch, J. B., 328
Bosch, John, 48, 60, 90, 327
Boschwitz, Rudy vs Dave Frederickson,
  268
Brandt, Bernie, 359
Brandt, Mrs. Phyllis, 291
Brannan Plan, 73, 75, 106, 140
Brannan, Charles - U.S. Secretary of
  Agriculture, 140
Bread For The World, 199
Breberg, Deb, 320
Brekke, Oscar, 33

Bremmer, Otto, 46
Brennan, Judge E. Thomas, 229
Bridges [Sen.] is either an unmitigated
  liar or dupe of schemers, 106
Bridges, New Hampshire Senator Styles
  and "red smears", 106
Brooks, Frank -41,000th MFU member,
  349
Brovold, Sanna, 162
Brown, George, 301
Brown, Henry, 358
Brown, Ted, 358
Brown's Valley Local, 343
Bruns, Fred, 359
Buddensick, Fritz, 358
Burdick, Quentin, 81, 98
Butcher, Dave, 300
Butenhoff, Earl, 192
Butz, Earl L., 204, 253
Byron, Jim, 301

C

Canby Oil Co-op, 34
Canby *Press*, 32
Capper-Volstead Act, 342
Capper Volstead Act of 1922, 43
Carlson, Elmer, 21
Carlson, Jerome, 213
Carpenter, Cy, 103, 139, 160, 191, 192,
  193, 194, 195, 207, 208, 238, 258,
  283, 305, 332
Carpenter, Frances Stauning, 191
Carpenter, Kris, 191
Carpenter, Richard, 191
Carson, Rebecca, 300, 329
Carstenson, Dr. Blue A., 306
Carter, President Jimmy, 252
Cassavant, William, 66, 359
Cattnach, Steve, 320
Cenex, 116, 239, 241, 255, 272
Cenex Convenience Stores, 129
Cenex Fast Lube, 129
Cenex/Land O'Lakes partnership, 128
Central Bi-Products, 323
Central Cooperative Association, 16
chartering convention, 16
Chase, Ray P., 87
Chervestad, Korydon, 255, 301
Childberg, Herman, 301
Chosen Valley Local, 329

Christen, Sue, 320
Christianson, Edwin, 66, 71, 104, 129, 143, 145, 146, 153, 154, 160, 207, 282, 332, 359
Christianson, MN Gov. Theodore, 20
Christianson, Nora, 145, 148, 152, 154, 197
Christianson, T. A., 66
Christopherson, Al, 274
Ciccarelli, Shari, 181
Clemens, Mike, 301
Clontarf-Tara Local, 329
Clough, B. Franklin, 60, 66, 71, 328, 359
Clough, B. Franklin Jr., 72
Clough, Minnie Bosch, 72
Cockshutt Farm Equipment, Ltd., 119
Cokato Farmers' Union Exchange, 18
Cokato Telephone Co, 17
Collins, D. D., 16, 33, 47, 59
Colorado Farmers Union, 312
Commodity Credit Corp., 102, 343
Commodity Exchange Act, 84, 102
Commodity Trading Co., 323
Communist Invasion of Agriculture, 158
communistic smear, 157
ConAgra, 269
Conrad, Marvin, 301
Consumers Cooperative Association of Kansas City. *See* Farmland Industries
Consumers Cooperative Wholesale, 119
Cooley, Harold, 141
Coolidge, President, 19
Co-op Bill of Rights, 342
Co-op Tractor, 49, 118, 119
cooperative grocery stores, 119
Cooperative Marketing Act, 342
Cormier, Shirley, 68
Corn Belt Committee, 21, 47
Cornbelt Rebellion, 60
corporate agriculture, 216
corporate farming, 75, 152, 154, 271, 274, 281, 288, 351
corporate take-over of agriculture in Minnesota, 216
Cory, William, 358
crop insurance, 84
Curtis, Carl, 98

D

Dalen, Junice, 66, 69, 296
Dandurand, Mona, 41
Daniel, F. B., 33, 35, 152, 192, 224, 229, 230, 232, 334, 336, 391
Daniel, Margaret, 213
Daniel, Mary Rose Vogel, 335
Danielson, Lois, 41
Davis, A. C., 45
Davis, Tom, 87
Dawes, Archie, 317
De Vries, Karen, 41
Dear Barney: How do you say 'Holy Cow!' in Polish, 238
Dechant, Tony, 76, 104, 139, 153, 339
Delvo, Jerome, 233
Deming, Leo, 360
DeNeui, Bud, 300
Des Moines River Township, 290
Des Moines Valley local, 22
Deutschmann, Elmer, 181
Deutschmann, Elmer, 291, 314, 319
developing 'Inner Space', 209
DeVries, Adrian, 300
Dewey Local, 17
Disaster Relief Act, 204
Doherty, Rumble and Butler, 43
Donnelly, Ignatius, 74, 326
Dostal, Steve, 320
Double H Farmers Union, 25
Dovary Township, 290
Drovers State Bank of S. St. Paul, 308
Dunn, Michael, 356
Durbala, Dan, 24

E

Edwards, Gladys Talbott, 327
Efterfield, Rodney, 42
Egerstrom, Lee -agri-businss writer, 195, 267, 274
Egley, Charles, 48
Eichorst, Louis, 314
Eischens, Joyce, 320
Eischens, Mervin, 320
Eisenhower, Dwight, 81
Eisenhower, Dwight -election of, 141
Eisert, Freda, 66, 72, 297
Eken, Betty Skaurud, 258
Eken, Kent, 258

Eken, Lee, 258
Eken, Loren, 258
Eken, Willis, 194, 210, 211, 254, 257,
    258, 260, 268, 283, 332, 355
Eklund, Marian, 283, 287
Eklund, Roger, 300
Ekola, Rev. Giles, 217
Eldevik, Alf, 358
Elevator M, 86, 87
Elgin, ND, 231
Emergency Farm Mortgage Act, 343
emergency moratorium on foreclosures
    for farm, home and small business
    mortgages, 261
Engler, Elder, 290
Engler, Eugene, 42
Equity Cooperative Exchange, 15, 46,
    49, 82, 102, 117, 321
Erickson, Jens, 242
Erp, Anna Stroh, 31
Erp, Donald, 32
Erp, John C., 16, 31, 47, 59, 61
Estenson, Noel, 272
Estenson, Noel K., 128
Evans, Tyndal, 359
Evans, W. B., 207
Evergreen Co., Colorado, 293
Everson, E. H., 38, 343
Extension Separation Bill, 348
Extension Service, 49

F

Family Farm Security Act, 258
Farm Board, 14
Farm Bureau, 16, 49, 147, 157
Farm Credit Act, 62
Farm Credit Administration, 84, 102,
    343
Farm Family Communications Center,
    213
farm G.I. Classes, 202
Farm Holiday, 21, 60, 72, 327, 344
Farm Holiday Association, 59, 61
Farm Holiday Movement, 90
Farm Market Guide newspaper, 15
farm numbers decline to 1.9 million,
    Aug., 1994, 356
farm population in Minnesota, 1925 =
    2.6 million, 18

Farm Prices Are Made in Washington,
    98, 104
Farm Resettlement Administration, 91
Farm Security Administration, 70, 85,
    91, 102, 343
Farm Strike, newspaper, 21
Farm Tenancy Commission, 344
Farm Unity Coalition, 210
Farmers Guild, 69
Farmers Home Administration, 85, 91,
    102, 220, 343
Farmers National Grain Corporation, 14
Farmers Union Central Exchange, 91
Farmers Union Central Exchange -
    becomes Cenex, 126
Farmers Union coffee, 119
Farmers Union convention dates
    conflicted with the Minnesota deer
    hunting season, 288
Farmers Union Co-op Tractors, 118
Farmers Union Education Service, 122
Farmers Union encyclopedia. See
    Daniel, F. B.
Farmers Union Grain Terminal
    Association, 91, 99, 344, 353
Farmers Union Group Legal Services,
    193
Farmers Union Livestock Shipping
    Association, 16
Farmers Union Marketing and
    Processing Association, 321
Farmers Union Milk Marketing
    Cooperative, 193
Farmers Union Mutual Life Association,
    312
Farmers Union Park Memorial, Polk
    Co., 354
Farmers Union Service Association, 293
Farmers Union Terminal Association, 63
Farmers Union Triangle logo, 25
Farmers Union-NFO livestock
    marketing agreement, 262
Farmers' Educational and Cooperative
    Union of America, 17
Farmland Industries, 129, 130, 272
farmland values, 18
farms that were victims of a national
    food system, 202
Farrell, Sister Lucy, 18
Featherstone, G. H., 314
Federal Crop Insurance, 102

Federal Emergency Relief Act, 343
Federal Farm Board, 14, 102
Federal Reserve System, 60
Federal Trade Commission, 121
Feed Grain Program, 109
Felt, Reuben, 76
Finn Hall, 287, 288
Finnish Relief, Inc., 288
First National Bank of Wadena, 306
Fisk, U. D., 290
Fitch, George, 288
Fitzsimmons, Lenore, 161
Fjeld, Dennis, 320
Flaherty, Michael, 227
Fletcher, Roy, 359
Flick, Anna, 41
Fluegel, August F., 359
Fly-In, 286, 298
FmHA. *See* Farmers Home
    Administration
Fog, Jorgen, 314, 319
Fogarty, Leo, 228, 323
Fogarty, Leo and Marion , plaintiffs, 233
Fogarty, Marion, 324, 327
Fond du Lac, WI, 293, 295
Food and Freedom, 106
Food for Freedom, 95
Food For Peace, 95, 107, 346
foreclosure sale, 62
Foreign Assistance Act, 346
Forsell, Dennis, 309
Fossum, Julius, 314
Frederickson, Dave, 254, 266, 267, 268,
    270, 272, 275, 318, 332, 338, 355
Frederickson, Donnell, 65, 275
Frederickson, Eric, 320
Frederickson, James D., 318
Frederickson, John, 65, 275
Frederickson, Kay Kennedy, 266
Frederickson, Rudolph, 65, 275
Frederickson, Theo, 65, 360
Freedom to Farm, 273
Freking, Leo, 25, 359
Fricker, Avis, 42
FUCE. *See* Farmers Union Central
    Exchange
FUMPA. *See* Farmers Union Marketing
    and Processing Association
funeral costs proved "excessive", 72

G

G.I. Bill of Rights, 202, 352, 353
G.I. Farm Training, 213
Geiser, Walt, 24
General Agreement on Agricultural
    Production and Prices (GAPP), 221
General Agreement on Tariffs and Trade
    (GATT), 221
General Andrews Rest Area, 308
Genereux, Sam, 317
Gennesse Local, 327
Germantown Township, 289
Gerten, F. L., 358
GI Farm Management Training
    program, 202
GI farm trainee school, 159
Giersdorf, Lloyd, 291
Giffey, Donald F., 225
Gilbertson, Walter, 359
Gilbery, Charles, 358
Gildersleeve, Jim, 300
Gilster, David, 301
Gjellstad, Markell, 286
Gjervold, Mrs. Art, 297
Good Guy vs Good Neighbor Awards,
    325
Gooden, Donald, 300
Gorham, Martin, 358
government student loans, 277
Grain Belt Federation of Farm
    Organizations, 21, 47
grain grading frauds, 58
grain prices, 18
Grain Terminal Association, 89, 94, 104
grain trade practices, questionable, 58
Grams, Lowell, 301
Grand Meadow Local, 329
Grange, The, 20, 49, 147
Granztke, Clarence, 25
Grasdalen, Marion, 299
Grasdalen, Ray, 299
Gravdahl, Otto, 358
Great Bend Township, 290
Great Depression, 35, 64, 90, 159, 201,
    299
Great Gully Cooperator. *See* Edwin
    Christianson
Green Thumb, 153, 154, 192, 305, 306,
    307, 308, 309, 314, 316, 350, 351,
    354

Green View, 212, 283
Green View, birth of, 304
Green View, Inc., 192, 258, 276, 303,
    304, 305, 307, 308, 309, 328, 332,
    339, 351, 356
Greenwood Encyclopedia of American
    Institutions, 58
Gregerson, Gary, 300, 301
Greseth, David, 41
Gresham, Isaac Newton, 59
Griesch, Joe, 320
Griffith, C. C., 40
Griffith, Fred, 358
Grothe, Walter, 314
Groundswell, 261
Grove, Alice, 297
growers routinely were being cheated,
    63
GTA. *See* Farmers Union Grain
    Terminal Association
GTA's "Daily Radio Roundup", 95
Gulbranson, Nels, 358
Gullickson, Sam, 360
Gully, 145
Gully Cooperative Credit Union, 149
Gundersonn, Elvern, 42
Gustafson, Wilt, 49
Guy, William L., 104

H

Haaland, Jewell, 48, 352, 357
Haaland, Sheldon, 42, 255
Hagen, Monophaye, 295
Hagen, Percy, 307
Hail Our Union, 292
Hakel, Milt, 82, 150, 197, 207, 217, 283,
    337, 338, 353
Hamann, Melvin, 217
Hample, Fred, 23
Handschin, author Robert "Great
    American Cooperators", 83
Handschin, Bob, 48
Hanson, Allen D., 353
Hanson, Harry, 162
Hanson, Hubert, 197
Hanson, James, 42
Hanson, Norma, 283, 356
Hanson, Ron, 320
Hardin, Clifford, 253
Hargens, Lowell, 235

Harris, Alvera, 297
Harris, Flossie, 41
Harris, Fred, 297
Harvest States, 90
Harvest States Cooperatives created by
    merger, 353
harvest yields, 18
Haslerud, Meredith, 217
Hauglum, Reuben, 358
Haugo, David, 217
Haugo, Olaf, 243, 245
Haugo, Oscar, 313
Haukos, Gale, 305
hay shortage situation, 206
Hayfield Local, 329
hay-lift, 29, 49, 205, 206
hay-lift coordinator F. B. Daniel, 206
Heaton, Lyle, 49
Hedeen, Dan, 319
Hedeen, Steve, 319
Hedman, Jon, 320
Hehn, Werner, 233
Heinen, Gloria, 320
hemp plant near Lake Lillian, 72
hemp production, government
    encouraged during World War II, 72
Hendrickson, Roy F., 87, 105
Henke, David, 320
Herzfeld, Bill, 301
Hess, Clinton, 285, 298
het/vanbelofte, 13
Heuer, William, 308
Hinds, H. K., 23
Hocke, E. E., 290
Hoekman, Martin, 301
Hoekstra, Robert, 301
Hoffman, Cleo, 291
Hofstad, Ralph - president of Land
    O'Lakes, 129
Hofteig, Cecil, 314
Hogs sold for 2 ½ cents a pound, 299
Holland, Ross, 359
Hollingsworth, Tom, 320
Holsum Foods, 353
Hoover, Herbert, 20, 248
Hortop, Duane, 301
House Un-American Activities
    Committee, 158
how big should a co-op be?, 124
Hubin, Merlyn, 255, 300
Huff, C. E., 16, 45, 47, 312

Huff, Charles, 63
Huff, Paul E., 312
Hulke, Bruce, 301
Hummel local, 22
Humphrey, Hubert, 71, 80, 141, 153
Hunt, Hugh, 300
Hunter-Heron Lake local, 25
HW: Try to work out, FDR, 90
Hyde, U.S. Agriculture Secretary
    Arthur, 61

I

immigrant trains, 57
Indian Lake Township, 300
Ingerson, Ralph, 49
Ingerson, Ralph - was called "creator of
    the Co-op Tractor", 119
Ingvalson, Lorene, 301, 329
International Federation of Agricultural
    Producers, 40, 193, 243
Isaacs, Alton, 313, 319
Isaacson, Roy, 68
Isanti Local, 284, 285, 286, 329

J

Jackson *Republic*, 22
Jacobson, Evelyn, 297
Jacobson, Larry, 300
Jacobson, Luella, 49, 297
Jaeger, Lori, 355
Jaloszynski, Alice, 300, 329
Jaloszynski, Roman, 283
Jandra, O. M., 17
Jardine, Agriculture Secretary J. M., 19
Jensen, Marv, 300, 323
Jeppeson, Ole, 25
Jerve, Audrey, 297
Jewett & Sherman Co., 353
Johannes, Helen, 320
Johannes, Paul, 320
Johansen, Richard, 48
Johnson, Andrew O., 23
Johnson, Arthur J. and Myrtle, 327
Johnson, Bud, 313
Johnson, Carol, 277
Johnson, Dave, 300, 301
Johnson, Ernest, 290
Johnson, Harold, 276
Johnson, Ladybird, 306

Johnson, Lyndon B., 105
Johnson, Milo, 314
Johnson, Mrs. Alfred, 297
Johnson, Mrs. Lewis, 18
Johnson, Mrs. Reuben, 284
Johnson, Otto, 17
Johnson, Roger, 233
Jones, Kenneth, 320
Jones, Veldo, 42
jord, 13
Jordahl, Leona, 181

K

Kandiyohi Rural Electric Association,
    40
Kandy Local, 328
Kansas State University, 49
Kausner, Joe, 309
Kelley, Oliver Hudson, 20
Kelly, T. J., 47
Kennedy, E. E., 69
Kennedy, John F., 80, 105
Kettle River Rest Area, 308
KIDS, Inc., 212
Kilen State Park, 25
Kirchenwitz, Ken, 300
Klaasen, Bob, 300
Kleberg, Mrs. Milville, 297
Kleven, Ione, 64, 69, 297, 345
Kliber, George, 359
Kliber, Mike, 301
Klima, Ben, 359
Kling, Lou Anne, 356
Kling, Wayne, 356
Klinnert, Leo, 76, 314
Klocow, Dennis, 320
Klose, Bessie, 297, 300, 327, 328, 329,
    353, 355
Klose, Jeffrey, 328
Klose, Nicole, 328
K-M Funeral Association, a cooperative,
    72
Knapp, Joseph -author- "Rise of
    American Cooperative Enterprise",
    85
Knudsen, Carol, 255
Knutson, Andy, 276
Knutson, Congresswoman Coya, 276
Knutson, Don -Exec. Dir., Green View,
    Inc., 303, 309, 310, 332

Koenen, Lyle, 300
Kohlhase, Orlyn, 359
Kompelien, David, 320
Kompelien, Steve, 320
Kopacek, Tom, 255
Koski, Bill, 288
Koskis, 289
Koster, John, 24
Kronebusch, Gerald, 360
Krueger, Arthur, 360
Kruschke, Myra, 297
Kruta, Jaroslav, 295
KSTP-TV, 240, 241
Kuehl, Art, 290
Kuivinen, Einar, 48, 66, 67, 77, 288, 359
Kuivinen, Florence Mattson, 68
Kuivinen, Jacob "Jake", 69
Kvam, Patty, 41
Kyllo, Orion, 300

L

Laabs, Herman, 32
Laase, Jack, 284
Ladies Fly-In, 40, 151, 328
Ladies' Camp, 208
LaFollette, Wisconsin Senator Robert M., 46
LaGuardia, Fiorello H., 96
Lake Elizabeth Local, 60
Lake Hansel, 308
Lake Sarah Farmers Union Park and Campground, 309
Lake, Stephanie, 356
Lakeside Local, 296
Lakeville, 293, 294
Lambert, George C., 48
Land O'Lakes, 16, 128
lande, 13
Landhuis, Joe, 301
landmand, 13
Langer, William L., 62
Langsev, Palmer, 300
Larson, A. M., 313
Larson, Arlene, 356
Larson, Blanch, 243
Larson, Darrell, 356
Larson, Harold, 359
Larson, Hollie, 355
Larson, Jerry, 243
Larson, Joe, 176, 242

Larson, Joseph, 241
Larson, Joseph E., 126
Larson, Wava, 300, 329
Latvalas, 289
Laurer, Kathy, 255
lawsuits against the USDA, 205
Lee, F. O., 23
LeFebvre, Nydia, 289
Legare, Orin, 300
LeVoir, Kevin, 301
Lewis, John L. (United Mine Workers), 69
license to farm, 281
Linder, Steve, 301
Lindsay, Ken, 320
Little Moscow. See Gully, MN
Livestock Exchange Building, 321
livestock populations, 19
Livingston, Frank, 47
Loftus, George Sperry, 46
Lohring, P. P., 290
Long, Glenn, 176, 321
Long, Laura, 297
Lorenzen, Ruth, 297
Lubbesmeyer, Bud, 300
Luckt, Marvin, 25
Lueders, Art, 358
Lundberg, Carl, 16, 33
Lundberg, E. A., 359
Lundquist, Edna, 286
Lundquist, Franklin, 281, 284, 285
Lundquist, Mrs. Franklin, 286
Lundquist, Mrs. Rufus, 297
Lupka, Carl, 360
Lyngen, Lloyd A., 66

M

maailma, 13
maamies, 13
Mackey, Don, 320
Madison Livestock Shipping Association, 33
Magnuson, George, 313
Malusky, Alfonse J., 235
Malusky, B. J., 103, 231, 235, 238, 353
Malusky, LaVon, 235
Manahan, James E., 49
Mankato Local, 349
Mankato State College, 293
Mann, George, 255, 290

Mansfield, Mike, 98
Maple Grove Union, 17
Marketing Act, 14
Marshall Plan, 74
Marshall, Ben, 359
Marshfield, WI, 293, 295
Masanz, Alice, 151
Matheson, Gordon, 218
Mathsen, Rodney, 320
Mattson, Harold A., 16
Mattson, Olaf, 16
McCabe, Ben C., 93, 121
McCarthy, Eugene, 81, 98
McClellan, Alvin, 41
McClellan, Erma, 41
McCollum, Dave, 320
McGovern, George, 81, 95, 98
McKay, John, 126
McLaughlin, F. J., 47
McNary-Haugen, 342
Meade, Claude, 291
Melbo, Maurice, 320, 348
Melbo, Maurice, 49
Mellum, John, 24
mercy wheat, 96
merger of three major Upper Midwest
    farm cooperatives, 353
Meyer, Jody, 309
MFU membership by year, 369
Midway Hospital, 197
Midwest Food Conference, 199
Milbrath, Alex, 25
Mile High Club, 317, 320
Mill City: nick name for Minneapolis,
    15
Miller, Martin, 23
Miller, Mrs. Arley, 297
Miller, Robert, 16, 64
Miller, Wendell, 143
Minneapolis Chamber of Commerce, 35,
    36, 49, 87, 121
Minneapolis Grain Exchange, 46, 232,
    336
Minnesota Agricultural Research
    Institute, 255
Minnesota Agriculture, 154
Minnesota Association of Cooperatives,
    105, 156
Minnesota Cooperative law of 1916, 43
Minnesota Department of Natural
    Resources, 304

Minnesota Department of
    Transportation, 304
Minnesota Family Farm Security Act,
    208
Minnesota Farmers Union lost original
    charter, 35
Minnesota Farmers Union Service
    Corp., 275
Minnesota Groundswell, 210
Minnesota is re-chartered - 1942, 360
Minnesota Nonpartisan League, Inc., 58
Minnesota Production and Marketing
    Administration, 49
Minnesota's charter is revoked -1937,
    360
Mitchell, C. J., 48, 130
Moen, Richard, 300
Mondale, Walter, 98, 137, 151
Montbriand, Bob, 206
Moore, Paul, 15, 48
Moore, Stanley, 38, 237, 316
moratorium on farm foreclosures, 72
Morse, Wayne, 98
Mosbeck, Mable, 297
Mosel, Darrel, 300
Moses, New Hampshire Senator, 98
Moseson, Darrell - Cenex President, 128
Motter, Wade, 320
Mundt, Carl, 98
Munsterman, Walter, 41
Murphy, Harold, 359
Murray, William, 317
Myhre, Nels, 300

N

Nash, Charles, 194
Nass, David L., 60
National Cooperative Business
    Association, 272
National Cooperative Refinery
    Association, 129, 345
National Cooperatives, 1933 main plant,
    Albert Lea, 343
National Defense Education Act, 277
National Farm Machinery Cooperative,
    119
National Farmers Organization, 21, 262
National Farmers Union Insurance
    Companies, 312

National Federation of Grain
  Cooperatives, 91, 344
national food budget, 198
National Producers Alliance, 59
National Tax Equality Association, 92,
  121, 240
Nelson, A. J., 290
Nelson, Ancher, 290
Nelson, Gaylord, 98
Nelson, Gilbert, 359
Nelson, Harry, 359
Nelson, Hemming, 16
Nelson, Hemming S., 33
Nelson, Jean, 300
Nelson, Miles, 359
Nelson, Sherman H., 359
Nerstad, Ron, 320
Nesland, Jeri, 320
Ness, Adolph B., 290
Neurer, Alan, 320
New Deal, 63
NFO. *See* National Farmers
  Organization
Niemi, Blanche, 289
Niemi, Brooks, 289
Niemi, Christopher, 289
Niemi, Donna, 289
Niemi, John, 289, 360
Niemi, Monica, 289
Nixon, Richard M., 81, 105
Noetzelman, Don, 320
Nolan, Congressman Richard, 208
Nolan, Joe, 48
Non-Partisan League, 15, 58, 75, 102
Nordine, Oscar, 313, 314
Nordstrom, Edgar, 66, 71
North Dakota Farmers Union, 38
North Pacific Grain Growers, Inc., 343,
  353
Northrup, Connie, 41
Northwest Committee, 26, 35, 45, 47,
  224
Northwest Cooperative Society, 120
Northwest Legislative Committee, 123
Northwest Organizing Committee, 342.
  See Northwest Committee
Norton, A. O., 17
Novak, Aton, 315
Novak, Raymond, 312, 315
Nygaard, Marcia, 213, 352
Nystrom, Violet Pratt, 300

Nystrom, William, 143, 197, 300

O

O'Connor, D. L., 49
O'Riley, Melvin, 360
Oelke, Paul, 33
Ogaard, Don, 314
Okeson, John, 358
Older Americans Act, 305
Olson, Alec, 307, 332
Olson, Curtis, 66
Olson, David, 24
Olson, Edgar, 255
Olson, Floyd B., 20, 62, 87
Olson, MN Lt. Governor Alec, 254
Olson, Sidney, 347
Onstad, Arnold, 217
Orville L. Freeman, 92, 102, 137, 146,
  151, 153, 282, 350
Otto, Vivian, 314, 319
Ouverson, Lola Erp, 32
Owens, Verna, 297

P

P. L. 480, 107
Palan, Craig, 314
Pampusch, Bob, 320
Patton, James G., 42, 76
Paul, Gene, 210
Pauley, Henry, 358
Paulson, George, 25, 300
Paulson, Harry, 359
Paulson, Irene, 66
Payment-in-Kind Program, 221, 260
peacetime price controls, 252
Pearson, Harold, 284
Pearson, Lulu, 297
Pederson, Nels, 16, 33
Peetz, Colorado, 312
Pehl, Ralph, 360
Petersen, Merten, 359
Peterson, Alan, 301
Peterson, Douglas, 255
Peterson, Harry, 255
Peterson, Nels A., 358
Peterson, Nordahl, 61
Peterson, Rep. Collin, 268
Pfeffer, Lucille, 306
Pigeon, Bob, 301

PIK. *See* Payment In Kind
Pikal, Leonard, 301
Piper, Albert, 290
Piper, Mrs. Stella, 291
Pockets of Poverty, 306
Point, Tex., 341
Polk, Kermit, 290
Polzine, Bobbi, 261
prairie populism, 58
Pratt, William C., 327
Presley-Johnson, Joyce, 320
Pribyl, Margaret, 292
Producers Alliance, 15, 36
pro-German, 59
Prom Center, 162, 292
Public Law 480, 348
Purfeerst, Clarence, 255, 359

R

Raatz, Verle, 358
Radniecki, Elaine, 206
railroads and the rail barons who owned
    and ran them, 57
Rains Co., Tex., 59
Raup, Philip, 219
red-smears, 74
Reeck, Lionel, 359
Reimnitz, Erwin, 313
Reimnitz, Lura, 38, 41, 42, 64, 71, 139,
    154, 197, 297, 313, 327
Reimnitz, Lyle, 255
Reimnitz, Oskar, 39, 206
Reitmeier, F. H., 127
Reitmeier, Ron, 255
Reno, Milo, 48, 60, 63
Rentschler, Ron, 301
Republican 4-H Club – Hoover, Hyde,
    Hard Times and Hell, 60
Resettlement Administration, 85, 343
Ricker, A. W., 36, 37, 45, 63, 82
Rickert, Robert, 23, 25, 197
Rimpy, Andrew, 360
Ripka, John, 300
Ripka, Robert, 359
Roach, Robert, 301
Rocky Mountain Farmers Union, 295
Roe, David K., 199
Roe, Larry, 308
Rolvaag, Gov. Karl, 153, 290
Ronholdt, Betty Jean, 42

Roosevelt, President, 62
rope nooses were dangled in front of
    foreclosure participants, 61
Rosedahl, Matt, 355
Rosehill Local, 294
Rosehill Township, 290
Rost Center Local, 22
Roth, Gordon, 290
Rousu, Ale, 288
Rousu, Emma, 287
Rousu, George, 288
Rubis, Shirley, 25
Rumble, Wilford E., 43
Rumsdahl, Steve, 301
Rumsey, Glensdale, 301
Runholt, Vernon, 359
Rural Economic and Community
    Development, 85
Rural Electrical Administration, 69
rural electrification, 156, 276, 288
Rural Electrification Administration,
    343
Rural Telephone Act, 347
Rural-Urban Action Campaign, 210
Russian billion-dollar grain purchase,
    252
Rustler Local No. 9, Wright Co., 17
Ruud, Russ, 181

S

Sagvold, Mark, 320
Salzwedel, Ray Sr., 25
Sanders, Betty, 41
Sandnesss, Claire - Cenex board
    chairman, 129
Sandstrom, Mrs. Deloren, 285
Sauk Rapids *Frontiersman*, 20
Save Our Co-op campaign, 293
Schaefer, Roger, 300
Scherbing, Gordon, 322, 356
Scherle, Republican Congressman Bill,
    253
Scherman, Joe, 317
Schluter, John, 23
Schmidt, Lyman, 358
Schmiesing, Randy, 300, 301
Schneider, Esther, 64
Schneider, George, 359
Schouville, Dennis, 289
Schouville, Marge, 289

Schouville, Tom, 289
Schreiner, Margaret, 206
Schultz, Vic, 255
Schuman, Glenn, 359
Schuster, Albert, 320
Schuster, Dean, 320
Schuster, Tom, 320
Schwandt, George, 289, 290
Schwandt, Russel, 25, 76, 139, 150, 251,
    289, 290, 358
Seelig, Mildred, 297
Sellman. Charles, 358
Shriver, R. Sargent, 350
Sikorski, Gerald E., 231
Silver King local, 22
Simonson, Ferne, 197
Simpson, John A., 63, 343
Sip-and-Scheme, 217
Sirjord, Hiram, 42
sisu, 68
Sjodin, Dennis, 181, 258, 263, 283, 286,
    300, 338, 356
Sjodin, Joanne, 283, 287
Sjostrand, Dan, 320
Sjostrand, Marie, 41
Slebodnik, Joe, 24
Slettom, Ed, 105
Slick, Grant, 300
Smith, E. L., 66
Smogaard, Ellsworth, 64, 255, 295
Social Security, farmers finally are
    covered by -1955, 348
Sohm, Roy, 359
Soil Service Centers, 126
Solberg, Joe, 358
Solberg, K. K., 20
Solberg, Loren A., 255
soldiers of the soil, 95
Solheim, Ron, 320
sons of the wild jackass, 98
South Dakota Farmers Union, 38, 194,
    259, 291, 293, 354
soybean checkoff, 283, 287
soybean embargo, 252
Soybrings, 289
Sparby, Steve, 300
Sputnik, 41
St. Anthony & Dakota Elevator Co., 47,
    92
St. Claire, Phyllis, 289
St. Lawrence Seaway, 109

St. Lawrence Seaway opens, 349
St. Paul Bank for Cooperatives, 231
St. Paul Grain Exchange, 46, 121
Stahl, Fred, 358
Star local, 22
Stedman, Alfred D., 43
Steichen, Thomas H., 47, 123
Stencel, John, 295
Stensrud, Larry, 41
Stickney, C. W., 49
Stoltz, Mildred K., 327
Stone, George W., 354
Storden Highwater Local, 290
Strack, Laurence, 360
Strawberry Lake, 287
Strobel, G. A., 358
Strommer, Glenda, 41
Sugar Bush Local, 287, 289
Sullivan, Jerry, 320
Sullivan, Jim, 317
Suomi, 13, 68
Suss, Ted, 258
SW Minnesota Hist. Center, 60
Sweep, Barb, 320
Swenson, Abner, 358
Swenson, Clifford, 291, 292
Swenson, Curtis, 42
Swenson, Donald, 42
Swenson, Donna Klatt, 72
Swenson, Elnora, 292, 294, 297, 332
Swenson, Leland, 194, 264, 265, 291
Swenson, Mrs. Lemuel, 297
Swensons, The "premier Torchbearer
    Family", of Cottonwood Co., 292
Swift & Co., 61
Syftestad, Emil A., 48, 116, 117, 120,
    240

T

Tacheny, Cletus, 66
Taft, Robert, 81
Takash, Pete, 356
Takle, Mrs. Clarine, 291
Talbott, C. C., 12, 37, 117, 122, 153, 327
Talbott, Glenn, 38, 153, 327
Tangborn, Verner, 358
Task Force on Corporation Farming, 217
Tauer, Fred, 192, 300
Tengwall, George, 319
Texarkana, Tex., 341

Thatcher given $500 to organize Minnesota, Wisconsin, North Dakota and Montana (1925), 86
Thatcher personally visited each of over 600 farmers... 27,000 miles over dirt roads..., 85
Thatcher, M.W., 38, 43, 46, 63, 67, 76, 80, 100, 123, 143, 235
Thatcher, Myron William. *See* Thatcher, M.W.
Thatcher-GTA-McCabe, 93
The President's Corner, 154
Thomas, Mrs. Clayton, 290, 297
Thompson, Jacob, 18
Thompson, Morley, 315
Thompson, Walter, 41
three-way merger, 218
timetable of Minnesota Farmers Union, 341
tin can campaign, 63
Tin Can Price Revolution, 89
Tisthammer, Art, 142, 313
Tobolt, Lyle, 358
Tomlinson, F. E., 17
Tongen, Robert, 217
Torchbearers, 41, 42, 255, 287, 291, 292, 293, 295, 355
Torchbearers, Swenson – Curtis, Donald, Inez, Kathryn, Kenneth, Leland, 293
Torgerson, Randall E., 354
Tosch, Elmer, 359
Townley, A. C., 58
Townsend, Ted, 292
tractorcade, 252
Transtrom, Robert, 192
Trelstad, Bert, 359
Truman, Harry S., 98, 105, 252
try to work out, 344
Tuttle, Therese, 279
Tuveng, Mrs. Ed, 297
Tvedt, Jerome, 131
Twin Cities Milk Producers, 16

U

U.S. Air Force, 293
U.S. Army, 242, 294
U.S. National Guard, 293
U.S. War Bond, 295
un-American, 59

United Nations Relief and Rehabilitation Agency, 96
Urtel, Wanda, 41
Uruguay Round, 221, 354
Utah Farm Bureau – libel suit against, 157

V

Van Overbeke, Gary, 300
VanDyke, Alice, 297
Velde, David, 308, 319
Vesecky, John, 344
Vesey, William, 359
Vikingstead, George, 358
Vincent, David, 66
Vogel, Frank, 360
Vogt, Roger, 192, 286, 300
Vollmers, Vere, 213, 300, 356

W

Wagner, Greg, 290
Wagner, LaVern, 359
Waisanen, John, 197
Wallace, H. C., 90
Wallace, Henry A., presidential campaign, 76
Wallace, Secretary of Agriculture, Henry A., 62
Walli, John Jr., 358
Wallin, Marvin, 286
Walnut Grove, 290
war on poverty, 306
we also had 1,800 farms in the county back then. Now, there's only 400 left, 251
Webster, Elroy - Cenex board chairman, 129
Wefald, Jon - MN state agricultural commissioner, 49, 206
Wegner, Curt, 181, 301
Welscher, Maynard, 300
Wendel, Charles, 22, 23
Wendt, John, 24
Werner, Gene, 301
Werre, Joseph, 287
Wertish, Gary, 301
Wertish, Jeanne, 300
West, Gale, 285
Westbrook Local, 290

Westbrook Park, 291
Westphal, Forrest, 285
Westphal, Myra, 283, 286
What's Green View got to do with
    Farmers Union?, 305
White House Conference on Education,
    293
Whiting, Ralph, 203, 213
Wieland, Ed, 239, 321
Willard Hotel stationery, 90, 102
Williams, Bob, 301
Williams, Gerald, 301
Willmar Farmers Union Oil Company,
    40
Windingstad, Harold, 300
Windingstad, Randolph, 313, 320
Windom Flaxters, 290
Wingren, Voyle, 41
Winters, Carl, 16, 33
Wiseth, Ethel, 138, 143
Wiseth, Ida Theresa Peterson, 135
Wiseth, Kenneth, 136
Wiseth, Robert, 138, 314
Wiseth, Roy E., 73, 76, 136, 359
Wiseth, Vance, 138
Witt, Jack, 313
Wobbe, Mike, 301

Wolfe, Frances, 297
Wolff, Laura, 297
Woodland, DeVon, 209, 262
World Hunger Action Coalition, 199
World War I, 29, 58, 83
World War II, 71, 72, 94, 96, 98, 109,
    118, 119, 154, 160, 193, 194, 219,
    242, 288, 295, 318
Wright, Cecil, 358
Wyum, Ivan, 314

Y

You can't buck the federal government,
    62
You're going to spend the rest of your
    life in the pen, 62
Young Farmer Credit Program, 213
Young, Milton R., 98

Z

Zak, Conrad, 301
Zimmerman, Kenneth, 359
Zumbrota, Covered Bridge Park in, 307
Zweber, Frank, 359

TABLE OF FIGURES

Figure 1 Horsepower in the 1920's   11
Figure 2 Farmers Union Logo   25
Figure 3 John C. Erp   31
Figure 4 L to R: M.W. Thatcher, A.W. Ricker, C.C. Talbott   35
Figure 5 A.W. Ricker   37
Figure 6 C.C. Talbott   38
Figure 7 Laura & Oskar Reimnitz   39
Figure 8 Wilford Rumble   43
Figure 9 Chas Egley, long-time Farmers Union Livestock
     Commission Co. mgr, organizer of many FU locals   51
Figure 10 Chas Barrett, Pres. NFU (1908-28); A key figure in
     "Amalgamation" of 1927   51
Figure 11 C. E. Huff. Kansan, NFU Pres. 1928-29   51
Figure 12 A. W. Ricker. Head of Producers Alliance, FU NW
     Committee Sec. , editor, Farm Market Guide & Farmers Union
     Herald, and an "Amalgamation figure.   52
Figure 13 C.C. Talbott. Organizer for Producers Alliance, later for
     Farmers Union; Pres. ND Farmers Union, member, Exec.
     Comm. FU NW Committee, 1st Pres., Board Chairman, FUCE   52
Figure 14 M.W. Thatcher. Equity Exchange auditor, chairman, FU
     Northwest Committee, Manager, FUTA, key staff member
     Farmers Nat'l Grain Corp., Pres. & Founder Nat'l Fed. of
     Grain Cooperatives, key FU legislation rep., & Gen. Mrg.,
     FUGTA   52
Figure 15 F. J. McLaughlin. Born in Dubuque, IA, moved to
     Dakotas in 1926, recognized as one of group responsible
     through the NW Committee for bringing Farmers Union
     movement into ND, MN, and WI. Employee of FU Terminal
     Association in 1929.   53
Figure 16 C. J. Mitchell. Organizer for Producers Alliance in SD,
     1st Credit Mrg., Farmers Union Central Exchange   53
Figure 17 Emil A. Syftestad. Succeeded M.W. Thatcher as mgr. of
     FUTA in 1931, 1st Gen. Mgr., FU Central Exchange, 1931-
     1957   53
Figure 18 Einar Kuivinen of New York Mills - 1st & 3rd MFU
     Pres. after rechartering.   54

Figure 19 Bob Handschin, dir. Econ. Research FUTA 1943-1976,
NFU legislative rep., 1938-43, editorialist Co-op Country
News & FU Herald. Died 1997   54
Figure 20 Richard Johansen. Staff at FUGTA, Press Sec. to M.W.
Thatcher. Skilled writer, authored many Thatcher speeches.   54
Figure 21 Ralph Ingerson, FUCE staff, developer of the Co-op
Tractor in 1930's which was made in St. Paul. Active developer
of oil co-ops.   55
Figure 22 Jon Wefald, MN Ag Commissioner who assisted FU hay-
lift campaign during 1976 drought, later Pres., Kansas State
Univ.   55
Figure 23 Einar Kuivinen   67
Figure 24 M.W. Thatcher, about 1926   80
Figure 25 Elevator M   86
Figure 26 Wheat protien sample can (1927-28 tin can campaign)   87
Figure 27 M.W. Thatcher at the microphone   95
Figure 28 Mercy Wheat kick-off in Climax, MN   96
Figure 29 Original Farmers Union / Equity St. Paul River Terminal   111
Figure 30 GTA's Superior, Wis. Terminals   111
Figure 31 1940's Farm Equipment   112
Figure 32 Roseau Farmers Union Co-op Oil c. 1930's   112
Figure 33 Yeild 30 bu./acre, harvest 50-80 acres/day   112
Figure 34 Farmers Union GTA Office, So. St. Paul, MN   113
Figure 35 Old Equity Exchange Building   113
Figure 36 The Co-op Tractor - From left to right are Alex Lind,
Ralph Ingerson and E. A. Syftestad   118
Figure 37 Combining Then   132
Figure 38 Combining Now   132
Figure 39 MN Gov. Karl Rolvaag, l., Thomas Steichen, a Farmers
Union Central Exchange General Manager   133
Figure 40 New York Mills FU Headquarters 1940's   133
Figure 41 Roy E. Wiseth   135
Figure 42 MFU Bus Group   142
Figure 43 Edwin and Nora Christianson   145
Figure 44 - The Latest: 1920's   165
Figure 45 25,000 Farmers Jam St. Paul Auditorium April 27, 1940   166
Figure 46 MFU Milk Petition. Front: Edwin Christianson, MFU
President, Minnesota U. S. Senators. Eugene McCarthy &
Walter Mondale   166
Figure 47 NFU Board Meets with President Harry Truman   166
Figure 48 1946 MFU Board & Office Members & Co-op
Managers. Back: Cassavant, Lyngen, Nystrom, Kuivinen,
Dwight Wilson; Center: Lura Reimnitz & Oskar, Junice Dalen,
Mitchell, Tom Croll; Front: M. W. Thatcher, Christianson,
Franklin Clough, Marv Evanson, Emil Syftestad   167

Figure 49 1947 MFU State Executive Committee. Back: Wm Nystrom, Wendell Miller, Roy Lindall; Front: Wm Cassavant, Einar Kuvinen & Freda Eisert   167

Figure 50 L to R: Wis. Sen. Gaylord Nelson, NFU Pres. Jim Patton, Orville Freeman, Mn Gov. & U.S. Sec. of Agriculture, ND FU Pres. Glenn Talbott, MFU Pres. Ed Christianson   168

Figure 51 Left to Right: Frank Livingston, Jim Jesson -Pres. Waseca Co. FU, Louis Huper -Pres., Faribault Co. FU, Nels Wongen, Freeborn Co. FU Pres.   168

Figure 52 L - R: MFU Pres. Ed Christianson, Orv Freeman, Wm Nystrom   168

Figure 53 Left: Paul Anderson, Dir., MFU Green Thumb, Right: Wendy Anderson, MN U.S. Sen., Gov. & State Sen.,   169

Figure 54 From Left: Glen Coutts, FUCE Bd. Member; Dennis Forsell, Dir., Green View Field Operations   169

Figure 55 Tony Dechant, NFU Pres., L., Ed Christianson, MFU Pres., R.   170

Figure 56 L to R: Ray Grasdalen, MFU Field Staff; Fred Gronninger, early organizer & GTA Fieldman   170

Figure 57 Axel & The Co-op Shoppers   171

Figure 58 Restored Co-op Tractor   171

Figure 59 L to R: Orville Freeman, MN Gov., U.S. Sec. of Ag; Wilton Gustafson, long-time MFU staff & bus driver   172

Figure 60 L to R: Norma Hanson, MFU Vice Pres., Exec. Comm., Co. Pres.; Harry Hanson, long-time Gully Co-op Bd.,   172

Figure 61 L to R: Hubert Humphrey, MN U.S. Sen. & Vice Pres. with MFU Pres. Ed Christianson   173

Figure 62 L to R: Milton Holtan; Leo Klinnert, MFU Bd., FU Ins. Agent   173

Figure 63 MFU Fly-In Group in Washington, D.C.   174

Figure 64 Ladies Fly-In Group Departing for Washington, D.C.   174

Figure 65 MFU Legislative Action in Washington, D.C.   175

Figure 66 Pres. Lyndon Johnson, a surprise guest at Minneapolis NFU Convention   175

Figure 67 Frank & Ethel Krukemeier, long-time MFU staff   176

Figure 68 L to R: Joe Larson, Climax, MN., Chm., FUCE Bd. Ch, NFU Ins. Bd.; Glenn Long, Manager, MFU Marketing Assoc.   176

Figure 69 Early group of Green Thumb workers reporting in.   177

Figure 70 MFU "old-timers" from 1945-1950. Photo taken 1993 L to R, Front: Mrs. Ann (Verner) Anderson, Irene Paulson, MFU Sec., Gwen (Paulson) Birch, Mr. & Mrs. James Youngdale. Back:Ione (Dahlen) Kleven, Edu. Dir., June (Smogaard) Carlson, Olaf Haugo, GTA Bd., Junice (Dahlen) Sondergaard, Edu. Dir., Verner Anderson, Ruth Lorentzen, Garland Birch, Regina (Mrs. Henry) Herfindahl, Alice (Van Dyke) Burlingame, Frances (Wolfe) Struck   177

Figure 71 L to R: George Mann, GTA Bd., MN St. Legislator; Ray Novak, Pres., NFU Ins.   178
Figure 72 L to R: Alec Olson, U.S. Rep., MN State Sen. & Lt. Gov; Gene Paul, NFO Pres., MFU Member   178
Figure 73 L to R: Orv Calhoun, Gully Co-op Mrg; Archie Bauman, cartoonist, MFU Sec.; Verner Anderson, life-long activist & organizer; Leo Klinnert, St. Bd. Member   179
Figure 74 L to R: Howard Peterson, GTA Staff, Co FU Pres.; Bob Rickert, MFU Sec., Bd. Member   179
Figure 75   L to R: Larry Roe, Dir., Green View; MN Gov. & Congressman Al Quie   180
Figure 76 Ale & Emma Rousu, Sugar Bush Local   180
Figure 77 MFU Foundation Bd meeting, 1997. L to R: Shari Ciccarelli, Sec., F.B. Daniel, Dennis Sjodin, Cy Carpenter, Orion Kyllo, Leona Jordahl, Russ Ruud, Elmer Deutschmann (seated), Curt Wegner. Not pictured: Brenda Velde, Harold Windingstad, Marion Fogarty   181
Figure 78 Typical Torchbearer Ceremony   181
Figure 79 L to R: Dave Roe, MN AFL-CIO Pres, Gordon Scherbing, CEO & Pres., FU Mktg.   182
Figure 80 Ferne Simonson, Sec. to Ed Christianson; Emil Syftestad   182
Figure 81 Honeymead, Mankato, MN   183
Figure 82 Refinery in Laurel, MT.   183
Figure 83 M.W. Thatcher   184
Figure 84 Mercy Wheat   184
Figure 85 Gully, MN Main Street on Meeting Day   185
Figure 86 Gully Co-op in the 1990's   185
Figure 87 MFU Managers at McPherson Refinery   186
Figure 88 MFU Campaign Yard Signs   186
Figure 89 L to R: Jerry Tvedt, Pres., FUCE; Nels Wongen, Pres. Freeborn Co. FU   187
Figure 90 MFU at the Minnesota State Fair   187
Figure 91 Bill Walker, MN Comm. of Agriculture, MFU member.   188
Figure 92 Ed Weiland, MFU Mktg Assoc. Gen. Mrg.   188
Figure 93 Robert Wiseth, Co. Pres., MFU Exec. Comm.   189
Figure 94 Devon Woodland, NFO President   189
Figure 95 Farmers Union Marketing & Processing Association Board, 1996. Seated L to R: Harold Eklund, Paul Symens, Harold Nearhood, William Johnson, Mathew Birgen, Palmer Pederson, Standing, L to R: David Eblen, Dwight Bassingthwaite, Marvin Jensen, Dennis Rosen, Robert Carlson, Dennis Sjodin, Frank Daniels   190
Figure 96 Cy & Frances Carpenter   191
Figure 97 The Haylift Action - 1976   205
Figure 98 B. J. "Barney" Malusky   236
Figure 99 Bob Bergland   247

Figure 100 Willis & Betty Eken   *257*
Figure 101 Dave Frederickson   *266*
Figure 102 Coya Knutson   *276*
Figure 103 Lee Swenson & Parents Clifford & Elnora Swenson   *292*
Figure 104 Clint Hess   *298*
Figure 105 Archie Baumann   *299*
Figure 106 Green View contract signing, 1969. L to R: Wm Heuer,
    Ed Christianson, Dale Wreisner, Richard Braun, Percy Hagen   *303*
Figure 107 Floyd Borghorst   *311*
Figure 108 James D. Frederickson   *318*
Figure 109 Marion Fogarty   *324*
Figure 110 Cy Carpenter   *333*
Figure 111 F. B. Daniel   *334*
Figure 112 Milt Hakel   *337*

AFTERWORD

By Cy Carpenter

The contribution which family farmers and rural communities have made and continue to make to the quality of life in America is one of this nation's great resources and one of its best kept secrets.

Peace, quiet, respect, neighboring, sunshine, clean air, breathing space and a strong work ethic are most abundant in rural America. These, together with freedom and an abundance of quality food and individual liberty have made this nation the envy of the peoples of all nations.

Far too many citizens today, if asked, could not tell us how and when these United States came to be, much less relate to the gallant struggles of past generations to found and to form a new nation which has been a world model for three centuries. How different would this nation be had it not been a "rural land," a land for the people, in those founding years?

Many, including rural residents, lack any desire to preserve rural America for the good of the nation, much less in fairness to those who, for more than a century, have provided abundance while being forced to accept substandard compensation and almost universal lack of respect for rural communities and their vital role in our society.

As family farms go, so will go rural America, its main streets, its job opportunities, its quality of life, indeed its very vitality. The family farm continues to be the only business in America which is denied the right to price its products on fair consideration of their contribution to society.

Nor are farm operators recognized as members of the "business community" even when the investment in their farm and the spectrum of necessary operating skills is several times that of some of their counterparts in other sectors of the economy

More serious perhaps. is the fact that after nearly 50 years of rural decline, more than half of our farm operators have never been willing to identify with a farm organization to consider or participate in the preservation of owner-operated farms and quality life in rural America.

History has demonstrated that the cities and towns in rural America are sustained by the number of operating farms in their respective trade areas as opposed to the number of producing acres and animals, a fact which has never been and is not now acknowledged in public policy.

To their eternal credit, the focus of the founders and *members of* Farmer Union has for 95 years, been the perpetuation of quality life for all Americans through the preservation of rural America. And in contrast, sad to say, most Americans seem to see all change as progress and they apparently see no problem with the continuing transfer of farm land from owner-operators to investor-operators or with tenant or absentee farming, or even with the annual blacktopping of thousands of acres of precious farm land.

Hopefully, in compiling and publishing this history, we will have paid fitting tribute to those who painfully blazed the trail for the good life we are privileged to share, will have made present and future generations more aware of their true heritage and have helped to bring proper focus *on the future of* our society. America will never be *what we* have known it to be and expect it to be without family farms and thriving rural communities.